W9-BTF-519

*f* **P**

ALSO BY KEN ALDER

*The Measure of All Things: The Seven-Year Odyssey and Hidden Error That Transformed the World*

*Engineering the Revolution*

# The Lie Detectors

## The History of an American Obsession

Ken Alder

*Free Press*
NEW YORK LONDON TORONTO SYDNEY

FREE PRESS
A Division of Simon & Schuster, Inc.
1230 Avenue of the Americas
New York, NY 10020



For information about special discounts for bulk purchases, please contact
Simon & Schuster Special Sales: 1-800-456-6798 or business@simonandschuster.com

Text Design by Paul Dippolito
Photo credits appear following the index.

Manufactured in the United States of America

1 3 5 7 9 10 8 6 4 2

Library of Congress Cataloging-in-Publication Data
Alder, Ken.
The lie detectors: the history of an American obsession/Ken Alder.
p. cm.
1. Lie detectors and detection—United States—History.
2. Polygraph operators—United States—History. I. Title.
HV8078 .A53 2007
363.25'4—dc22 2006052468

ISBN-13: 978-0-7432-5988-0
ISBN-10: 0-7432-5988-2

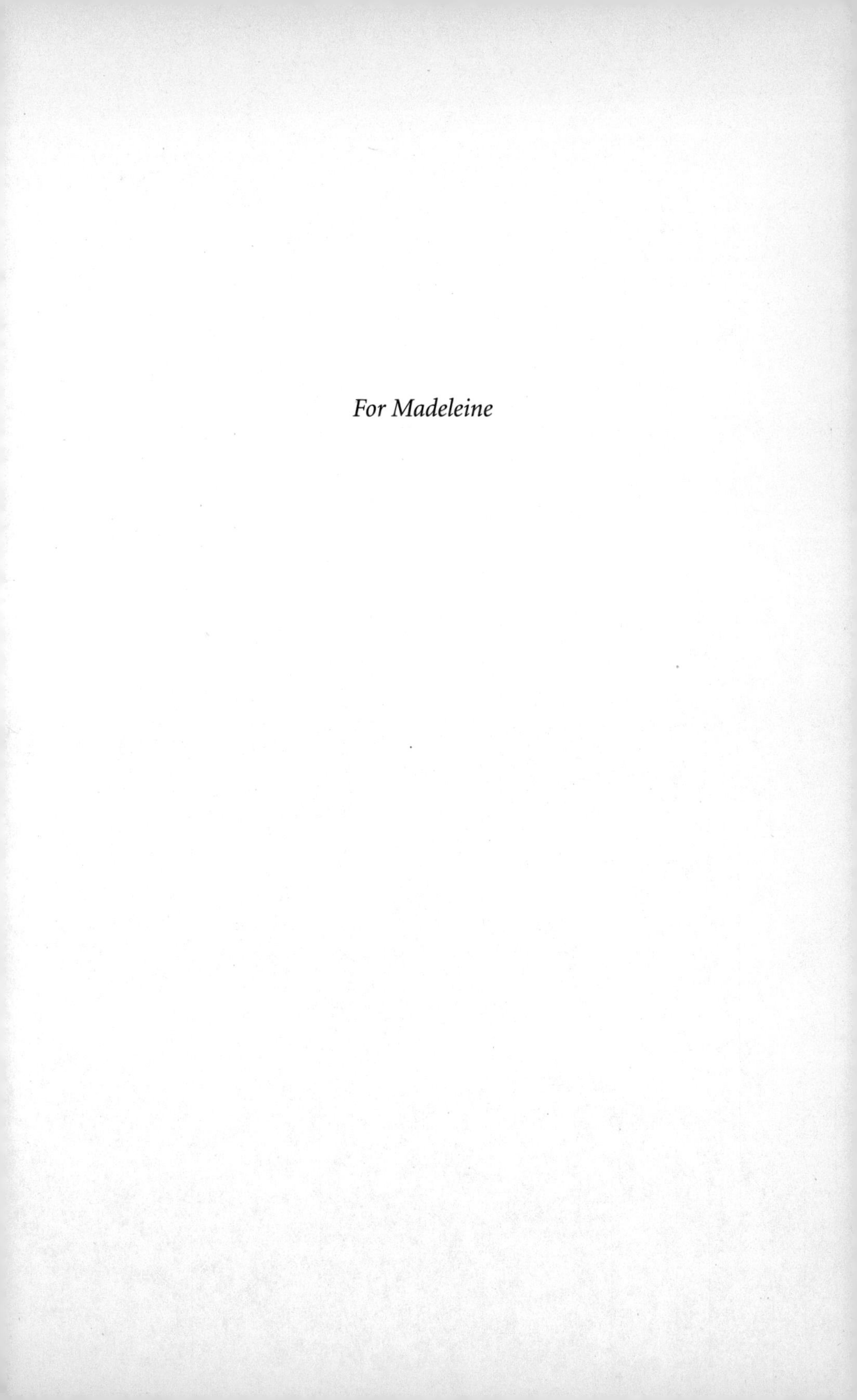

*For Madeleine*

*All we got on him is he won't tell us nothing.*

—RAYMOND CHANDLER, *THE LONG GOODBYE*

# Contents

# Preface

WHO WOULDN'T BE TEMPTED BY A DEVICE THAT LET YOU read the thoughts of your fellow citizens? And who wouldn't hesitate to be hooked up to such a device?

In the early part of the twentieth century two Americans announced that they had solved the age-old quest for a reliable method of distinguishing truth-tellers from liars. The momentous announcement came from Berkeley, California, where the two men were working as disciples of the town's police chief, himself the nation's leading advocate of bringing scientific methods to police work. The first disciple, John Larson, was the nation's first cop with a doctorate, a forensic scientist of restless integrity. The second, Leonarde Keeler, was a high school–age enthusiast, with less integrity but considerably greater charm. Working first together, then at cross-purposes, they fashioned an instrument—and a technique of interrogation—that was hailed by the public as history's first "lie detector."

To a nation obsessed by criminal disorder and political corruption, the device seemed to light the way toward an honest society. Adultery, murder, conspiracy, espionage—the bright lamp of the lie detector would pierce the human opacity which allowed these secret vices to flourish. On the strength of this utopian ambition, millions of Americans have been hooked up to a polygraph machine that monitors their pulse rate, blood pressure, depth of breathing, and sweatiness—and on the basis of their physiological responses during a brief interrogation, have been declared either sincere or deceptive.

Is America more honest as a result?

This book tells the story of the lie detectors and their creation: a device which seemed to mimic the actions of the human soul, until, like Frankenstein's monster, it threatened to supplant its creators, even while terrifying—and enthralling—a nation. This is also the story of our secret selves. Saint Augustine long ago defined a lie as the gulf between the public utterance of one's mouth and the secret knowledge of one's heart. Everyone has felt it: the skip in our hearts when we tell a big fib, the effort it takes to breathe evenly. As the poet Joseph Brodsky has observed, self-consciousness does not really begin until one has told one's first deliberate lie. To deceive is human.

The problem is, human beings are opaque. Some have considered this a design flaw—perhaps the original flaw. According to the ancient Greeks, the first human being owed his existence to a competition staged by Zeus to reward the most inventive of the gods. It was to be the world's first science fair, with Momus, the god of criticism, to serve as judge. Each competitor tried to outdo the rest. Athena constructed a magnificent dwelling; Poseidon built the first bull; and Prometheus made the first man. Momus didn't think much of any of the entries, but he was particularly scathing about man. Athena's dwelling was so grand it was unmovable; what if its inhabitants quarreled with their neighbors? Poseidon's bull had horns on either side of its head; they would have been more effective up front. And as for Prometheus's man, he lacked a window in his breast whereby others might look in and see "all the man's thoughts and wishes . . . , and whether he was lying or telling the truth." In later years Momus grew so infuriated by the duplicity of this second-runner-up invention that he plotted mankind's destruction—for which sacrilege he was driven from heaven.

Despite this warning, the search for Momus's window has continued down the centuries. The Greeks developed a science of physiognomy to assess people's character from their facial features and gestures. On the assumption that anxious deceivers generated less saliva, suspected liars in ancient China were asked to chew a bowl of rice and spit it out. Judges in India scanned for curling toes. One pious Victorian physician suggested that God had endowed human beings with the capacity to blush so as to make their deceptions apparent. Today, you can pick up the basics of body language for a few bucks on almost any library resale table—"Who's Lying to You and Who's Lusting for You!"—along with guides for spotting tricksters when you travel abroad. Popular manuals, updated with the latest findings of neuroscience, advise you how to track the eye movements and hand gestures of your spouse, boss, and stockbroker.

Yet experts on deceit—the sort of psychologists who regularly ask Americans to lie to one another in laboratories—tell us that the vast majority of us are very bad at detecting deception, despite our confidence in our own powers. In 2006, one review of the available research concluded that people can successfully sort truth-tellers from liars only 54 percent of the time, or about as well as blind guesswork. Surprisingly, the more intimately we know the deceiver, the worse we do. Even cops, judges, and psychologists—those citizens professionally licensed to sort truth-tellers from liars—don't get it right much more than half the time.

That is why, in the early years of the twentieth century, a coterie of American psychologists set out to decipher the operations of the human mind by peering beneath the skin. They recorded the body's involuntary tremors: its secret pulsations, hidden pressures, and suppressed gasps. In doing so, they drew on a new theory of the emotions that declared that human emotions were nothing more than a set of physiological responses inherited from our animal ancestors. Each act of deception, they suggested, produced a divided self, a disjuncture between the heart, fearful that its secret feelings would be exposed, and the mind, desperate to suppress the body's betrayal.

The first person to attempt this feat of detection was an artful young Harvard psychologist and lawyer named William Moulton Marston, later famous as the creator of the cartoon character called "Wonder Woman" (the embodiment, he said, of all the psychological principles behind his technique of honesty testing). But the lie detector did not truly come into its own—or even acquire its name—until the 1920s, when John Larson and Leonarde Keeler adapted Marston's method to the interrogation of criminal suspects. There was much that Larson and Keeler shared; for instance, both men first met their wives while interrogating them on the lie detector. But soon after they both moved to Chicago to prove their methods in the American capital of crime and corruption, they become rivals and eventually enemies. In the end, each man paid a personal price for his obsession, led by the device to mistrust the people closest to him, including one another. The machine that launched their careers poisoned their lives.

By then, however, they had established the machine in the heart of American culture, transforming the device from a detector of lies into a monitor of loyalty. What began as a way to confirm honesty in precinct stations, office towers, and government agencies became a way to test the credibility of Hollywood movies and Madison Avenue advertising. By mid-century, the device was being used to safeguard nuclear secrets, assure the political fidelity of scientists, and purge homosexuals from government jobs. By the 1980s, some

5,000 to 10,000 polygraph operators were testing 2 million Americans each year. The lie detector had become America's mechanical conscience.

Today, the lie detector is still used to interrogate criminal suspects, expose fraud, safeguard nuclear secrets, and combat terrorism. Yet no country other than the United States has made use of the technique to any significant degree. And even in America, the lie detector has been consistently banned from criminal courts and discredited by panels of illustrious scientists, from the Congressional Office of Technology Assessment to the National Academy of Sciences. Surveys of studies of role-playing games conducted in labs suggest tremendous variability in how accurately the technique detects guilty reactions: from only 35 percent to nearly perfect for standard polygraph procedures, with most in the range of 60 to 90 percent. As for field studies, their accuracy is just as variable, as best anyone can determine. Indeed, according to a review by the noted psychologist David Lykken, in real-life "field situations," when results were graded on the basis of the polygraph results alone, the innocent were called truthful only 53 percent of the time, which is to say, hardly better than guesswork. Despite all this, the lie detector lives on.

Given this history, this book does not attempt to expose the scientific pretensions of the lie detector. Not only would such an exposé be redundant; it would hardly achieve its purpose. Instead, this book addresses the obverse problem: Why, despite the avalanche of scientific denunciations, does the United States—and only the United States—continue to make significant use of the lie detector?

Of course, the machine's proliferation in twentieth-century America shows that the public believed the tests served some purpose. But in the case of the lie detector something additional was required, because persuading Americans of the machine's potency was itself a prerequisite for the machine's success. As its proponents acknowledged, the lie detector would not distinguish honest Americans from deceptive ones unless those same Americans believed the instrument might catch them. In short, the lie detector depends on what medical science has dismissively termed the "placebo effect." At the same time, as its proponents also acknowledged, the lie detector did not test whether people were actually telling the truth so much as whether they believed they were telling the truth. So either way, America's obsession with the lie detector poses the most troubling question of all: What do we believe?

PART 1

# *The Athens of the Pacific*

---

*There are two kinds of liars the kind that lie and the kind that don't lie the kind that lie are no good.*

—GERTRUDE STEIN, *A NOVEL OF THANK YOU*, 1925

CHAPTER 1

# "Science Nabs Sorority Sneak"

*Her eyelids drooped. "Oh, I'm so tired," she said tremulously, "so tired of it all, of myself, of lying and thinking up lies, and of not knowing what is a lie and what is the truth. I wish I—"*

*She put her hands up to Spade's cheeks, put her open mouth hard against his mouth, her body flat against his body.*

—DASHIELL HAMMETT, *THE MALTESE FALCON*, 1929

THE CASE HAD ALL THE SIGNS OF AN INSIDE JOB. ONE OF the ninety young women in College Hall was a sneak thief. For several months, someone had been filching personal possessions from the rooms of her dorm sisters: silk underthings, registered letters, fancy jewelry, cash. It was the springtime of the Jazz Age in 1921, and young women were returning to the boardinghouse on the campus at Berkeley to find their evening gowns spread out on their beds, as if someone had been sizing them up. A sophomore from Bakersfield had been robbed of $45 she had hidden inside a textbook; a freshman from Lodi lost money and jewelry valued at $100; and Margaret Taylor, a freshman from San Diego, could not find her diamond ring worth $400—though she wondered whether she had simply misplaced it.

Unable to wring a confession from any of her boarders, the housemother turned to the Berkeley police department, famous for introducing modern scientific techniques into crime-fighting. But Jack Fisher, an old-time cop on the force, didn't have much to go on. He learned that on March 26, Ruth Benedict had put $65 in her purse before going down to dinner at six; when she returned at six-thirty, the money was gone. One boarder,

Alison Holt, had been seen watching Benedict hide her purse, and had not come down to dinner immediately. This made her Fisher's prime suspect, especially as she was "one of these big baby eyed types [who] cannot remember what took place on any given date and answers all questions with the big innocent baby stare." The other girls thought her "queer."

Also, at that same meal, another young woman, Helen Graham, had carried a plate up to a Miss Arden, sick in her room. Officer Fisher was plied with various rumors about Miss Graham, a tall, well-proportioned woman with deep-set eyes, dramatic eyebrows, and an intense manner. Her roommate told him Miss Graham spent money out of proportion to her modest Kansas background; also, she wore a diamond ring and a pendant with big stones. She was a bit older than the other young women and had trained as a nurse. "She is of the highly nervous type," Fisher wrote, "and has been suspected of being a hop head." She also had more experience when it came to men, and her dorm sisters seemed to resent her for it.

Then there was Muriel Hills, who had been seen in the vicinity of another theft: a "very nervous type, the muscles of the eyes seem to be affected, the eyes moving all the time, and she . . . has to hold her head sideways to see who she is talking to."

So far Fisher had a baby-faced queer girl, a high-strung bad girl, and a jittery nervous girl, plus other suspects. He did not see, amid these female intrigues, how he would ever solve the case.

Then the housemother began to worry that repeated visits by the police would give the house a bad reputation. College Hall was the sole sanctioned residence at the University of California, filled with respectable young women of eighteen and nineteen from good families. The housemother asked that the investigation be wound down.

So Fisher called in his colleague John Augustus Larson, the nation's first and only doctoral cop: a twenty-nine-year-old rookie who had earned a Ph.D. in physiology from the University of California. Larson was a solid man of medium height, who led with his forehead, his blond hair pasted firmly to one side. A man with something to prove. He was currently working the four-to-twelve downtown beat like any other rookie, but he was not much of a cop in other respects. For one thing, he was almost blind in his right eye and was the worst shot in the department. For another, he was just learning to drive and had recently wrecked two squad cars in a single day. Meanwhile, he was still toiling in a university lab looking to bring new scientific methods to police work.

Only a few weeks back, Larson had read an article entitled "Physiological Possibilities of the Deception Test," by the lawyer-psychologist William Moulton Marston. In experiments conducted at Hugo Münsterberg's famous emotion laboratory at Harvard, Marston had discovered that he could determine which of his fellow students were spinning tall tales and which were giving an honest account. All he had to do was track the rise in their blood pressure as they reached the climax of their story. Larson wondered: might this method be applied to the dirty business of police interrogation?

As a trained physiologist, however, Larson saw several ways to improve Marston's technique. He began by reversing Marston's procedure. Whereas Marston had taken intermittent blood pressure readings while his subjects told their tall tales, Larson decided to take continuous readings while his subjects answered specific questions. With the help of a lab technician, he assembled an apparatus that registered a subject's systolic blood pressure and breathing depth, and recorded these values permanently on a roll of smoke-blackened paper. Though the machine would record the relative values of a pulse-pressure amalgam, and not the absolute value of the blood pressure, as Marston's cuff method did, its great advantage was that the automated device minimized the examiner's judgment in taking the readings, thereby fulfilling one criterion of the scientific method, which was to "eliminate all personal factors wherever possible." This was particularly important in cases where the examiner might be led astray by his own feelings about a test.

In another sense, however, Larson's procedure was hardly new. For more than half a century physiologists had used this sort of automatic recording device to track bodily processes beneath the skin. Some had even tried to correlate these interior reactions with subjective feelings. As early as 1858, the French physiologist Étienne-Jules Marey had built a device that simultaneously recorded changes in blood pressure, respiration, and pulse rates while his subjects experienced nausea, sharp noises, and "stress." By the late nineteenth century, the American psychologist William James had come to *define* emotion as bodily changes that occurred in response to the cognition of an exciting stimulus. Larson was simply proposing to read the body's emotional script for signs of deception.

The resulting device, which Larson dubbed the "cardio-pneumo-psychograph," was a bulky Rube Goldberg contraption, and this is no idle comparison. Reuben Lucius Goldberg, class of 1904, the Berkeley engineering school's most famous graduate, had recently achieved renown for his cartoons skewering the American credo that all of life's problems had a mechanical solution. And what was Larson's "cardio-pneumo-psychograph" if not a mechanical solution to one of life's oldest mysteries: What is going on inside the head of another person?

At the same time, in keeping with police procedures, Larson swapped Marston's analysis of invented stories in favor of a "controlled" comparison of answers to yes-no questions, some irrelevant to the matter under investigation, others of an accusatory nature. One of the great challenges of lie detection was to match the human ingenuity for deceit. Dissembling comes in many guises: Machiavellian lies are disseminated by the strong; defensive lies are woven by the weak; and white lies keep the social machinery running smoothly. As the essayist Michel de Montaigne once observed, "If a lie, like truth, had only one face, we could be on better terms, for certainty should then be the reverse of what the liar said. But the reverse side of the truth has a hundred thousand shapes and no defined limits." By narrowing the range of possible deception postulated by Montaigne, Larson could calibrate the device for each person. He also insisted that all the questions be identical for each suspect, and that all be posed in a monotone.

Finally, instead of testing his technique against Marston's role-playing games, Larson found an experimental setup more in tune with the world of practiced deceivers. In the past decade psychologists had largely given up trying to derive the universal qualities of mind by introspecting within their own roomy consciousness, and had instead begun to deduce human behavior by testing the outward responses of ordinary individuals. As the most convenient source of ordinary individuals, university undergraduates had become the preferred subjects of laboratory psychology; they also had the advantage of being relatively homogeneous, healthy, able to follow simple directions, and unlikely to complain. Larson found a way to preserve these advantages and still investigate a serious matter. Larson set out to investigate crime in Berkeley's sororities. In this real-life test the examiner could not know the identity of the guilty party in advance. He might also never know whether he had truly solved the mystery.

For just this reason, Larson planned his protocol with care. He readily secured permission from the housemother and the young women to run the test; after all, he noted, anyone who refused would have appeared guilty. He devised his list of yes-no questions: first a set of innocuous questions to define the student's "normal" bodily response, to be followed by a set of questions pertinent to the crime. Then he invited five young women—two victims and three suspects—to the physiology lab on the Berkeley campus for "a preliminary or sparring examination." Of these, four produced records of sufficient ambiguity to justify retesting: big-eyed Holt; worldly Graham; sickly Arden; and even Ethel McCutcheon, one of the young women who claimed to have been robbed. Larson attributed the poor discrimination in his results to the fact that he had peppered the subjects with questions too rapidly; in the future he would allow more time for tension to build. At last, on April 19, 1921, he turned to the main event: a full-scale test on the same young women, plus nine presumably innocent women from the house, who would serve as his controls.

Larson began with Margaret Taylor, the freshman who had lost the $400 diamond ring. Not that he doubted her word; she had been one of his and Fisher's "confidential informants" inside the dorm. But a policeman must always be skeptical, as many a complaint was faked and many a victim embroidered her tale of woe.

While the other young women waited in the antechamber, he invited Margaret Taylor into his lab and seated her alongside the elaborate machine. She was a blue-eyed, fair-haired specimen of the California southland, an eighteen-year-old native of San Diego with honest-to-goodness golden ringlets cascading to her shoulders. Larson wrapped one of her bare biceps in a cuff to calibrate her blood pressure, then strapped the automatic blood-pressure gauge to her other bare arm and pumped its cuff until it gripped her firmly. He wound a rubber hose with its leather brazier tight around her chest to measure the depth of her breathing, then told her to hold her body perfectly still, lest the least muscular movement be mistaken for a guilty reaction. Then he turned the instruments on. The drums began to revolve, the black recording paper turned, and the long rubber hoses swelled and subsided to the rhythm of her body's organs, while a pair of long sharp needles scratched out her body's message against the black recording paper, as if tracing a silhouette of her thoughts. After a short preamble, he began.

1. Do you like college?
2. Are you interested in this test?
3. How much is 30 x 40?
4. Are you frightened?
5. Will you graduate this year?
6. Do you dance?
7. Are you interested in math?
8. Did you steal the money?
9. The test shows you stole it. Did you spend it?
10. Do you know where the stolen money is?
11. Did you take the money while the rest were at dinner?
12. Did you take Miss Taylor's ring?
13. Do you know who took Miss B[enedict]'s money?
14. Do you know who took Miss S[chrader]'s hose?
15. Did you at any time lie to shield yourself or others?
16. Are you accustomed to talk in your sleep when worried?
17. During the past few nights do you remember having dreamed when you might have talked in your sleep?
18. Do you wish at this point to change any of your statements regarding the thefts?

Each test took no more than six minutes. Much longer than that, and the pressure cuff became "uncomfortable and painful." Larson worked his way systematically through the list until he came to Helen Graham, the full-figured student nurse with dark eyes and eyebrows.

No sooner had he brought up the subject of the diamond ring and stolen money—"The test shows you stole it. Did you spend it?"—than Graham's record showed a precipitous drop in blood pressure before beginning what looked to be an alarming rise, along with skipped heartbeats and an apparent halt in her breathing. Then, as Larson leaned forward to calibrate her blood pressure, the young woman exploded with rage. Ripping off the restraining cuffs, she leaped to her feet and ran over to the rotating drum to read the squiggly lines that traced her body's reactions. Larson's police report describes what happened next:

> We forcibly prevented her from going near the drums and upon going outside she told Miss Holt that if I had not had her tied down she would have smashed Officer Fisher in the face and told another girl that she felt like tearing up the record. Just before leaving the room she told all of us that the questions asked were perfectly atrocious and that she agreed with [the housemother] that such things should not be allowed.

Whereupon she rushed back to College Hall to accuse her roommate of betraying her, stormed out of the dorm in a rage, and "promptly spent the night with her lover."

None of the other students, according to Larson, posed any objections to the test. Nor, he reported in a scientific paper, did their tests show any anomalies—although the record of Alison Holt had not been entirely untroubled. Overnight, Larson learned more about Helen Graham from Miss Taylor and the other good women of College Hall. Apparently, she had doped herself before the preliminary test and slept heavily that afternoon; this may have accounted for her serenity the first time around. Moreover, she had admitted to friends that she had once stolen a notebook to cheat on a high school exam, had conducted more than one "affaire d'amour," and had once taken quinine to induce an abortion.

The next day, a distraught Helen Graham came to the police station, demanding to see her record. For twelve hours, Larson and Fisher bombarded her with questions until she "broke down and had an attack of sobbing." She continued to assert her innocence, but admitted she might have taken the items "in her sleep or in some possible mental disorder." She even offered to replace the ring and money if that would end the investigation. Larson, playing the good cop, told her she ought not to make restitution if she was innocent. Fisher, playing the bad cop, told her that if she was guilty, she would be prosecuted to the full extent of the law. Then she was sent home.

Every day that week she stopped at the police station to demand an appointment, but by prior agreement, the police refused to speak with her. Only when she threatened suicide did Larson meet with her again. Again she insisted on making restitution with the understanding that the case be closed. Again he refused to accept the money unless she admitted her guilt. A few days later, she returned with a substitute diamond ring; the original, she said, had been "lost." But as this new ring was of lesser value, Miss Taylor insisted on another. The next day, the police tailed Graham when she took the ferry across to San Francisco to meet her lover, Roger Harvey, with whom she went to Morgan's jewelry store to pick out a suitable replacement ring. That evening, she presented the ring to Larson, swearing that the stone was similar to the original, while admitting that the original setting had been melted down. The next day, she was followed again, but this time she "made" the tail and eluded him.

The denouement came on April 30, when Larson arranged an interrogation in the time-honored manner: "Officer Fisher played the role of 'hard-boiled cop' with his usual adroitness, and I was her friend." After several hours, Fisher stormed out of the room, telling her that when he returned he would show her she had been "booked for San Quentin." While he was gone Larson got Graham to admit taking the money and the ring, plus some hose off the line—though she denied stealing underwear. She then signed this confession in Fisher's presence, agreed to make restitution, and gave Larson and Fisher a version of her life history. After that she moved into a hotel, withdrew from the university, and prepared to return to Kansas, where she would wait for Harvey to come and marry her.

It was the first real-life crime solved by the "lie detector," though some time would pass before hard-boiled reporters, the sort of men who judge a thing by the end it serves, would give Larson's cardio-pneumo-psychogram that name—to his perpetual irritation. Of course, Larson's rigged assembly had not itself exposed the guilty party. Instead, the instrument had nabbed Helen Graham by indirection: heightening her sense that she had been marked out as guilty; confronting her with the jagged evidence of her guilt; and then tightening the emotional screws until, in a climactic scene, she broke down and confessed. Its success owed less to the modern science of experimental psychology than to archaic rituals of guilt and absolution.

For Larson, the College Hall case did more than launch the American lie detector; it turned his life inside out. For one thing, it flung him on a scientific quest that would consume his efforts until the day he died, an old man obsessed with the device he had unleashed on the world. "Beyond my expectation," he would write shortly before his death, "thru uncontrollable factors, this scientific investigation became for practical purposes a Frankenstein's monster, which I have spent over 40 years in combatting." In the interim, his machine-brought-to-life would commandeer America's police forces, its business establishment, the national security apparatus of the U.S. government, and the public's imagination.

The College Hall case also changed Larson's life more intimately. One year after strapping her to his instrument, Larson married Margaret Taylor, the freshman victim of the College Hall thief. Immediately after the ceremony, a raiding party of college cops handcuffed the newlyweds together, packed them into a paddy wagon, and abandoned them in the countryside as a prank. As one of Larson's assistants would later acknowledge, "It was an odd way to begin a romance." A thirty-year-old Ph.D. cop married a nineteen-year-old Californian coed whose diamond ring he had recovered. The first time he met his wife-to-be he strapped her down and probed her innermost thoughts and feelings. Years later, he still had the record of their first meeting in his files, the zigzag trace of her heart as he asked her, "Are you interested in this test?"

The meet-cute story certainly proved irresistible to the boys in the press room. "INVENTOR OF LIE DETECTOR TRAPS BRIDE," read the headline above their oval portraits on the front page of the *San Francisco Examiner.* According to the newspaper, Miss Taylor was so grateful for the return of her ring that she volunteered to play the role of "criminal" in further scientific tests of the detector. This, of course, involved asking personal questions. Then, one day, Dr. Larson was inspired to take their relationship to a more intimate level:

> Fixing the "criminal's" blue eyes with his own, the psychologist sternly asked: "Do you love me?"
>
> "N-no," murmured Miss Taylor.
>
> And the wings of the ... "lie detector" trembled, fluttered, waved a frantic "S.O.S."
>
> "You lie!" cried the scientist.
>
> And Miss Taylor didn't deny it.

Though Larson derided the newspaper dialogue as "pure hooey," he privately acknowledged that the story contained a germ of truth. He had been trying, he said, to eliminate all those factors, aside from criminal guilt, which might have influenced the young women's responses, when it dawned on him that some of them "might have been reacting to the questioner, not the questions." So the Ph.D. cop brought back the attractive young Miss Taylor to test this proposition on the machine; first by asking her to lie to him, then by asking her out.

The instrument's allure was irresistible that way. Given a chance to peer into the soul of a colleague, a friend, or a (potential) lover, who would not be tempted to pose a few personal questions?

Yet Larson soon had reason to doubt that he had actually solved the College Hall case. Was Helen Graham guilty, or had she merely *felt* guilty? After all, she had been subjected to a month of intense pressure and surveillance by the police, not to mention by her dorm sisters and housemother. They had turned Graham inside out, but what did anyone really know about her?

As Larson honed his technique that year on a dozen more sorority cases, he became increasingly convinced that even an innocent person could be tripped up. Physicians had long been aware that certain physical signs were altered by the medical examination itself. The act of taking patients' blood pressure, for instance, raised their blood pressure, and insurance examiners even factored in this test anxiety. By asking innocuous questions, Larson was able to define each subject's "test normal." But in the context of a police interrogation, a question like "Did you steal the ring?" was surely more stressful than "Do you like math?"—whether or not the subject was guilty. And there was the rub: guilty of what? As Larson quickly discovered, even people who had not committed the crime in question were troubled by "complexes" brought to the fore by interrogation. These clusters of emotions had to be cleared away before the subject could be cleared of the crime; and this in turn meant delving into their personal history, getting them to confess to unacknowledged "crimes," some real and some imaginary, with no sure way to distinguish between them. In the course of his sorority investigations, Larson unmasked midnight poker games, petty shoplifters, pregnancies, and attempted abortions, often without solving the original crime itself.

In another case of petty theft, when Larson put the supposed victim on his machine, she confessed to being pregnant and having gonorrhea, and threatened to commit suicide. A physician found no trace of gonorrhea or pregnancy, but he sent her to the Pacific Coast Rescue and Protective Society for psychiatric observation. In his effort to solve a petty crime, Larson had opened up a greater mystery. Larson, who had been thinking of attending law school, decided to study forensic psychiatry instead.

As for Helen Graham, as part of her police confession Larson extracted a version of her life history, with particular attention to her sexual past. "My first knowledge of sex matters came at the age of 7 years; we had a man working for us on the farm . . . [who] taught me all the things that a girl should know and used to play with my parts." She had sexual intercourse at fourteen. A long-term sexual relationship began at fifteen. "I taught him the things that was taught to me." She then pursued an affair with a medical student before coming to California and meeting Roger Harvey.

All this gave her an acute sense of shame. Among the litany of sins she confessed to Larson: she had once been caught stealing a notebook to cheat on a test in high school. "As the town was small, I always thought that everyone knew about it and that made me very unhappy. . . . I think I have never told you that I can hear voices in the air, and I firmly believe that the trees speak." Larson began to suspect that Graham's confession was the product of an overactive sense of guilt. No sooner had she returned home to Kansas than she wrote to say her confession had been obtained by trickery, and only out of fear that her affair would be exposed. The episode had precipitated a "complete nervous breakdown," she said, and she had even contemplated suicide. In his write-up of the case, Larson acknowledged that Graham presented "all the indications of a psychopath, in all probability of a manic-depressive type." Indeed, the evidence strongly suggests that Helen Graham was singled out mainly for her sexual transgressions, much like her contemporary Carrie Buck, whose sterilization, supposedly for "eugenical reasons," was upheld by the Supreme Court.

For his part, Larson wrote Graham a letter of consolation. He pointed out that she had much to live for. If she was guilty, she had been treated leniently. If she was innocent, why had she told so many fluctuating and contradictory stories? He told her not to lose her faith in men. "I am very sorry that you have been feeling blue and wish that I could do something to make you feel better." A year later, she wrote a more upbeat letter to August Vollmer, Berkeley's chief of police. Though Harvey had never turned up in Kansas, she had met a charming Irish architect who was working on her parents' home. As for John Larson, she wrote, "Dr. Larson is indeed a wonderful scientist and truly a Man. The department was indeed fortunate in securing his services." Yet she still insisted on her innocence. "This," she wrote "is the closing 'chapter' of my case."

But for the lie detector, it was the opening. The Berkeley police ballyhooed the machine's victory over deception. Chief Vollmer himself was the first to tell the story for the general public, in a soft-core version he published that year in the *Los Angeles Times.* Casting himself in the role of Sam Spade, the Chief wallowed in the hothouse sexuality of the all-female dorm. "Listening in on the heart beats of fifty charming, impulsive, romantic university coeds to discover which one was a thief and save an innocent pretty girl from unmerited disgrace was a job big enough, if not impossible, for the average police department." For what "average criminologist" could possibly stand up against the collective judgment of "forty-nine giggling, thoughtless, loving embodiments of budding flapper exuberance"—especially when they had already fingered the aloof Marjorie Small as the thief? According to the testimony of Georgia Long—a "magnetic, gorgeous creature" with "revealing eyes of purple velvet"—Miss Small had been seen entering another girl's room and removing the stolen book. But Dr. Larson's machine quickly discerned the "wild reactions" beneath Miss Long's cool front. "Her face was like a mask now, cold, composed. The dark lashes lifted to discover hard, steady eyes. The smile was gone. She leaned forward slightly and gritted her teeth as if determined to betray no more emotion." Yet against her will the machine read her body's message. Haunted by her guilt, she broke down in tears and confessed to having stolen the book. Yes, appearances can be deceiving—especially in a tight sweater and pearls—but the lie detector could not be seduced or bamboozled. In his two years with the Berkeley police, Larson would investigate some two dozen sorority cases, and very few fraternities.

In an America beset with gangland murders, industrial sabotage, bootlegging, and political corruption, these trivial sorority cases dramatized the lie detector's potential as an instrument of justice. Its proponents—and editors quick to see the hot angle to a story—took advantage of the age-old misogynistic assumption, applied first to Eve, that women are subtle, deceitful, and collusive, if only to confer the contrasting virtues on the machine's operator. Where cops were distracted by appearances, the lie detector probed beneath the skin. Where institutions were corrupt, the machine could not be bribed. Where men were slaves to their emotions (and to the organs those emotions aroused), the machine recorded the emotions of others, so that its operator could remain dispassionate. Like the antiheroes of Dashiell Hammett's new style of crime fiction, the lie detector operated in the name of disinterested justice.

This was a man's justice: skeptical, mistrustful, objective—with women cast as creatures of guile and temptation. It's the oldest dichotomy in the nature lore of the West—masculine science investigating feminine mystery. And it played to a lurid sort of voyeurism. As one polygraph examiner later admitted, "I sometimes feel like a window peeper."

Yet Graham was hardly the naive test subject Larson portrayed in his scientific write-up. Even on her first exam—the original encounter between a human subject and the modern deception machine—Graham had apparently taken medication in advance to mask her physiological reactions. Several decades would pass before psychologists—who ought to have known better—would realize they could not treat their objects of study as other scientists do, as brute phenomena of nature. When Einstein inscribed above his fireplace the motto, "Nature's God is subtle, but He is not malicious," he acknowledged as a corollary the possibility that people might be malicious, if also sometimes subtle. Yet even polygraph operators—who surely knew better—seemed not to consider the myriad ways their subjects came forearmed. Countermeasures to lie detection are as old as lie detection itself.

All through 1921, long after Graham had returned to Kansas, the petty thefts continued in the College Hall residence. In retrospect, Larson wondered whether some of the young women had conspired to distract him during the exam. He bemoaned the way his investigation had been hurried. He should have been allowed to test the chambermaids, he said, not to mention the housemother and her family. It would become a familiar regret; get hold of a lie detector, and who knew whom you could trust?

CHAPTER 2

# Policing the Polis

*In a certain sense, a large part of the criminalist's work is nothing more than a battle against lies. He has to discover the truth and must fight the opposite. . . . Utterly to vanquish the lie, particularly in our work, is of course, impossible, and to describe its nature exhaustively is to write the natural history of mankind.*

—HANS GROSS, *CRIMINAL PSYCHOLOGY,* 1898

BERKELEY TODAY STILL CONJURES UP IMAGES FROM THE 1960s, when protesters smeared the town's police officers as "pigs." Yet in the first half of the twentieth century, Berkeley was world-famous as the seedbed for a new kind of police officer: technologically sophisticated, respectful of the law, and closely allied to social work. This reputation was the legacy of Chief August Vollmer, often considered "the most significant individual in the annals of American law enforcement." By 1920 he was the nation's most famous cop, and Berkeley was his experimental proving ground. The lie detector was an integral part of Vollmer's program to regenerate the morality of both the police and Berkeley.

There was no limit to Larson's admiration for Vollmer. To say that Vollmer was the father he would have wanted is to reduce to a pop-psychological commonplace what Larson expressed in far loftier terms. He dedicated his first book to "the genius and altruism of Chief August Vollmer, humanitarian, scientist and criminologist." Vollmer, a man who never passed the sixth grade, replied with characteristic verve: "First of all I am not a humanitarian, I'm a 'cop.' Secondly, I am not a scientist, I am a good guesser. And thirdly, I am not a criminologist because a criminologist has recently been defined as one who studies crime and knows nothing about it."

August Vollmer knew something about crime. He was born in New Orleans in 1876 to German immigrants. His father died when he was eight, and his mother settled in Berkeley in 1891, when Vollmer was fifteen. He grew to be a tall young man with swimmer's shoulders and a long face with pale lips and clear gray eyes. He could appear stern and unyielding, but he was a keen observer of human foibles and comfortable with people from all walks of life. Everyone wanted to please him.

When he was twenty-one Vollmer sold his feed store and volunteered for the Spanish-American War, during which he served in the military police and ran river patrols against Philippine guerrilla groups, adapting his tactics and cutting deals with locals. By his own account it was the formative experience of his life. On his return to Berkeley he worked as a letter carrier until 1905, when Friend W. Robinson, publisher of the *Berkeley Gazette* (and future governor of California), recruited him to run for town marshal. According to the paper, not only did Vollmer possess the requisite "mental acuity and sagacity"; he had the "physical strength to cope with any criminal."

Vollmer was swept into office in a three-to-one landslide, along with a slate of other Republican "good government" candidates. Twenty-nine years old, without formal experience in law enforcement, he relied on his own rigorous integrity, judgment of character, and military-minded ability to match means to ends. Immediately he persuaded the city board of trustees to replace the two part-time deputies with six full-time policemen. It was the start of something extraordinary.

Berkeley has always been two towns joined at the hip. First came the commercial district of warehouses, working-class housing, and rough saloons clustered along the Bayshore. Soon after, the trustees of a small college in Oakland created a campus at the base of the Contra Costa hills, a public university to outshine the universities back east, partially funded by selling panoramic lots to middle-class householders. The prospect through the Golden Gate led one trustee to recall the line of Bishop Berkeley: "Westward the course of empire takes its way." Thus did Berkeley, famous for the proposition that objects exist only in the mind of God, give his name to what its founders hoped would be the "Athens of the Pacific." By the early twentieth century, the town of Berkeley was approaching its Athenian promise, without fully disengaging from its hard-luck neighbors.

Vollmer's first move as marshal was to clean out the Chinese gambling dens. His logic was strategic; gamblers' payoffs threatened to corrupt town politics. In his second year in office, the San Francisco earthquake struck. Vollmer organized an auxiliary force of 1,000 men (nearly every adult male in Berkeley) to maintain order among tens of thousands of refugees. He was reelected by an even larger majority. Then in 1909 he was appointed police chief, a post he held for the next twenty-three years, with only brief leaves of absence to transplant his methods to other cities. During those decades he introduced the various features of the "Berkeley system," the core tenets of American professional policing.

Though he was a workaday police chief in a university town rather than a social theoretician, Vollmer had a well-thought-out view of the police as guardians of democracy. He sought a middle ground between the narrow Anglo-Saxon view of the police as crime-fighters and the European continental view of the police as regulators of social life. Though he would have preferred his men to focus on property crime and personal violence, Vollmer came to recognize that it was more efficient to prevent crime, even if this meant inserting the police into the community's messy life. The challenge was that most working-class Americans feared the police for their brutality, while the well-to-do considered them fools or knaves.

Vollmer's lifelong goal was to dispel this blend of fear and contempt by raising the social, intellectual, and moral stature of police officers until they got the respect they deserved. This explains Vollmer's embrace of scientific police work, his vaunted program of professionalization, and his rejection of police violence, corruption, and favoritism. It also meant doing something about America's scandalous crime rate.

In the early twentieth century the rate of violent crime was four to ten times greater in America than Europe. Vollmer insisted that the police could not be blamed for this, yet he was determined to do better. The main obstacle was "politics," by which he meant the spoils system of the municipal machine: the way the police were hired, fired, or promoted on the basis of pull and patronage rather than competence. This explained why the police enforced the law violently and selectively; it gave them leverage to extract bribes and favors. Thomas Byrnes, New York's notorious cop, is said to have coined the term "third degree"—perhaps a pun on his name—for his violent interrogations; and his colleague Captain Alexander Williams once boasted that there was more law in the end of policeman's nightstick than in all the decisions of the Supreme Court.

Vollmer agreed that police officers had to exercise discretion; he just wanted them to enforce the law fairly and efficiently. Building on the reforms of Progressives like New York's police commissioner Theodore Roosevelt, Vollmer urged a managerial revolution in police work, one analogous to the revolution in "scientific management" then transforming American business. This included the centralization of command and communication, specialization of tasks, and the deployment of scientific know-how. In 1906 Vollmer was the first to outfit his entire force with bicycles; and in later years he was the first to outfit it with squad cars, although, oddly, Vollmer himself never learned to drive. In 1907, he installed the nation's first citywide network of signals, enabling a dispatcher to contact a police officer anywhere in town; and in later years he installed its first two-way radio system. He also created his own filing system to organize crimes by fingerprints and handwriting, cross-referenced with his own classification system based on the criminal's modus operandi, and from these he compiled statistics to map the intensity of crime and assess which police methods best reduced it. His identification system became the model for California, and then for the nation as a whole under the aegis of J. Edgar Hoover, who applied many of Vollmer's methods.

These technical innovations, which made Vollmer the darling of the press, ought not to obscure his drive to create a new kind of police officer. Unsatisfied with the level of police training, he established an in-house police school in 1908, with members of the university faculty teaching courses on evidence law, forensic methods, crime-scene photography, and medicine. In 1916, he hired a chemist from the pharmacy school as a full-time criminologist. As for his own men, he announced that he would hire and promote officers solely on the basis of merit. In 1919 he began to recruit college graduates, vetting applicants with the intelligence tests developed during World War I, along with a battery of psychiatric exams.

Only this new kind of officer, Vollmer believed, would be able to balance proactive policing against the danger of becoming the enforcer of some unobtainable ideal. Vollmer himself always took a paternal interest in doings throughout the city. He encouraged citizens to turn to him or his officers for help with domestic disputes and unruly teenagers. One disciple recalled Vollmer telling recruits, in somewhat contradictory fashion: "You're not to judge people; you're just to report what they're doing wrong. Better still, you can prevent people from doing wrong. That's the mission of a policeman."

Around 1920, in a provocative address to his fellow police chiefs entitled "The Policeman as Social Worker," he turned his attention to the problem of juvenile delinquency. Vollmer had hired the region's first female police officer, a woman with training in psychology, to work with local schools to identify potential delinquents. On a map in his office, each problem child in town was represented by a color-coded pin, so that the Chief could track these children's development. Fourteen years later 90 percent of the "problem" children had been placed in institutions of one sort or another. In the noisy debate then raging over the root causes of crime—was it the fault of the individual, or due to biological and social forces?—Vollmer finessed the issue by considering crime a disease and suggesting that the solution was a comprehensive effort to improve the nation's moral hygiene.

It helped that Vollmer was himself neither a moralist nor a hypocrite. Like many policemen, Vollmer delighted in gossip. He made no secret of being a ladies' man. His first wife was an opera singer, who apparently grew tired of his philandering ways. He thought Prohibition a disastrous diversion of police resources and handled the town's liquor laws with tolerance. He even frequented speakeasies. He played the guitar and enjoyed practical jokes, especially if they exposed pseudoscientific pretension. Once Vollmer invited a phrenologist to demonstrate his ability to read people's characters from the shape of their skull, then planted one of his own ace detectives, a chemistry graduate, in a jail cell for the expert to assess. The phrenologist pronounced the man a confirmed criminal.

It helped too that Vollmer achieved results. His statistical surveys not only helped determine which methods best helped fight crime, but also convinced the stingy city council of his success. On his watch burglary rates declined by 50 percent even as Berkeley's population increased, the city employing half as many cops as were typical in other towns its size. In later years he led the town's efforts to collect every citizen's fingerprints—voluntarily—as the way to ensure a well-policed community.

As he approached middle age Vollmer maintained his trim bearing, high forehead, hawklike nose, and creased laugh lines. Visitors invariably commented on his clear gray eyes. They were less windows into his soul than instruments of discovery. William Dean, one of the first "college cops," and later a winner of the Congressional Medal of Honor, compared Vollmer to generals George Marshall and Douglas MacArthur in that respect. "All three had the capacity to look through you and you'd think they knew exactly what you were thinking. You didn't feel uncomfortable but you felt you'd better not try to tell anything but the whole truth when you spoke to them." Some people speculated that Vollmer cultivated this piercing gaze because he had some secret of his own to hide. If so, no one ever learned what it was.

Central to Vollmer's strategy for ensuring respect for the law was making his police themselves law-abiding. For Vollmer, this meant treating citizens with the presumption of innocence and disavowing those coercive police interrogations known as the "third degree." His officers left political radicals alone. Vollmer personally denounced the death penalty as ineffective. And he hired Walter Gordon, the first black football player at the University of California, who in later years, as governor of the Virgin Islands, recalled Vollmer's contempt for racism and insistence that his officers use persuasion instead of physical force. Another disciple recalled Vollmer's telling his officers never to "strike any person, particularly a prisoner, except in extreme self-defense; and then he said, if you ever do, you have just resigned." To some extent, this "by the books" image was a myth cultivated by his disciples. In practice, Vollmer's paternalist, preemptive approach to police work meant that he sometimes took a more direct hand in resolving conflicts. In his first year on the job, the Chief "mildly" whipped three youths for stealing $75 from a quarryman. And though he forbade violence, he never renounced the use of psychological pressure to extract information. Today, we associate this approach to policing with the rulings of the Warren Supreme Court of the 1960s. This is no accident. Earl Warren began his career in the 1920s as a district attorney in Alameda County (which encompasses Berkeley), and his approach owes much to the lessons he learned while working alongside August Vollmer.

No wonder Vollmer was enthusiastic about the College Hall case. Larson's technique seemed to extract information from suspects without tempting his cops to resort to brutality. It also promised new insights into the criminal mind and—who knew?—perhaps even its cure.

Before he OK'd the instrument for general use in Berkeley, Vollmer asked Larson to prepare a personal demonstration. When he arrived at Larson's lab, however, he immediately turned the tables, making Larson strap *himself* into the machine. At that instant, Larson recalled, he anticipated the exact topic of Vollmer's interrogation. Two weeks earlier, the rookie had been off his beat, sharing a malted milk, when a hardware store had been robbed—and the Chief, who knew everything that happened in Berkeley, had undoubtedly been informed. So by the time the anticipated question came—"Were you off your beat the night of the Sunset Hardware Store burglary?"—Larson didn't even have to answer; his body's response was so dramatic that "the needles went off the drum." At which point, Jack Fisher, the old-time cop, slammed his star on the table and offered to resign rather than go on "that Goddam thing." "I don't need the machine for you," laughed the Chief. That December the Berkeley city council approved the construction of a new device. For the next two years Vollmer gave Larson carte blanche to try his technique on hundreds of cases in Berkeley. It was the chance Larson had been waiting for.

When Dr. Larson, a freshly minted Ph.D., joined the Berkeley police force in 1920, it was actually a logical career move. The son of an Indiana quarryman of Nordic stock, John Augustus Larson was born in Shelbourne, Nova Scotia, on December 11, 1893. At a young age he moved with his family to industrial New England, where his parents separated. At Boston University he studied biology as well as classical and modern languages, earning his way through college by working at a variety of odd jobs: busboy, paperboy, stock boy, elevator operator, stonecutter at the quarry, and fireman on the graveyard shift, plus two summers on a hospital ship. For extra cash he caught stray cats on Tremont Street, three or four to a suitcase, which he toted down Boylston Street to be boiled in preparation for zoology demonstrations.

During the school year he tutored wealthy coeds, conscious of his poverty among the college swells. He was just as uneasy among the laboring classes. Working as a summer trolley conductor, he was so assiduous in his ticket-taking—conductors customarily pocketed one fare in three—that his comrades mocked him as an honest fool. His mother told him to quit, "fearing," he recalled, the "antisocial conditioning." Hardworking but impulsive, physically vigorous but clumsy, compulsively honest but self-conscious, Larson set great store by doing the right thing. He wanted to become a criminologist.

Was it fate or upbringing? The family heirloom, signed on the flyleaf by generations of Nordic Larsons, was an early Reformation vilification of perjury: how it was a moral sin to break an oath.

In 1915, at his professor's suggestion, Larson wrote his master's thesis for Boston University on fingerprint identification, a technique that was finally being recognized by the courts. Larson created his own classification system, but his greater ambition was to make fingerprinting a predictive science, on the premise that both prints and criminality were heritable. He failed to find any familial pattern amid the swirls, loops, and whorls. His taste for forensic science then took him to Berkeley, where in 1920 he completed his Ph.D. in physiology, examining thyroid deficiencies, then widely believed to be a leading cause of emotional and criminal deviance.

If the old-timers on the force didn't appreciate the new college cops, they found the doctoral cop unbearable. Larson was running himself ragged: writing a book on his fingerprint system, continuing his lab experiments, auditing courses in criminal psychiatry—all while working the four-to-twelve beat. Vollmer later conceded that the force's hazing had been particularly cruel. But Larson was not cowed. He may have looked all wet, but he had a fierce sense of honor. When Officer Henry Villa, the department's self-styled ace detective, ladies' man, and prizefighter, needled Larson one time too many, Larson put him in a chancery lock so hard he nearly broke the cartilage in Villa's Roman nose.

The suspects Larson tested in Berkeley between 1921 and 1923 exhibited the full range of American lies: burglars, forgers, bootleggers, arsonists, murderers, blackmailers, gamblers, men, women, students, teachers, juveniles, vagabonds, musicians, housewives, whites, blacks, Chinese, Mexicans. Berkeley aspired to a utopian ideal, but it was still an American town of 60,000. Some subjects were mentally disturbed, others feebleminded, still others drug addicts. The sex offenders were kinked every which way: homosexuals, masturbators, exhibitionists, a medical student accused of disseminating a picture of an erect penis, and two students caught in flagrante delicto. The machine served as a police tool, a screening device, a marriage counselor, a priest. The lie detector became whatever the circumstances called for.

In the hands of the Berkeley police, the situation was most often criminal, and the crime a bike theft (a category in which Berkeley still leads the nation). As with Helen Graham, the lie detector often exposed petty crimes by eliciting confessions: a restaurant chef pleaded guilty to stealing silverware, a paint-store employee to robbing the till, a custodian of the Unitarian Church to pocketing a purse and watch. Alcohol infractions were especially common, with Vollmer generally filing away a written confession and issuing an oral warning.

Some subjects, the machine revealed, were victims of crimes that never happened. When Larson questioned a young soldier who claimed to have been bound, gagged, and robbed, the young man hurled invectives: "You go to—! This is an outrage on an American soldier!" Larson, shamefaced, lent the soldier his overcoat and money to travel across the bay. Months later, he had to petition the soldier's commanding office to get the coat returned. In the interim the soldier had admitted spending the money on a woman friend. For an expert on lying, Larson was a trusting soul.

Among the sex cases, some were criminal, and others offered a glimpse into the intricacies of domestic life. Concerning one marital dispute, Larson wrote: "Mrs. Simons accused of masturbation by husband. Had *puritis ani* [itchy anus]." The organist at the local movie theater who fondled boys in the dark was exiled to Los Angeles to begin a new life; a month later he wrote to "Friend Larson" to kindly inquire, "How is the test?"

When he ran low on cases, Larson asked his fellow college cops to round up hoboes in the train yards. With such vagrants Larson went on "fishing expeditions," asking them if they were being sought by the army, the navy, the police, etc., then showing them a map of the United States and asking if they hailed from this state or that. Not only did he catch military deserters this way; the machine acted as a deterrent. Larson boasted that vagrants soon learned to give Berkeley a wide berth. This confirmed one persistent gripe about Vollmer's methods: that he had not so much reduced crime as driven it into neighboring towns.

In these early years the lie detector had yet to acquire its aura of infallibility, and as word spread of the machine's prowess, some locals considered it a challenge. One medical student had heard his professor scoff at the notion of a lie detector. When he was accused of stealing a bicycle, he tried to beat the machine by controlling his breath and tensing his fingers. Writing up the case for publication, Larson claimed that the young man's record showed "disturbances due to a guilt complex." But the original police file indicates that the student was released after obtaining a "very smooth" record. Only after the student consulted with his father and a lawyer did he confess to having stolen the bike.

Then there is the case of the two bunco men caught dealing a trick deck of cards on an Oakland-bound train. At the Berkeley station house their fingerprints identified them as confidence men with long records. One agreed to sit for a lie test. Despite his stony poker face, Larson read his record as an invisible "tell." But in their cell, the hustlers plotted their revenge. On the wall above their bunk, they drew a subversive cartoon of themselves standing before the solemn bench of justice; and underneath, they scrawled this doggerel on scientific interrogation:

> "Have you ever been in jail before?"
> "Did you give us your right name?"
> "Is it true you got a sucker's coin
> In a 'smoker' poker game?"
> "And was the game on the up and up?"
> "Or did you use marked cards?"
> "And did you think you could get away
> By running through back yards?"
> "Please answer each question
> By saying 'yes' or 'no';
> Don't wiggle there while in the chair,
> And don't answer quite so slow."
> These are some of the things they ask—
> If they think that you are green—
> In Berkeley's super city jail
> In front of the lying machine.

In two years Larson tested 861 subjects in 313 cases, corroborating 80 percent of his findings by post-exam confessions or subsequent (unspecified) checks. In total, 218 criminal suspects were identified and 310 exonerated. It was an impressive achievement, and from it Larson deduced several principles. He discovered that the citizens of Berkeley were overwhelmed by a sense of guilt, at least when interrogated by the police. He noted how easily he obtained confessions. And he found that when he retested suspects after confession, their records appeared similar to those deemed innocent of the crime. He also found that he could train his fellow cops to conduct these tests.

In those same two years Vollmer's initial skepticism turned to unbridled enthusiasm. The lie detector eased the administration of justice and supplied a physiognomic portrait of the town to match his map of colored pins. And Vollmer was determined to let the world know that Berkeley had defeated the age-old problem of criminal deception. In a ten-part silent movie serial, *Officer 444*, Vollmer played himself—"one of the world's leading criminologists"—calmly marshaling his scientific police force against a criminal scientist who exploited science for evil purposes. Filmed in Berkeley, the popular serial was meant by Vollmer as an antidote to the hapless Keystone Kops, whose antics he despised. By contrast, Officer 444 was brave and efficient and even got the girl after solving the crime with the help of the "'lieing machine'—a modern marvel of criminology, which records a crook's guilt even while he is denying it." And the machine was proving successful in the real world too. When a murder suspect who had been cleared by Larson's test had his innocence confirmed by an unimpeachable alibi, Vollmer told the press that this was the "most convincing case" yet. "So far we have never made a mistake with our machine. I will not say that it is infallible," he informed the *San Francisco Examiner.* "But thus far, it has proved so."

CHAPTER 3

# A Window on the Soul

*Then the Officer began to spell out the inscription and then read out once again the joined up letters. "'Be just!' it states," he said.*

—FRANZ KAFKA, *IN THE PENAL COLONY,* 1919

THE NEWSPAPERS BAPTIZED THE LIE DETECTOR; THEY named the device, launched its career, gave it its purpose. The machine made great copy, great pictures, great drama. During the 1920s nothing moved print better than tales of true crime, and here was a new angle on noir stories of depravity: an instrument that let readers peer directly into the criminal soul. The machine's judgment could be delivered to the morning doorstep, a front-page deus ex machina that resolved the mystery of whodunit—at least until the next morning's paper. This was voyeurism by retail, and San Francisco's papers, locked in furious competition, begged Vollmer to try his Berkeley instrument on the city's bold-headline crimes. With the International Association of Chiefs of Police due in town for their annual convention, its president-elect, Vollmer, decided to showcase his new methods of crime-fighting and sent John Larson across the bay.

The summer after Larson's triumph at College Hall, a priest had been abducted from the San Francisco diocese by a mysterious stranger in a car. In response to calls from the newspapers, thousands of grief-stricken citizens combed the city streets in search of his body. Yet not until a reward was offered did a mysterious drifter named William Hightower lead a reporter from the *San Francisco Examiner* to a foggy beach where they found the priest's body at midnight, buried under a billboard ad for flapjacks. The

*Examiner* was the first paper in William Randolph Hearst's empire, and with it he had inaugurated a new style of page-one investigation, showing how dastardly criminals would have eluded justice but for "The Invincible Determination of the *Examiner*."

In this instance, the *Examiner*'s intrepid reporter persuaded the police to hold Hightower in secret all night so he could file his morning exclusive. Over the next few days, as bigger and bigger headlines splashed across the city's front pages, reporters and the police competed to gather circumstantial evidence against the suspect. Reporters found experts who matched the ransom note to his typewriter; police discovered a custom-built machine gun in his room, as well as a .45-caliber revolver whose bullets might have killed the priest; reporters unearthed half a canvas tent at the grave site, the other half of which was in Hightower's room. The prosecutor thought he had Hightower dead to rights. But despite the prosecutor's threats and a howling lynch mob outside his cell, Hightower admitted nothing. As the reporter described it, "William A. Hightower is unshaken in his surface assumption of innocence. But in his mind is an area like a large bruise. Whenever a question touches upon the edge of that bruise he winces, pauses and maintains silence. Possibility of confession is remote. . . . His high, curving forehead is bland and unwrinkled. His hands do not tremble, nor do his lips quiver."

Against this effrontery, what could justice do? Thousands of avid readers wanted to know. Then a rival paper, the *San Francisco Call and Post*, scored a coup of its own. Its editors secretly arranged for Dr. John A. Larson to put Hightower on his scientific "soul test"—with exclusive rights for the *Call and Post* to publish the findings. The previous night, the prisoner had been unable to sleep, telling his jailer, "My dreams are raising hell with me." Summoned from his cell just after midnight and told that a scientist wished to take his blood pressure, Hightower was strapped to the apparatus and warned, "If you lie to us on a single question we will detect it." The next day, under the headline "SCIENCE INDICATES HIGHTOWER'S GUILT," the *Call and Post* filed its exclusive. "Science penetrated the inscrutable face of William A. Hightower today, revealed that beneath an unruffled exterior is a seething torrent of heart throbbing emotions, and that these emotions indicate strongly that he was the murderer of Father Heslin of Colma."

> Nothing could have been more dramatic, more dispassionately heartless than the manner in which science dissected Hightower, felt his heart beats, his pulse, examined his breathing, looked beneath the flesh for indications. And nothing could have been fairer.

And there, unfurled across the page, were the telltale traces of Hightower's jagged heartbeat and herky-jerky breath, with ominous black arrows to point out the "explosive" reactions confirming what the public wanted to believe: that behind even the stoniest criminal mask there lurked a conscience, aware of its sins, and that science could track the soul to its lair.

Inside the paper, Larson described his improvements to Marston's deception test, already "100% accurate," and Chief Vollmer announced that the graphical record left "no question" that Hightower was guilty. Not only did the Chief think the results reliable enough to be admitted into court; he called Larson's instrument a forerunner of still better methods, "a mechanical instrument of the future [which] will prove, beyond a question of doubt, the guilt or innocence of the accused." The press rested its case and dubbed the instrument "the Lie Detector." A few weeks later a jury (needing no lie detector) sentenced Hightower to life imprisonment in San Quentin.

Less than a year later, as the city geared up to welcome the International Association of Police Chiefs, the association's president August Vollmer gave Larson a chance to demonstrate his technique before 300 police chiefs. "No longer can we hope to compete with criminals," Vollmer announced in his presidential address, "unless we discard antiquated and obsolete equipment and strengthen our force with the recognized and desirable tools of our profession." That evening, on the roof of the Saint Francis Hotel, Larson demonstrated his device, with San Francisco's police chief, Daniel O'Brien, in the hot seat, telling whoppers.

As luck would have it, a more grisly test of the lie detector was simultaneously under way. One week earlier, the San Francisco police had invited Larson to investigate a crime that had horrified readers across the nation: a husband accused of complicity in his wife's murder. On Tuesday evening,

May 30, 1922, Henry and Anna Wilkens had been returning from an outing in the Santa Cruz mountains with Henry, Jr., age eight, and Helen, age three, when another car forced them to the curb, and a gunman jumped out, stuck a revolver through the driver's-side window, and demanded cash. Wilkens docilely handed over three $100 bills. But when the bandit reached across him to grab his wife's diamond engagement and wedding rings, the husband reacted with pardonable fury. "Haven't you enough?" he said, and reached for his own gun in the vehicle's side pocket. Unfortunately, the bandit shot first and struck dear Anna in the chest, before speeding off. Outraged headlines played to the nation's growing obsession with car bandits. At the inquest, Henry, Jr., touched the raw nerve of the tragedy "My daddy loved my mother—she died to save the bandit's bullet from hitting him."

But the police, always suspicious, soon directed their attention toward the grieving husband. Two days after the murder, two hardened ex-cons, Walter and Arthur Castor, sons of a San Francisco police officer killed in the line of duty, had been questioned by the police after trying to buy gas with a $100 bill. Wilkens failed to recognize either man in the police lineup. But no sooner had the Castors been released than the police learned that Wilkens had actually employed Walter Castor in his auto shop four years earlier. By then, the Castor brothers had vanished. So the police asked Wilkens if he would be willing to submit to a lie test.

In the city's Hall of Justice, Larson set up his modified assemblage. The device (now stored in the Smithsonian) was even more Rube Goldbergian than before. The reporter for the *Examiner* described it as "a combination of a radio set, a stethoscope, a dentist's drill, a gas stove, an aeronoid barometer, a time ball, a wind gauge, and an Ingersoll watch." In short, it was the sort of "mystic apparatus" one might find in a "thought laboratory."

In fact, on its six-foot-long plank, the mechanical-electrical device resembled nothing so much as a mechanized human body flayed open for inspection, like a cubist anatomy lesson. Each instrument was the end product of decades of physiological research. Each transposed the activity of an internal organ onto an external mechanism, which mimicked its action in a way that was measurable. Each was animated by the living force that drove the subject's bodily functions. The pulsating rubber hose extended the swelling arteries to record the blood pressure. The rubber-drum tambour rose and fell with the lungs to measure the depth of breathing. The clock-wound motor drove the mechanism like the muscles of the constrained subject. The electronic circuit, poised like the nervous system, reacted the instant a puff of air signaled speech. Finally, and most visibly, looping between the two upright drums was the blackened sheet of paper—broad as a human torso—where needles scratched out the instruments' response, as if scoring their message on the subject's skin.

One organ, of course, was conspicuously absent. The instrument lacked any correlate for the brain. You are nothing but a body, the apparatus implied, a mechanical cadaver laid out on the table—with this one difference: the cadaver on the table was not quite dead. So long as the machine was hooked up to the living, breathing subject, it became a kind of Frankenstein's monster, an artificial human given life by the subject to whom it was intimately attached. No wonder Wilkens was reluctant to undergo the ordeal. But the police gave him the veiled threat they always give in such cases; they told him that if his "hands were clean," he had nothing to fear. He agreed—with a sullen expression and gloomy black eyes.

So while the intense, high-minded Dr. Larson hovered over his creation, Wilkens's sleeves were rolled up to reveal a surprising musculature and a navy tattoo. The rubber "bands" were wound tight around his biceps and pumped full of air, and the rubber-and-leather "brazier" was strapped tight around his chest. After a short preamble Larson posed his questions at one-minute intervals, beginning with innocuous "controls": Did Wilkens like the movies? (Yes.) Had he had a nervous breakdown? (Yes.) Had he suffered from heart trouble? (No.) Did he walk in his sleep? (No.) Did he fear going insane? (No.) Then, once Wilkens's nerves had settled, Larson began to probe the possibility that Wilkens had plotted to murder his own wife.

Several hours later, the interrogation done, Larson gave his exclusive verdict to the *Examiner.* Although Wilkens had been extremely nervous during the interrogation, Larson's "hasty examination" suggested that he was not implicated in the crime. Around the bay, the newspapers announced that Wilkens had given the lie detector a "run for its money and emerged with flying colors."

This was not quite correct. As Larson privately acknowledged, Wilkens's record was "doubtful." When Vollmer reviewed the records that evening, he doubted Larson's interpretation—though publicly he seconded his officer's judgment.

Nor were the San Francisco police entirely convinced, and with good reason. As Wilkens left the Hall of Justice, supposedly cleared of any wrongdoing, a detective tailed him to a secret meeting with Robert Castor, brother of the men he denied knowing. The detective observed them exchange money and overheard Wilkens boast about his experience on the lie machine. A week later, on the same day that August Vollmer urged his fellow police chiefs to adopt national identification files, criminal psychology, and routine use of the lie detector, Arthur Castor was arrested in Eureka and escorted in chains back to San Francisco. The next day, John Larson offered the chiefs a summary of millennial progress in the detection of human deception, from the medieval ordeal and torture to the "third degree" and the scientific innovations currently underway in Berkeley. The police chiefs left town with the distinct impression that science would help them win the war against crime. But the cops on the Wilkens case weren't so sure.

A month after the chiefs left town, the jailed Arthur Castor publicly confessed that Henry Wilkens had paid him to murder Anna. In short order, Wilkens was arrested and interrogated by the police in the traditional manner—to the point that he had to be hospitalized for "appendicitis." Larson was appalled by the cruelty, yet the widower still would not admit his guilt.

Then the story took a still more grisly turn. On August 2—a week before Dr. Larson married Miss Taylor—Walter Castor, still hiding out in San Francisco, was betrayed by his lover (his brother Robert's wife), and in a dramatic shootout, killed her and a policeman before succumbing to a hail of bullets. The same day the *Examiner* announced the wedding of Dr. Larson and Miss Taylor, an article two columns over revealed that Wilkens had been the lover of his wife's sister and that prosecutors were investigating his other infidelities as well as his history of wife-beating. With headlines surging to new heights, the police became more determined than ever to make this wife-killer and cop-killer pay.

During a contentious monthlong trial, the prosecutors exposed Wilkens as an admitted liar on matters ranging from the fate of his wife's rings to his affair with her sister, while the defense challenged the honesty of Castor's confession, for which Castor had been given immunity. There was a brief moment when Larson believed that he and his lie detector evidence might be summoned before the court on Wilkens's behalf, but the defense never did call him, presumably because Wilkens had already admitted lies the machine had failed to catch. The jury deadlocked, six to six. In a second trial, Wilkens was acquitted, much to the consternation of the judge and the public.

Was Wilkens guilty? Could the lie detector be beaten? The old-style cops were furious that science had bungled the case. Larson heard rumors that the San Francisco police had vowed never to use the instrument again. The city's captain of detectives stood on the podium at the next meeting of the International Association of Chiefs of Police and declared that lie detectors "would not be countenanced."

For his part, Larson blamed the press. He had been duped by the papers, he decided, and not for the first time. Already he had encountered the sort of unscrupulous editor who begged to sit in on an interrogation in return for a promise to keep the story under wraps—only to run the scoop in the morning. The newspapers fed all manner of misconceptions, stirring up wild fantasies of instantaneous justice. The Sunday supplements had started running stories about "How the Electric Detective Catches Criminals," invariably illustrated with "a pretty girl attached to an imposing array of apparatus." One crank had begun writing to Larson obsessively, contending that he too could read people's minds. Larson made a vow: he would refuse to talk to the press about ongoing investigations, except when cases were already sensationalized. That way at least he would do no harm.

But there was still one question Larson couldn't stop turning over in his mind: Was Wilkens guilty or not? It's the question everyone always wants to answer: Whodunit? Larson also had his scientific results to clear up, not to mention his reputation. In his frustration, he tried an unusual approach; he befriended Wilkens. He congratulated him on his acquittal, helped him find a job in Oakland under an assumed name, and drove him around town while he went house-hunting. Then Larson persuaded Wilkens to sit for another round on the lie detector. After all, Wilkens could not be tried twice for the same crime. But the second results mirrored the first in their ambiguity, and again Wilkens insisted on his innocence.

Unsatisfied, Larson tried yet another tack. The lie detector was not the only new solution to the hoary problem of extracting the truth from recalcitrant witnesses. Since the publicity given to Larson's exploits, Dr. R. E. House of Ferris, Texas, had announced his own method: "truth serum." This was the drug scopolamine, used by obstetricians to induce a "twilight sleep" during childbirth and ease the memory of pain. House now wished to promote it as a memory aide, coaxing drugged subjects to recall events honestly. One year after Wilkens's acquittal, when House came to San Francisco to present his work to the American Medical Association and test his drug on prisoners at San Quentin, Larson asked him to try his serum on Wilkens.

Wilkens was game. "I am going to take every step to establish my honesty in the community." For several hours on the evening of June 25, 1923, a semi-somnolent Wilkens reiterated his previous claims of innocence while Larson, Vollmer, House, and several psychiatrists posed questions. But Larson was disappointed. Wilkens, he noted, had managed to tell lies under the influence of the drug, falsely denying, for instance, that he had had sex with his housekeeper. Worse, Larson decided, truth serum was unethical because suspects' words were not under their conscious control, so they could not be held responsible even for words that were true. At a minimum, this form of unwilled self-incrimination could never be accepted in a courtroom.

By then, Larson was preparing to resume his training in criminal psychiatry. At the end of 1923, with Vollmer's blessing, he was to leave Berkeley for Chicago to attend Rush Medical School and work part-time at the Institute for Juvenile Research, the nation's leading center for behavioral approaches to crime prevention. There Larson would seek to transform his technique into a method for curing crime as well as detecting it.

But in the meantime, the Wilkens case had raised questions: Was it appropriate to introduce lie detector evidence in court? Should it be permitted if the suspect had agreed to take the test voluntarily? What sort of truth did the lie detector coax from the body? Larson addressed these questions when the American Bar Association met in San Francisco in 1922. He entertained the hope that its graphical results might one day be shown to juries, as they "are so striking that they could be easily recognized." However, these presentations would have to be made by qualified experts and then only after "careful standardization . . . , much cooperation and experimental work."

But there were others, more impatient. The publicity surrounding the Wilkens case had attracted the notice of William Moulton Marston, the Harvard-trained psychologist whose work had inspired Larson's work in the first place. Marston still hoped to validate his own brand of lie detection, though he understood (as Larson did not) that the lie detector was not antithetical to publicity, but fed on it, even magnified it. Marston was in Washington, D.C., having used his method to exonerate an accused murderer, and he hoped to take the case all the way to the Supreme Court. He wrote to Larson in the hope that they might pool their efforts.

CHAPTER 4

# Monsterwork and Son

POLONIUS: *See you now,*
*Your bait of falsehood takes this carp of truth,*
*And thus do we of wisdom and of reach,*
*With windlasses and with assays of bias,*
*By indirections find directions out.*

—WILLIAM SHAKESPEARE, *HAMLET*

IN JULY 1922—ONE MONTH AFTER LARSON FIRST TESTED Wilkens—a young African-American named James Alphonso Frye was visited in his jail cell in Washington, D.C. by William Moulton Marston, the Harvard-trained psychologist and lawyer who had inspired Larson's research and who still had ambitions for his own lie detector. Frye, a twenty-five-year-old veteran of the Great War, had been arrested on August 27, 1921, for a car heist in collusion with four other black men. A few days later the police had induced him to confess to the unsolved murder of a prominent black physician, which had taken place almost a year earlier. Then a few days after that—to the consternation of his attorney—Frye had retracted his confession, having made it, he said, in exchange for a policeman's promise to drop the robbery charge and for a half share of the $1,000 reward. Frye now claimed that he had agreed to this implausible deal because he knew he would never be convicted, thanks to an airtight alibi. But when the witnesses to his alibi refused to come forward, Frye's new lawyers hunted for a way to counteract the confession. Coincidentally, Dr. William Marston, teaching at nearby American University, had run extensive lie detector tests on Negro suspects during the war.

Marston's goal in testing Frye was to create a legal precedent, thereby realizing the ambition of his mentor, Hugo Münsterberg, to usher in a new era of scientific justice. In this he would succeed, though not as he intended. Frye's case would effectively end the chance for the lie detector to have a career in the criminal courtroom, even as it gave the device free rein outside the formal procedures of the law.

The justice system has long wrestled with methods for gauging honesty. Criminal activity, almost by definition, cloaks itself in just the sort of falsehood society wishes to uncover. There have been three major phases in the development of such tests in the European west, each with its preferred methods of extracting the truth from recalcitrant human beings—and each increasingly in tune with prevailing assumptions about how to extract the truth from inanimate nature. Proponents of lie detection hoped to introduce a fourth and final phase in the administration of justice in which the methods for humans and inanimate nature would finally converge.

In the first phase, the medieval trial by ordeal, the community obliged the accused to confront a physical challenge so that God might judge the outcome. Sometimes this ordeal involved mortal combat. At other times, the faith of the accused was tested by an act of endurance. For instance, presumed liars were asked to lick a burning hot poker; if God wanted to commend their honesty, their tongues would not be scorched.

The second phase, which lasted from the twelfth to the eighteenth century, came in two distinct forms, one on the European continent, the other in England. Though the continental system of inquisitorial justice would be later denounced as cruel and unusual (see the Fifth and Eighth Amendments to the U.S. Constitution), it actually operated according to its own impeccable logic. The goal of inquisitorial justice was certainty of judgment—that, and the reintegration of the guilty into the kingdom of the saved. In an age when almost all crimes were punished by death, no Christian magistrate dared substitute earthly judgment for God's say-so, at least not without proof "as clear as the sun at noon." Hence, the judge calculated guilt according to an arithmetic rule. The sworn testimony of one reliable eyewitness constituted a half proof; that of two independent eyewitnesses constituted a full proof. Unfortunately, as few crimes were witnessed by even one person, let alone two people, magistrates were authorized to torture the accused—confession being considered the "queen of proof."

In theory, judicial torture was justified to save the soul of the accused, lest the person die having violated a sacred oath of innocence and suffer the far greater torments of hell. It was also thought to be conducive to truth-telling. Early modern jurists believed that the truth was less a production of the will than a spontaneous utterance, and hence that bodily pain, by crushing the will, would release the truth, much as the pain of childbirth would induce an unmarried woman to honestly name her infant's father. Yet the jurists also recognized what their Roman predecessors had long known: that many people resisted torture to the end, and that others, to end their agony, simply told interrogators whatever the authorities wanted to hear. For just these reasons the Roman jurist Ulpian was often quoted as saying that torture was "weak and dangerous, and inimical to the truth."

To mitigate against such deceptions, the jurists generally authorized torture only on the basis of circumstantial evidence of the sort we would consider sufficient to convict. Moreover, jurists introduced safeguards to ensure that torture was not abused. Examiners were forbidden to ask suggestive questions; the confessor had to supply corroborative information that only the guilty party could know; and the confessor had to repeat the confession after the torture had ceased. The torture itself was executed according to strict procedures. Judicial torture was not wanton sadism, but a formalized practice for obtaining the truth, one which matched the most up-to-date understanding of how to gain knowledge by putting nature "on the rack."

In England, by contrast, the common law actually retained many features of the trial by ordeal, except that parties contended in verbal rather than physical combat, with local jurymen posing questions and passing final judgment, while law judges, appointed by the king, served as neutral adjudicators. In the common law, torture was forbidden and no man was bound to accuse himself, although in practice the defendant did have to answer the charges or risk almost certain condemnation. Still, there were those in England who longed for the rigor of the continental system.

These currents clashed in the writing of Francis Bacon, often called the godfather of experimental science. Bacon was also the last Englishman to direct treason trials where torture was ordered by the king's Privy Council. Bacon criticized the scholastics for merely touching nature "by the fingertips," and urged a more aggressive line of interrogation. As he put it, "the nature of things betrays itself more readily under the vexations of art than in its natural freedom." He urged investigators to bind nature in chains, driving matter "to extremities," causing it to "turn and transform itself into strange shapes" until—coming full circle—"it returned at last to itself." Bacon acknowledged that experimenters who forced nature in this way risked misconstruing its qualities. "When bodies are tormented by fire or other means, many qualities are communicated by the fire itself . . . whence strange fallacies have arisen." Yet he was confident that God's benign superintendence would "return" nature to its true properties. Experimentation was less a form of torture than a struggle, from which both nature and experimenter emerged purified. Bacon's natural philosophy found an uneasy echo in his jurisprudence. He condoned judicial torture, but only to discover corroborative evidence, not to extract confessions as evidence of guilt. As he acknowledged: "By the laws of England, no man is bound to accuse himself." The queen, he said, would not "make windows into men's souls."

The resolution of this contradiction emerged later in the seventeenth century, when the third and current phase of judicial inquiry took shape. It was then that elite jurists and natural philosophers modestly agreed that justice could depend on judgments short of certainty. Henceforth, judges and adversarial lawyers were to probe the statements of witnesses in cross-examination, with judgment on their veracity—and the guilt of the accused—to be assessed in probabilistic terms. We can still hear the echo of this new approach to proof in such phrases as "beyond a reasonable doubt" and "a preponderance of evidence." And it was under this probabilistic banner, as much as under Enlightenment humanism, that the continental states gradually abandoned torture. Not coincidentally, natural philosophers at this time began to accept the notion that scientific knowledge of nature was provisional and probabilistic.

This grand accommodation, however, did not find favor with those who preferred swift and certain justice. Built into this new probabilistic justice were two contradictory trends, at least in the Anglo-American legal system. In a democratic vein, the courts increasingly granted lay jurors unfettered license to weigh the truthfulness of testimony. In an authoritarian vein, the courts increasingly put their confidence in expert-interpreted circumstantial evidence because it could not be dissembled. The ambition of advocates of the lie detector was to resolve this contradiction—in favor of the experts.

Only gradually over the past three centuries have Anglo-American judges allowed jurors to hear the defendant and witnesses for the defense testify under oath in criminal cases. Judges thought they had good reasons for this hesitation. Some worried jurors might be reluctant to convict defendants who swore oaths of innocence. Others feared defendants would readily perjure themselves, damning their souls and providing an excuse for further jeopardy. As late as the eighteenth century, certain sorts of people were thought so prone to lie on the stand that they were not allowed to take the oath: children, the spouse of the accused, those with a financial interest in the outcome, slaves, and the defendant. It was only after the American Civil War that U.S. courts finally allowed defendants to be sworn in when they took the stand in their own defense (and then mainly to avoid the accusation of hypocrisy as newly freed slaves were allowed to testify under oath). Only then, with all parties sworn in, did jurors have full freedom to weigh the testimony of all witnesses in an evenhanded manner, assessing their words and demeanor: the way they moved, blinked, blushed, sweated, or spoke. The institution of the jury may be 1,000 years old, but only in the past century have jurors acquired an unfettered license to distinguish truth-tellers from liars.

Yet over this same period, the justice system increasingly sought to corroborate human testimony with circumstantial evidence beyond the power of human beings to dissemble. Because lay juries and magistrates were themselves unable to assess such evidence, the courts increasingly turned to expert witnesses. The courts already drew on the testimony of skilled practitioners of specialized crafts—surveyors, physicians, trades workers—regarding technical matters of which they had direct knowledge. But only during the eighteenth century were adversarial lawyers in England permitted to call experts who extrapolated from special experimentation using general scientific principles to give their opinion regarding the case at hand. With the growing reach and authority of science in the nineteenth century, reformers hoped these experts would underwrite the reliability of a justice system otherwise in the hands of a lay jury.

The problem was that adversarial lawyers proved adept at finding adversarial experts to make diametrically opposed arguments. By the latter half of the nineteenth century, quarreling among experts in the courtroom had become a subject of general mockery. One legal aphorism of the period had it that there were three kinds of liars, "the common liar, the damned liar, and the scientific expert." It didn't help that a vast array of practitioners of new sciences clamored for admission to the courtroom. How would the law decide whom to admit? The dream of certainty dies hard.

In the early decades of the twentieth century a coterie of reform-minded Americans set out to inaugurate a new phase in the administration of justice: the scientific interrogation of accused persons and other witnesses. The time had come, these reformers proclaimed, for the law's archaic methods of assessing human honesty to give way to the new science of psychology. Just as the discovery of X-rays now allowed expert radiologists to peer into a patient's body—and, according to some physicians, perhaps the mind as well—so might these new instruments allow expert psychologists to peer into a witness's body to infer whether the conscience was disturbed. The human body, suggested these investigators, could serve as a kind of circumstantial evidence for the mind.

This challenge to the law first issued from the laboratory of Marston's mentor, Hugo Münsterberg of Harvard. In the years before World War I, Münsterberg, a German émigré, founded "brass instrument" psychology. Lured to Harvard by William James, but despised by him in later years, Münsterberg was the first scientist to lay out the rationale for a science of lie detection. The seventeenth-century polymath Gottfried Leibniz had famously asked whether we would understand human consciousness any better if we could visit the machinery of our minds as if taking a factory tour. At the turn of the twentieth century Münsterberg created such a factory over Harvard Yard: a lab filled with students transforming their inner feelings into the kind of public phenomena we call science.

Münsterberg hoped to explore the deepest questions about the human mind. What was consciousness? What was emotion? What were honor, loyalty, and love? To learn the soul's mute truth, he suggested, one had to listen to the body, precisely because it was not subject to the caprice of the human will or the vicissitudes of language. "We must bring man before a registering apparatus to find out . . . whether sunshine or cloudiness prevails in his mind." As one student at Radcliffe recalled: "[O]ur pulses strapped to his recording needles and cylinders, we registered irrevocably our susceptibility to patriotism, romance, horror, joy and a dozen other influences of daily life." Converted into graphical tracings—the new universal language of science—the qualities of subjective feelings could at last be accumulated, compared, and controlled.

In this, Münsterberg was pushing William James's own theory of the emotions to its limits. The tone of our feelings, James proclaimed, derived not from our minds but from our bodies. As he put it, in one of those overly vivid examples that caused him so much grief, when we see a bear, we do not run because we are afraid; rather, it is our running that *makes* us afraid. Subtract bodily sensation from our feelings, James suggested, and no emotion-stuff remained.

Though James's theory drew on Darwin's evolutionary account of the expression of emotion in animals and humans, it marked a departure as well. So eager had Darwin been to refute the Victorian theologians who believed that every human characteristic served some divine purpose—such as the pious physician who called the blush God's way of assuring honesty—that he had ascribed emotions to the persistence of animal instinct. This recast emotions as part of the natural human endowment without assigning them any social function. James emphasized that emotions were a physiological response honed by evolutionary pressures to serve a social role, even when they ran counter to the promptings of our conscious reason.

As a crude slogan, this view met with mockery. Surely, said his critics, there must be cognition before emotion; we must decide that a bear is threatening (and not, for instance, a circus act) before we fear it. Surely our bodily reactions cannot uniquely determine our emotions; we shiver from cold as well as fear, weep from joy as well as sorrow. But partially revised, James's theory took these features into account. More to the point, his theory could be investigated experimentally.

There was just one hitch; James hated the tedium of lab work. This made Münsterberg a godsend. James wrote his brother Henry that Münsterberg was "the ablest experimental psychologist in Germany." Luring him to Harvard would make its psychology lab the nation's "unquestioned first." After an extended courtship, Münsterberg settled in America. "Now I am yours forever," he wrote to James.

But it was not long before James realized that in hiring Münsterberg he had unleashed a single-minded monster who would supplant his work. After all, James's theory had preserved a small domain of action for the human will. It was not much more than the capacity to direct one's attention, nearly a will-o'-the-wisp will. But it underscored James's conviction that human beings could choose their sense of themselves and perform the tasks of democracy. Münsterberg's "action theory" left no room for such wishful thinking. Münsterberg believed that our experience of conscious choice was an illusion: nothing more than our memory of often repeated bodily acts. Even when you feel you have exercised your free will—such as when you raise your right hand to take an oath—all you have really experienced is your body once again preparing your biceps, triceps, and lats to raise your arm. Münsterberg preferred the autocratic institutions of his homeland. His experimental protocols sought to recover the automatic responses beneath our public acts.

Of course, to be brought before Münsterberg's instruments was to become the sort of person Münsterberg described. Another student at Radcliffe, the incomparable Gertrude Stein, spent several semesters in his lab. She described how under the gaze of his apparatus she became an "automaton" on which others could work their will; or as she put it in her journal, "One is indeed all things to all men in a laboratory." In other words, wasn't a subject who could be obliged to tell the truth sufficiently pliable to tell her interrogators what they wanted to hear?

In her characteristic third-person voice Stein described how she felt herself divide in two while she watched her classmates watch her thoughts being registered on the revolving cylinder. "Strange fancies begin to crowd upon her, she feels that the silent pen is writing on and on forever." While her body is imprisoned, her mind is displayed for public amusement. "Her record is there she cannot escape it and the group about her begins to assume the shape of mocking friends gloating over her imprisoned misery."

There is some debate as to whether Stein's prose was influenced by her encounter with induced disassociation and automatic writing. (She thought so.) But not only did Münsterberg's program challenge Romantic notions about human creativity and autonomy; it also challenged the law's methods for assessing human beliefs and personal responsibility. As a scientific mandarin, Münsterberg did not hesitate to speak out in public. His "scientific conscience," he said, would not let him do otherwise. In a series of muckraking articles in magazines like *McClure's,* Münsterberg demonstrated how eyewitness testimony—even confessions—could be mistaken; how false memories could be planted by police interrogations; and how difficult it was for students to recall staged crimes accurately. On a positive note, he explained how psychology was finally turning the tables on human duplicity. In just the past decade Cesare Lombroso, an Italian who asserted that criminals constituted a distinct race, had interrogated suspects while their hand was encased in a special glove to register changes in blood pressure, on the assumption that their physiological reactions would betray their state of mind. Carl Gustav Jung, the Swiss psychiatrist, had adapted his word-association test to probe the patient's psyche, on the assumption that a hesitant response to words relevant to the crime meant that the subject was deceiving the analyst. Jung had also tracked emotional changes with a galvanometer that recorded the conductance of current across the sweaty surface of the subject's skin, and had even induced one young man to confess to a crime while questioning him on this device. And all these methods were akin to the efforts of Jung's mentor, Sigmund Freud, who had tried inducing a quasi-hypnotic state in his patients so as to press them to recall repressed memories of sexual abuse, before he changed his mind and decided they were just fantasizing.

Then, in 1907, Münsterberg got a chance to try these truth-testing techniques in the most prominent trial of the time. Prosecutors invited him to Idaho to assess the honesty of Harry Orchard, a political assassin whose confession to the murder of the state's governor had implicated western labor leaders. Millions of readers across the country were following the trial of "Big Bill" Haywood in what was billed as a titanic struggle of labor against capital. In the seclusion of the prison outside Boise, Münsterberg sat Orchard down for a battery of tests designed to "pierce his mind." The most telling of these, said Münsterberg, was the word-association test. When prompted by the word "confession," Orchard took only eight-tenths of a second to respond "true," the brevity of the lapse indicating that this answer was as innocent as his pairing of "river" with "water." After two days of gathering such data, the psychologist left the prison assuring prosecutors that he had "not the slightest doubt" Orchard was telling the truth.

Regrettably, on his return to Boston, he confided this conclusion to a reporter. By morning his verdict had been wired to every newspaper in the country. Editorials excoriated the presumptuous expert who dared to usurp lay justice. Journalists began to refer to Münsterberg as "Monsterwork." Jurists denounced his pandering to the newspapers as "yellow psychology." And his colleagues wondered how "Dr. Münsterberg can have the face to ply the American public with these shallow, platitudinous half truths." Yet the professor was not one to retreat from a good *Kulturkampf*. He did moderate his views after Haywood was acquitted. (In his draft essay on Orchard, he wrote, "My nerves protest against twelve jurymen in rocking chairs, each one rocking in his own rhythm. . . . [A] few hours of experimenting were more convincing than anything . . . in all those weeks of the trial.") But he still insisted publicly that the time would soon come when "the methods of experimental psychology cannot longer be excluded from the court of law." And it wasn't long before other scientists echoed his hope that "truth-compelling machines" would soon be adopted by the courts. In 1911, an article in the *New York Times* asked readers to look forward to the time when all judicial questions would be decided by impartial machinery.

> There will be no jury, no horde of detectives and witnesses, no charges and countercharges, and no attorney for the defense. These impedimenta of our courts will be unnecessary. The State will merely submit all suspects in a case to the tests of scientific instruments, and as these instruments cannot be made to make mistakes nor tell lies, their evidence would be conclusive of guilt or innocence, and the court will deliver sentence accordingly.

As a college junior working in Münsterberg's lab, William Moulton Marston was captivated by this vision—and the career it promised. Marston was a jaunty young man from a good Massachusetts family, in step with the rhythm of the new century. He had already written a prizewinning "photoplay" entitled "Jack Kennard, Coward," based on the real-life story of a football player at Harvard, crushed by gambling debts, who wins back the love of his girl. And Marston had a knack for trumpeting his achievements. As he informed one reporter during his senior year: "This study of the psychophysics of deception is going to prove a great help to me when I begin to practice law."

Marston was determined to cut his own path. As he informed Professor Münsterberg, he had discovered that some of his fellow students enjoyed lying so much that their response times actually contracted on the word-association test. How could Münsterberg's favorite test distinguish these eager liars from confident truth-tellers? At this impasse, Marston recalled, he told his adviser he would take the advice of a young woman friend from Mount Holyoke: he examined his heart. He married Elizabeth Holloway, "the girl from Mt. Holyoke," and together they proved that by keeping intermittent track of a storyteller's blood pressure they could pick out 96 percent of the liars, whereas ordinary student-jurors were fooled half the time. On the promise of this method, he enrolled for both a J.D. and a Ph.D. in psychology.

Münsterberg died just before America entered the Great War against his homeland. But Marston proved an apt disciple, and wartime was propitious for testing deception. Robert Mearns Yerkes, the chief of the psychological branch of the National Research Council—the same outfit that designed the IQ test for the mass army—agreed to sponsor Marston's research so long as he followed new, more rigorous protocols. Applying his methods to twenty detainees in the Boston municipal court, he identified 100 percent of the liars, compared with 75 percent identified with a method of tracking breathing patterns recently proposed by an Italian researcher. Yet Marston's over-the-top success only made his seniors more suspicious.

They were not reassured by his performance in his first and only real-world case. When surgical instruments and a military codebook vanished from the U.S. Surgeon General's office in Washington, D.C., intelligence officials suspected one of the building's eighteen African-American messengers. Asked to find the culprit, Marston identified one man "with very strong consciousness with regard to something he had done," but he also admitted that he was not certain of the man's guilt. The problem, he explained, was that blacks, perhaps because their will was more "primitive" than that of whites, seemed to respond differently: "The factor of voluntary control which, with white men, seems to make a deception rise regular and almost an absolute one, apparently is almost altogether lacking in negroes, so that, tho the change is really even more sharp and extreme, it is vastly more difficult to estimate norm plus excitement."

The racist assumptions on display here highlight the methodological vulnerability at the core of lie detection. No psychologist doubted that bodily changes often accompanied mental activity. The psychologists did doubt, though, that a specific change could be read as a sign of a specific mental act, like lying. In 1915 the physiologist Walter B. Cannon had published animal experiments to refute James's theory of bodily emotions, proving that diverse stimuli could produce identical physiological responses. Dogs salivate from more than one cause. Humans cry from joy as well as sorrow. Then too, some people are more sorrowful than others. To reverse the inference, as Marston did, and assume that a specific mental state had caused the rise in blood pressure, was to abandon the norms of inductive reasoning in favor of what the philosopher Charles Peirce called abductive inference: the effort to reason one's way backward from consequences to cause. Although such reasoning is a valuable way to generate novel ideas for testing, it is generally held suspect without further corroboration because it leaves so much scope to the prejudices of the reasoner.

Marston's brilliant successes had simply cast suspicion on the experimenter himself. His new thesis adviser admitted that while the young man had "always proven entirely trustworthy," he was also "slightly overzealous in grasping opportunities, which causes him to take the corners a little sharply." That said, the professor couldn't help admiring Marston, who was ambitious, resourceful, and "very much of a man." Now he just had to prove he was a scientist—in other words, that other researchers could replicate his success.

The army sent Marston to Camp Greenleaf, Georgia, to see whether other interrogators could master his technique. In a new series of deception games these interrogators were able to separate honest men from liars 74.3 percent of the time—better than they did without the technique, but nowhere near Marston's 94.2 percent success rate. So that even as the tests confirmed Marston's skill, the psychologists could not decide whether his achievements were due to science or personal ability. Although the war ended before he had convinced his fellow psychologists, Marston did not give up. He and his wife returned to Massachusetts, where they both took the bar exam and resumed joint research on deception. He received his Ph.D. in 1921 and found a temporary academic job in Washington, D.C. while he angled for a grander stage.

It was a stage the law would deny him. In order to create a legal precedent in the Frye case, Marston first had to qualify as an expert witness. To this end, Frye's lawyers supplied Judge William McCoy with a weighty stack of scientific reports: on emotion by William James and on lying by John A. Larson, plus Marston's Ph.D. dissertation. But by his own admission, Judge McCoy spent only five minutes perusing these tracts before he ruled the next morning that he would not allow Marston to take the stand.

It was true, he acknowledged, that courts often heard from fingerprint experts and other experts, including alienists who reported on the mental states of suspects; so why not Marston, a qualified Ph.D. in psychology? The difference, he explained, was that lie detection was not yet "a matter of common knowledge." Of course, science had made stunning progress in the past generation: airplanes, radio, telephones. A time might come when lie detectors were as common as telephones, whereupon trials would be conducted on the basis of a "mere record" of the defendant's truthfulness. By that time, however, he expected to be dead (a sentiment expressed so as to suggest it was his preference). Until then, "the jury looks at the witnesses, hears what they have to say, compares their statements with other statements, and so forth; and then does what human beings do out of Court when they determine whether or not a man is telling the truth." That, he said, "is what the jury is for."

On July 20, 1922, the jury did what it was for and found Frye guilty of second-degree murder, and soon after, Judge McCoy, doing likewise, sentenced Frye to fifteen years. (The relatively light verdict, Marston claimed, was a sign that his exonerating test had somehow reached the jury.)

Even so, Frye's lawyers appealed. But in a terse opinion, which cited neither precedent nor authority, the court of appeals of the District of Columbia upheld the lower court's rejection of the test. In doing so, it laid down a rule that would ban the lie detector from criminal courts for the rest of the century and set criteria for the admission of all scientific evidence for the next sixty years.

The ruling made no mention of the objection that had haunted Judge McCoy's refusal and had been raised in the prosecutor's appeal brief. "If such tests ever are adopted," McCoy warned the court, "it is probable that the jury system will have to be abandoned." Nor did the ruling turn exactly on whether lie detection was "common knowledge," as Judge McCoy had proposed. What mattered, the court of appeals declared in 1923, was whether the science was widely deemed acceptable by the relevant scientific experts. The "Frye rule" was simple and flexible. Instead of having judges evaluate new science themselves, it made them into black-robed pollsters of scientific opinion—while retaining control over who was to be polled.

> Just when a scientific principle or discovery crosses the line between the experimental and demonstrable stages is difficult to define. Somewhere in this twilight zone the evidential force of the principle must be recognized, and while courts will go a long way in admitting expert testimony deduced from a well-recognized scientific principle or discovery, the thing from which the deduction is made must be sufficiently established to have gained general acceptance in the particular field in which it belongs.

In the Frye ruling, in the absence of any poll, the court of appeals took judicial notice that the lie detector had yet to gain "general acceptance in the particular field to which it belongs," and upheld the verdict. For the next seventy years, despite various challenges, the ruling acquired the status of a Supreme Court precedent, both with regard to the lie detector and with regard to scientific evidence in general.

Three years later, the eminent law professor Charles McCormick did poll the group he deemed the most "relevant" scientific experts, his colleagues in academic psychology. Of the thirty-eight scientists who responded to his inquiry, eighteen partially favored presenting the technique to judges and juries, thirteen opposed its introduction, and seven were of "dubious classification." Among the enthusiasts were William Marston; Robert Mearns Yerkes; and Walter Dill Scott, a business psychologist who was the president of Northwestern University. The naysayers included luminaries like the behaviorist John B. Watson, who declared it "a thing for the laboratory for another 25 years," and others who ascribed

Marston's and Larson's success to their personal skills at interrogation, rather than to the science of psychology per se. Several also worried that the scientific trappings of the technique would be overly persuasive to a jury.

In short, these academic psychologists dismissed the lie detector because it simultaneously challenged and exploited their discipline's newly won scientific authority. This authority was founded on the presumption that sophisticated experimenters had the upper hand on naive human subjects, and could treat mental properties as natural objects, not unlike those studied by physics, chemistry, and biology. Watson's ascendant school of psychological behaviorism took this attitude so far as to deny that subjects had minds worthy of investigation. Thus, to accept the core premise of the lie detector—that some subjects willfully deceived investigators—would have forced psychologists to confront the fact that their work was unlike the work of their scientific colleagues; that their experimental subjects approached psychological tests with their own agenda; and that some subjects even carried the day. Already, one researcher at the Harvard lab had discovered that subjects who discerned the goal of the experiment could escape detection; indeed, sophisticated subjects might even "prevent detection of their sophistication." Marston himself, working with his wife, not only had begun to recognize the diversity of liars—male and female, black and white, eager and clumsy—but now announced the shocking discovery that subjects' reactions depended on the qualities of the examiner: for instance, whether the examiner was a man (Marston himself) or a woman (Marston's young wife). It was a radical thought; psychology—Marston was saying—was not like other sciences.

Today we are less surprised by the "discovery" of something so obvious than by the unwillingness of these psychologists to acknowledge it. But on reflection, their unwillingness is not so hard to understand. To acknowledge their subjects' trickery would have obliged them to counteract these deceits with deceptions of their own. Not for another three decades would psychologists embrace experimental dishonesty to preserve subjects' naiveté. For the time being, to dismiss the lie detector was to affirm the highest scientific value they knew: professional honesty.

Though Larson was not included in the survey, its author consulted closely with him. In the interim, Larson had come under conflicting pressures. On the one hand, he had been scolded by the California State Medical Association for abusing medical techniques. On the other, he was under pressure from Vollmer to validate these techniques. Under these constraints, Larson took a stand that squared his scientific principles with his ambitions as criminologist. The time had not yet come for the lie detector to be used in court, but the technique remained a legitimate tool in police investigations.

Larson's position dovetailed neatly with the Frye rule. It allowed the machine to flourish in the largely unregulated world of police investigation, even as it denied the lie detector entrance to the courtroom. It acknowledged that psychology had turned its back on the technique, even as it left open the possibility that the technique might one day be validated by scientific opinion. Larson hoped this day was not far off. He told McCormick of an innovative new machine for lie detection then being assembled by his disciple Leonarde Keeler. He even hoped Keeler's more sensitive and reliable apparatus might resolve the Wilkens murder.

CHAPTER 5

# The Simple Home

*A detective official in San Francisco once substituted "truthful" for "voracious" in one of my reports on the grounds that the client might not understand the latter.*

—DASHIELL HAMMETT, "FROM THE MEMOIRS OF A PRIVATE DETECTIVE," 1923

THERE ARE CONFLICTING STORIES OF HOW LEONARDE Keeler first got hooked up with the lie detector. To hear his father tell it, his father deserved the credit. According to Charles Keeler, Berkeley's most famous poet, he was visiting his dear friend August Vollmer when the Chief showed him a "psychological chart." Intrigued, Charles asked the Chief to show the chart to his son, Leonarde, then a high school student recuperating from an illness. The Chief had known Leonarde since Vollmer's days as village mailman, when he had hauled his two-wheeled cart up to the family's isolated redwood house on the barren slopes of Berkeley. "Send him down," the Chief responded. Soon young Keeler was working on the machine to the exclusion of his studies.

In Chief Vollmer's version, told as part of his eulogy for Leonarde Keeler, it was the Chief, not Charles, who advised the young man to take an interest in lie detection. Young Keeler—whom the Chief had come to regard "with the affection that a father would a son"—had been feeling listless and apathetic, so Vollmer let him watch John Larson extract a confession from a burglary suspect. "From the moment that Nard saw [this], his life took on new meaning, and his mind and body seemed to be charged with energy."

But there was another episode which put Keeler in Vollmer's office, one that neither his real nor his adoptive father had any wish to recall. Apparently, Leonarde and a frolicsome high school buddy, Warren Olney, had been grilled by the Berkeley police after an eyewitness saw the two youths jump out of a Cole Aero-Eight waving a gun and run into a store in Oakland's Chinatown. Quick detective work enabled the police to nab Olney, who admitted to a mustachioed tough cop, Frank Waterbury, that he and Leonarde Keeler had indeed borrowed his father's car and gun. Brought in for questioning, with bright lights shining in his eyes, Keeler confessed that the boys had been organizing a high school circus for which he would perform a magic disappearing act punctuated by pistol shots to misdirect the audience. They had driven into Chinatown to purchase some blanks for Olney's father's gun—Olney's father being at the time an associate justice on the California supreme court—and to buy exotic incense for Keeler's sister's performance as a snake charmer. This was absurd enough to be credible, and the boy's parents demanded that Vollmer reprimand the arresting officers. This the Chief refused to do, though he did dismiss all charges. After all, no one in Chinatown had filed a complaint.

In later years Warren Olney became legal deputy to Alameda County's rising young district attorney, Earl Warren, and followed him to Washington to serve as the head of the Justice Department's Criminal Division. Leonarde Keeler also slid around to the other side of the interrogation table. Olney always got comical satisfaction from the fact that in his youth he and his pal Keeler, acknowledged master of lie detection, had been grilled by a cop so suspicious that no amount of contradictory evidence could shake his theory. And all his life, Keeler continued to practice sleight-of-hand tricks and amateur magical acts.

But the tough cop, Frank Waterbury, never forgot the tale either. Forty years later, when Keeler was dead, he wrote up his version for an elderly, embittered John Larson, his fellow veteran of the Berkeley police force. Back in the bad old days, he told Larson, he had once arrested that "snake charmer and rotten liar" Leonarde Keeler for pulling a gun on a Chinaman on the corner of Fourth and Franklin. It was a story that confirmed everything Larson had come to believe about Keeler, his first disciple and his greatest regret.

As for Leonarde Keeler, in his version of how he met his daemon, he strode jauntily into Larson's lab confident that he could beat the machine "easy"—and got caught the first time out.

In fact, the first thing Leonarde did with the lie detector was what any high school student would do, at least in the liberated 1920s: he dug into his friends' love life. His sister recalled how he interrogated her best friend "Chickie" in the dank police lab in the basement of City Hall. When he asked her if she loved Curtis or Harry, the answer was twice no. But when asked, "Do you love Charlie?" the needles lurched, the onlookers laughed, and Chickie blushed bright red. "It's true," she confessed. "I do have a kind of crush on Charlie. But I never dreamed anybody would find out." The machine was so potent that when Leonarde put his sister on it, she fainted. "I imagine I was afraid you might detect certain hidden thoughts," she told him. Everyone knew that Eloise adored her older brother—worshipped him, really. Both children were subject to periodic fainting spells. Eloise, after years on the psychiatrist's couch, came to blame her attacks on their father. She wanted so badly to please him and always failed. Keeler never hazarded a guess about his own tendency to black out, though he too spent half a year in psychoanalysis with a lie detector strapped to his ankle and the topic often reverting to his father.

Charles Keeler fancied himself a scientist as well as a poet and the founder of a new world religion. His greatest contribution, however, was a new aesthetics of living, which would come to define California's bohemian bourgeoisie. Charles's romantic sensibility conceived of no boundary between art and nature, mind and body, morality and necessity. While still in college he dabbled in spiritualism and mind reading. At twenty-two he published a Darwinian treatise that explained the beautiful plumage of male birds as the outcome of an alignment of inner striving (avian exuberance at mating time) with outer necessity (competition for females).

Charles Keeler was himself a male bird of fine plumage: six feet tall, trim and handsome, with dramatic eyes, hollow cheeks, and glossy black hair that fell like curtains on either side of his forehead. He hated barbers. Even for an ordinary commute on the ferry across the bay, he affected a black cape and a gold-headed cane. He had traveled to Cape Horn, Alaska, and the South Pacific, and he cultivated the friendship of older artists and intellectuals: the landscape painter William Keith, the naturalist John Muir, the architect Bernard Maybeck.

In the year of his Darwinian treatise, Charles married one of Keith's pupils, a sixteen-year-old Berkeley girl named Louise Mapes Bunnell, in whom he saw a chance to consummate his moral and aesthetic ideals. She would illustrate his poems with beaux arts woodcuts. He would pose for her in Athenian robes, a wreath of flowers in his hair, while declaiming his poems in a booming voice which "could fill stadiums." One of his allegorical works, "The Truth," began:

> I crave the truth, stark naked, unashamed:
> And should it smite me, let me face the pang,
> Aye, turn the other cheek, and cry, again!

Keeler's next step toward this new mode of living came when he gave Bernard Maybeck his first private commission: a house on an unspoiled hillside just north of the university. As the antithesis of the "wedding cake" houses back east, this house would be "simple and genuine," conforming to the site and clime. Under its generous peaked roof the interior walls and rafters were of unadorned California redwood. Persian rugs trailed across the floors, Chinese lanterns hung from the ceiling, and the bookcases gleamed with leather. And lest neighbors spoil his vision (and his view), Keeler induced his friends to commission Maybeck homes in the vicinity. In his manifesto of 1904, *The Simple Home,* Keeler wrote that a residence must not just mirror nature out-of-doors but offer "a genuine reflection of the life which it is to environ." A home, like its inhabitants, was to be honest, direct, and beautiful, dedicated to "the conviction that we must live art before we can create it." Its finest work of art was the family.

Leonarde Keeler was born into this home on October 30, 1903. His father named him after Leonardo da Vinci, he later explained, in the hope "that you would grow up to be one of the great contributors to the thought of this wonderful generation in which we live." But as young Keeler grew up, he preferred to be known less grandly as "Nard."

Calamity struck when Leonarde was two. His father managed to snatch him to safety when the San Francisco earthquake nearly brought Maybeck's chimney down on his head, but the strain of caring for the earthquake's many refugees took a severe toll on Charles's wife. Despite hypnotism and mental cures, she was dead within a year. Charles Keeler was never the same afterward.

He rented out his ideal home, packed his children off to their maternal grandmother on the opposite side of Berkeley, and set out on a two-year round-the-world poetry tour. In 1913 he returned briefly to Berkeley to discover ten-year-old Leonarde stealing money from his grandmother to buy candy for his neighborhood gang, to whom he was in thrall. Charles consulted his friend August Vollmer, and resumed the old technique of plying the boy with hypnotic suggestions while he slept so as to substitute "courage and initiative" for "subservience to the mob and cowardice." Night after night, he sat beside his sleeping son, repeating the words: "Leonarde, you hear me speak. You are brave, you have great courage. You are independent. You make your own plans and work them out for yourself." Charles believed that truthfulness could be "indoctrinated." Only later did he concede the contradiction.

That fall Charles brought his children back east, placing Leonarde in the Hackley School in Tarrytown, New York, while he tried to make his name as a poet in Manhattan. But America's taste for enchanted verse was wearing thin amid the slaughter of World War I, and Charles's inheritance was nearly spent. In 1916 the family returned to Berkeley and moved into a Maybeck-inspired studio in a secluded canyon on the southern edge of town. To pass under the Japanese torii gate and cross the bridge between the redwood trees was to enter a California of early settlement. In a shed out back, Leonarde kept a menagerie of reptiles, as well as a wireless set. On the hillside below Charles performed his neo-Athenian poetry—to the acute embarrassment of his children—until the great influenza epidemic closed the theater, and Charles, most piteously, caught the flu. He was nursed back to health by an assistant principal at Leonarde's progressive school. They married in 1921. An adulator of women who adulated him, he never strayed far from home again.

Flamboyant, fusty, and self-important: Charles Keeler was a would-be prophet in an age that mocked idealized harmonies. In the early 1920s, he found two new outlets for his energy. His Cosmic Religion was a Bahai-like creed that sought to unite all the world's prophets and peoples under the divine trinity of "love, truth and beauty." Its devotional meetings consisted of Chautauqua-style talks on parenting, evolution, eugenics, birdwatching, invention, and the arts and crafts—all given by Keeler. If successful, he predicted, it would become "the greatest single contribution to humanity of modern times," though he admitted that he did not expect this to happen during his lifetime. Attendance soon dropped from sixty to ten, and he was disappointed when neither Palo Alto nor Hollywood established a chapter.

At the same time, in desperate need of money, Keeler took a position as secretary for the Berkeley Chamber of Commerce, placing his prolific pen in the service of "the most beautiful city in the world." From this position, he helped his friend Chief Vollmer block a merger with Oakland that would have unraveled Berkeley's police department. The idealist poet and the pragmatic cop were unlikely allies, but they were united by a common vision of Berkeley: the dream of an honest polis. While Charles led a moral and aesthetic regeneration in the Berkeley hills, Vollmer set out to police the unruly flats. Their most famous collaboration would be Leonarde.

In his youth Leonarde alternated between feats of physical prowess and bouts of disabling illness, leaving him little time for school. He was active in the Boy Scouts, making forays along the Russian River, building barracks, and organizing Liberty Loan drives. Pictures of him at age fourteen show a handsome, self-assured youth, wiry and tanned, with tousled hair, gazing frankly at the camera. Then at the age of fifteen—the year of the "great influenza"—he suffered a severe streptococcus infection which affected the valves of his heart and the lining of his brain. In the hospital he heard the nurses say he was at death's door. His father smuggled in a Christian Scientist who ordered Leonarde to rise—the first steps, Leonarde later recalled, in his recovery. He convalesced at home, caring for his snakes and photographing birds—though vulnerable to fainting fits. In his senior year at University High, while on an expedition with the John Muir Club, he developed acute appendicitis and was again hospitalized for several months. During his recovery, he became entranced with Larson's machine and its power to compel the body to be honest.

It was Vollmer's fond ambition to have his two young disciples collaborate on the lie detector. While Larson validated the method scientifically,

Keeler would handle the practical business of redesigning the instrument. All that would be necessary, Vollmer believed, was that they keep one another apprised of their progress in their respective domains.

For a time Keeler persisted in his other pursuits. He led expeditions of young women into the high Sierras for pay and pleasure. Women were charmed by his resilience and vulnerability, as well as his physical grace. The future dancer and choreographer Agnes de Mille was one of these. The party flirted, clowned for the camera, drew sketches. On the final day Keeler led a smaller party above the timberline. "This is my land," he told Agnes as they looked down on the granite basin. "This is the California I love. I have to come back at regular intervals. I know it is here, and I draw my strength from it. Whether I see it or not. It is here."

Keeler was comfortable in his body's lithe strength. He had wide-set eyes, a broad face, and a playful smile. He did not talk much, but he was such a good listener that a girl hardly noticed. For the rest of her life, Agnes de Mille recalled Leonarde Keeler as "the best of all my dance partners."

> My God, how this rough boy could move on a parquet floor! Leonarde Keeler was the finest partner I ever stood up to in my entire life. He was not fancy, but he led so strongly, so securely, so rhythmically, that I followed as though borne by a tide. His sense of timing was exact and the sense of latent power in his body was inviting and reassuring.

Adventurous but vulnerable, sociable but guarded, Leonarde Keeler was determined to make his mark. Raised to be an individualist, he was happiest in a group. Raised to be a second Leonardo da Vinci, marrying science and art, he took up Vollmer's practical badge of science and crime-fighting. Born in a redwood house consecrated to domestic fidelity, he built his life around a mechanical "box" predicated on mistrust. The box would become his personal obsession and his point of entry into human relations. He would use it to wring confessions from hardened criminals, help managers check the loyalty of their employees, and spy into troubled affairs of the heart.

CHAPTER 6

# Poisonville

*And don't kid yourselves that there's any law in Poisonville except what you make for yourself.*

—DASHIELL HAMMETT, *RED HARVEST,* 1929

BEFORE TOUGH-MINDED CRIME-FIGHTERS WOULD AGREE that the lie detector fought crime, the technique would have to prove its mettle in a tougher town than Berkeley. So when a secret delegation from Los Angeles approached August Vollmer in 1923 to see if he would be willing to transplant his scientific methods to the nation's fastest-growing metropolis, he agreed to consider it, as long as he would not have to "battle politicians or political influences either within or without the department." Vollmer knew, of course, that he would be confronting politics at every turn—that was why he encouraged Keeler to join him.

That the city would approach an outsider like Vollmer for police chief shows how deeply city leaders feared their metropolis was on the verge of a violent unraveling. Vollmer's name had originally been placed in candidacy by the civilian Crime Commission, founded by the great oligarch himself, Harry Chandler, publisher of the *Los Angeles Times.* The commission wanted more efficient police to protect private property, combat union radicalism, and support all those organizations dedicated to 100 percent Americanism. With the city in the throes of a vicious crime wave and the police department "honeycombed with crooks," city leaders feared vigilante violence.

Yet the final approval to hire Vollmer came from the very people inciting this populist revolt. For the past few years, the reverends Gustav Briegleb and "Fighting Bob" Shuler, pioneer radio evangelists and founders of the modern fundamentalist movement, had thundered against crime and licentiousness, as well as the corrupt police and judges who let vice besmirch America's one "white spot." "Is it true that thugs and thieves, libertines and degenerates, dope sellers and other criminals are operating on our streets with apparent security and if so, who is responsible for such a [*sic*] conditions?"

Briegleb and Shuler believed that justice in Los Angeles was so hard to come by that parishioners sometimes had to take the law into their own hands. Shuler defended a gang of "White Knights" who had nearly flogged to death a man they (falsely) believed had assaulted a teenage girl. He defended leaders of the Ku Klux Klan on trial for attacking a family of Spanish-Mexican "bootleggers" in Inglewood, even after one of the 200 night riders—fatally wounded by the night marshal—turned out, under his mask, to be the local chief constable. The ensuing trial showed how deeply the Klan had permeated the Los Angeles police. Even Police Chief Louis Oaks was found to be a member of the Klan, as were some 10 percent of police officers and any number of prosecutors and district attorneys. In Anaheim, the Klan controlled the city council, and police officers there were asked to patrol the streets in Klan regalia.

Then, when one of Oaks's captains had the temerity to insult Shuler's flock as a "semi-intelligent" mob, Shuler and Briegleb led their howling congregants into Mayor Cryer's office to demand a crackdown on vice and to lay down an ultimatum: if Chief Oaks was not deposed, the two ministers would petition to recall the mayor. Kent Kane Parrot, the mayor's patronage chief, had little reason not to comply; cracking down on petty vice dealers enabled him to extract a bigger protection fee from their bosses. But no sooner had Chief Oaks tried to crack down on Parrot's main ally inside the police department than the chief was conveniently discovered in the backseat of a car with a half-naked young woman and a half-empty bottle of whisky, and promptly fired. It was enough to make the *Los Angeles Times* wonder "WHO RUNS THE CITY?"

In the early 1920s, two true-blue American solutions to the uncertainties of the law—the lynch mob and the third degree—were converging on Los Angeles. The lynch mob, a legacy of southern slave patrols, swarmed directly to the vengeful heart of justice and strung up its victim in an ecstasy of righteousness. The third degree, a legacy of northern urban patrolmen, was more measured, a beating delivered by pros, so that the suspect could be handed over to formal procedures of the law with his confession tied around his neck. In the early 1920s, as America's second wave of the Ku Klux Klan surged across the nation, the lynch mob seemed to be coming to urban America, with the third degree as its station-house auxiliary. The *Los Angeles Times* warned that the city had reached a crossroads. "Business has continued to flow in this direction very largely for the reason that this community has always been a place where law and order were supreme—where the courts were respected and where mob rule had no roots." Vollmer's scientific policing offered a way out of this abyss, providing certainty in law enforcement while evading populist violence and a lawless police force.

So it was a desperate mayor who sent a second delegation to Berkeley to vet August Vollmer, a delegation consisting of Kent Parrot, the Reverend Bob Shuler, and Reverend S. T. Montgomery (head of the Anti-Saloon League). Vollmer agreed to place himself on loan to Los Angeles for one year.

Instead of a university town of 40,000, Vollmer now policed an urban center of 800,000. Instead of two dozen cops, he now commanded nearly 2,000. But Vollmer had not risen to be police chief of Berkeley without knowing how to play the game. On his arrival he immediately announced his intention to "divorce" police work and politics, then spent six hours a day meeting the public, publishing articles in the *Los Angeles Record* (the city's reform paper), and giving regular interviews to the *Los Angeles Times*. He even addressed Shuler's congregation, promising to clean up the department if they would cease heckling the police.

Vollmer offered himself up as a "new type of police chief," unlike the two-fisted chieftains of yore. One early newspaper profile established the new chief as the modern executive: decisive, factual, all-seeing. "Somehow one feels, after looking Chief Vollmer in the eyes while he is talking, that is exactly the man to do the job, no matter what it is."

Vollmer set out to make the Los Angeles Police Department (LAPD) honest and efficient. He created a records department as the "hub of the police wheel," with a tabulating machine to map crime by place and modus operandi. He divided the force by function, segregating tasks such as vice, traffic, and detective work so that local police stations could adapt to shifts in population and crime. He created a 250-man "flying squad" of "crime crushers" equipped with prowler cars for quick pursuit to wipe out gambling dens and end the scourge of automobile thefts. He pushed through a $1.6 million bond issue and a bigger police budget to fund new stations, hire 500 more officers, and build more jails. "I am going to strip all the mystery and hokum from police work and place it on the basis of efficiency," the Chief proclaimed.

As in Berkeley, the key was better-quality personnel. This entailed insulating officers from the spoils system of municipal politics. Vollmer got the mayor to fire the commissioners after they refused to dismiss two cops accused of taking bribes. In his year as chief, 115 of his officers (nearly 7 percent) left under "pressure," that is to say, for such peccadilloes as stealing, neglect of duty, intoxication, suspicion of murder, and suspicion of rape. At the same time he won civil service protection for the police leadership, and he made all officers take intelligence exams to qualify for their posts. When quibblers insisted that he take the exam too—expecting a man with only a sixth-grade education to flounder—he outscored all his junior officers (albeit with a formula that added points for police experience and military service). He established a police academy and obliged his ranking officers to attend classes in criminal psychology, criminology, forensic science, and collection of evidence.

And Vollmer announced that he would not tolerate police brutality or vigilante justice. When thugs hired by industrialists destroyed the Wobblies' union hall, he denounced the lawless action and refused to send the police to bust strikes. In his annual report he included an article by the local head of the American Civil Liberties Union denouncing the third degree. That said, Vollmer was not a by-the-books officer. He hated the "foolish worship of technicalities," which undid arrests of vagrants. He sometimes expressed a desire to see his officers shoot a few criminals to set an example. And to resolve these contradictions, he promised the city a new scientific tool to sort out true testimony from lies and good cops from bad, and to end the need for populist violence or police brutality:

the lie detector, which he grandly referred to as "a modified, simplified, and humane third degree."

Keeler had just begun his course work at the University of California-Berkeley in the fall of 1923—without notable success—when fire swept through the Berkeley neighborhood of his infancy. The lingering smoke affected his breathing, and with his Berkeley grades already in the tank, he decided to transfer to the university's "southern branch" (as UCLA was then known) and join his mentor.

It is doubtful that he ever attended class. The city was itself an education: the idle rich waiting to move into their cemetery plots; the Idaho farmers dreaming of getting rich; and the "morbid, painted-up" women, so interesting "from a physiological point of view." In the auction houses he watched hucksters stoke the bidding far beyond the items' worth. "You can get more knowledge of the inside workings of people's brains in one of those sales in five minutes than you could from a year's study of some dry professor's book." Keeler thought Los Angeles a "crook's paradise" with a "hodgepodge of riff raffs that could be easily dropped in the Pacific Ocean and yet not have the country feel the loss."

Two weeks after Vollmer gave Keeler the go-ahead, the *Los Angeles Times* announced the machine's "first victim" on the front page. Bert T. Vernon—a "giant negro"—had quarreled with a fellow tenant and shot him before a multitude of witnesses (who were presumably thought to lack credibility because they were black). After some preliminary questioning on the machine, Vernon confessed to the deed. The newspapers recorded more victims that winter. When a suspect in a botched bank job reacted with agitation on Keeler's lie detector, further interrogation led to a full confession. When a suave bank bandit was told that his record on the machine showed "great inward emotion," he confessed. When $10,000 in jewelry was pilfered from the screen stars Jack Pickford and Marilyn Miller, Keeler interrogated the servants.

The most telling victory came when Keeler got a cop to confess to having stolen a pistol taken as evidence in a case of sexual assault. While some police officers denied the instrument had solved the crime itself, the message was clear: the box could be used on them. But the hard boys at the

LAPD would have none of it. They refused to give Keeler an official role on the force. Even the money to build his new machine came out of the Chief's discretionary funds.

To build it, Keeler had enlisted the aid of Hiram Edwards, an assistant professor of physics at UCLA who was a family friend. Edwards housed Keeler on his arrival in Los Angeles, and helped him build a more sensitive and reliable tambour, the crucial link in the instrument that transferred the pressure in the rubber tubes into the movement of the pen. Their relationship broke off, however, when Edwards wanted to exploit the device commercially, or so Keeler informed Larson. "It seems to me that this work is altogether too much in its infancy to start anything in a commercial way; and besides my interest is more in the results obtained than in the actual mechanics."

Writing from Chicago, Larson approved of his disciple's high-mindedness. "You did right to keep out of the commercial proposition of Edwards, for I think it would ruin you scientifically." Before he could take his machine to market, Keeler had at least three problems to solve: how to register blood pressure fluctuations in quantitative terms, how to combine physiological measures on a single scale, and how to make the device portable. Only then, noted Larson, would Keeler have "something very good and worth getting a patent on at once."

Keeler was aiming in just that direction. With two pals he moved into a one-room shack on an empty lot in Hollywood, where he bootlegged enough electricity to try new designs. One evening in March Keeler came back to find the shack in flames. He told his father that the lie detector had been saved. The truth, he admitted to Larson, was that the machine had been incinerated, and he had begun work on an improved version.

Again, Keeler got expert assistance, this time from Charles A. Sloan, crime editor for the *Los Angeles Times,* yet another in Keeler's long series of protective friends. Sloan was not just a publicist; he was also a skilled machinist, and that winter Keeler moved in with him so they might work in his machine shop. Their new device, the first to be driven by AC power, included a second pressure reducer and a more reliable recording mechanism. Sloan had come to appreciate Keeler and his machine while covering the disappearance and murder of the wealthy mine owner George E. Schick. Though Keeler's "uncanny" machine had been unable to crack the case, Sloan's articles in the *Los Angeles Times* described Keeler as a twenty-year-old whiz kid whose "emotiograph" never erred. Thanks to the lie detector, Sloan wrote, "the inhumane methods of the third degree will be to future generations as unknown as the tortures of the Spanish Inquisition are to the present school child." In fact, thanks to Sloan, Keeler discovered that the machine need not wean the police from third-degree methods so much as give a scientific edge to coercive interrogation. Although Keeler was not noted for his modesty, this was one case he never publicized.

In the early twentieth century, the village of Los Olivos, thirty-seven miles north of Santa Barbara, was a dusty, isolated hamlet, five miles off the coastal highway, whose 500 residents had a life as insular as the residents of any Appalachian settlement. Sometime around 1916 a seventy-year-old itinerant blacksmith named John J. McGuire began feuding with his neighbors, especially when he was drunk, as he often was. Apparently the local children tossed rocks onto the tin roof of his shop while he was shoeing horses. McGuire quarreled most violently with his neighbor William H. Downs, the town's postmaster, barber, and headman. Their enmity turned ugly after McGuire went around town saying that he'd caught Downs practicing cunnilingus on his wife and forcing his children to fellate him. Then, one afternoon, according to Downs, McGuire barged into his store, drew his gun, and told the storekeeper to say his final prayers. Though McGuire left without firing a shot, Downs decided to take action. For starters, he called in the law. The constable sent old man McGuire down to the Santa Barbara jail for thirty days to let the quarrel cool off. No such luck. Three days after McGuire returned to Los Olivos, at three o'clock in the morning on December 16, 1923, a huge explosion ripped through his shack. Although the blast was heard as far away as Santa Barbara, none of the townsfolk got out of bed. Instead a nearby telephone crew had to carry McGuire's mangled body to a Santa Barbara hospital, where he died a day later—though not before damning Bill Downs to hell and blaming him for the bombing. Suspicion was further aroused when the county sheriff uncovered a freshly dug trench—ideal for a fuse?—running under the fence from Downs's property to McGuire's.

Downs denied any knowledge of the trench, and flaunted his airtight alibi: he had spent the night of the explosion with his family in a hotel in Hollywood. No one in town admitted having any foreknowledge of the blast (though some witnesses later testified that the entire town had actually drawn lots to decide who would do the deed). For four months the investigation was stalled. Then the district attorney of Santa Barbara hired a new assistant, Williard Kemp, who happened to be the King Kleagle of the local chapter of the Ku Klux Klan.

Kemp had heard reports of Keeler's lie detector and decided to set an unorthodox trap. He told Downs that the best way to avoid prosecution was to get in good with law enforcement officers by joining the Klan. To this, Downs eagerly assented. Kemp then secretly arranged for a fake Klan initiation, with the lie detector to play the starring role. In mid-April, Keeler and Sloan drove up from Los Angeles to don Klan regalia and join Kemp, a court stenographer, and twenty other fake initiates, all actual Klan members. Downs was escorted into the initiation room, strapped into the machine, asked a few harmless questions, and then told: "In the solemn secrecy of this room you may answer the questions put to you without fear or favor, for we must ask you, that the records of this organization may be kept clear of all unworthy names, questions pertaining to the death last December of John J. Maguire. Have you any objection to this test?"

A startled Downs answered, "No."

Of the thirty questions that followed, twenty-five named bombing suspects, including Downs himself and his father. Like the map test Larson had used to identify the home state of vagrants, this "peak of tension test" is today known as the "guilty knowledge test." In all cases, Downs denied the charges or kept silent.

No matter. Keeler determined that Downs had specifically reacted to seven names, and a posse quickly rounded up five men for interrogation. After thirty-six hours of interrogation without break, punctuated by sessions on the lie detector, Keeler discovered that two—Harvey Stonebarger, owner of a local machine shop; and William Crawford, a wealthy rancher—reacted strongly when asked if Downs did the killing. Sloan, reporting for the *Los Angeles Times,* fanned rumors that, thanks to the lie detector, confessions would soon be forthcoming. Finally, after an additional thirty-six hours of uninterrupted interrogation on the "mechanical truthseeker," Stonebarger fell to his knees, sobbing, and confessed. Downs and his father had planned the bombing, he confessed; they had bought the explosives, laid the charge with Stonebarger's help, and left town. Then Crawford had lit the fuse with his cigar. The prosecutors jailed all four men. But at the trial Stonebarger recanted his confession on the witness stand and accused the prosecutors of forcing him to tell falsehoods by threatening him with "inhuman third degree" techniques. The men were acquitted.

The lie detector, it turned out, was not so much a thing-in-itself as a mirror that magnified the context in which it was used. Change the context, and the meaning changed. The machine could just as easily amplify the intimidating mob violence of the Ku Klux Klan as the sanctioned investigation of the model station house. Like any mirror (like any placebo), it was not so much an agent in itself as a question: What do you believe?

Keeler believed that the accusations of intimidation in the case were overblown. As subsequent cases would show, he was willing to do what it took. It was not just the lie detector that took its cue from the context. The young man who operated the lie detector could conform just as easily. Larson, by contrast, was appalled. Furnished by Sloan with a behind-the-scenes account of the case, Larson likened the interrogation to torture. Indeed, in his book of 1932, ostensibly cowritten by Leonarde Keeler, he obliquely condemned Keeler for resorting to such methods.

In the short run, however, Keeler and Vollmer were eager to convince the police that their methods got results. In the months after Vollmer's arrival in Los Angeles, the arrest rate increased sevenfold, the crime rate dipped slightly, and citizens acquired a new respect for the police—or so the newspapers reported. Behind the scenes, progress was less encouraging. Many of the Chief's public supporters were undermining him.

Outwardly, Vollmer had the support of the business elite. They stood by him when he moved against the "fixers" who funneled money from African-Americans to the political machine. They were less pleased when he forbade harassment of the Wobblies. And they were furious when he attacked corruption at its source by setting up a covert gambling den to nab policemen and politicians on the take. Through back channels the "kings of the underworld" warned him that they had financed Mayor

Cryer's campaign, and unless he left gambling alone, "the hounds would immediately be unloosed." Soon city councilmen were denouncing the Chief's high-handed ways.

In the meantime, Vollmer lost the support of his own department. Someone had leaked to the newspapers the results of his IQ survey, showing that only 27 percent of officers had received a grade of A or B. This implied that three-quarters of the force were incapable of carrying out their tasks and 7 percent were less capable than a fourth-grader. To add to the insult, the public was given the score of every cop by name. Soon one newspaper was describing Vollmer as "tired, jaded, his voice nervously sharp."

As the end of his year in Los Angeles approached—and the renewal of his appointment began to be publicly debated—a quip circulated through the town: "The first of September will see the last of August." In fact, August Vollmer did not outlast July. His enemies arranged for a sex scandal to escort him out of town. The setup was the oldest form of entrapment, but as Vollmer acknowledged, "Sometimes a frame-up can be staged so perfectly as to seem almost true." On the front page of the *Los Angeles Times* were the coy brown eyes and ample décolletage of the twice-divorced "concert singer" Charlotte Lex, who had filed a $50,000 breach-of-promise suit against the Chief, alleging that they had been lovers since his arrival in Los Angeles. Among her accusations was that Vollmer's lovemaking resembled that of "a cave man."

Vollmer's response to these imputations would become a staple of American scandal management: he offered to test his word against that of Mrs. Lex on the lie detector. "If such a test were made," he challenged her in the press, "the suit would never appear in court."

Needless to say, Mrs. Lex spurned the ordeal. On aquamarine hotel stationery she accused the Chief of ungentlemanly tactics, while she angled for a settlement.

By then Vollmer had rushed back to Berkeley to marry his longtime friend Millicent "Pat" Gardner, who publicly rebutted the foul characterizations. As for Mrs. Lex, Vollmer declared her "either mentally unbalanced or . . . a tool of the underworld." Indeed, the two men who seconded Mrs. Lex's cause were none other than the ministers Shuler and Briegleb. By then Vollmer's contacts had dug up enough dirt on her, and he had her suit quashed.

No wonder Vollmer's assessment of his time in Los Angeles was tinged with bitterness. In his summary report of 1924—the foundational text of professional policing—Vollmer laid out what it would take for Los Angeles to complete the job. But printed alongside his blueprint was its negation: his own police captains jawing about what they thought it would take to enforce the law in Los Angeles, including (1) an end to trial by a jury of one's peers, (2) an end to marriage for "diseased or mentally deficient persons" (with compulsory sterilization), (3) an end to the presumption of innocence, and (4) an end to the existence of whole peoples. ("The Chinese have no excuse for existence. They are gamblers and dope fiends. They are a menace to our community.") The captains ascribed crime to dope, prostitution, Mexicans, and citizens who blamed the police when the courts let crooks off scot-free. Captain R. Lee Heath, appointed Vollmer's successor, warned that the police could not enforce the laws in a city of "poorly assimilated races" without recourse to physical force.

"[I]t is my opinion," Vollmer wrote nine years later, "that under the present system and laws the police department of Los Angeles will never be separated from politics, nor be free from the vicious and frequently unfounded attacks made upon the department by some of the newspapers and preachers of Los Angeles." Vollmer expected that progress would take at least two generations. Indeed, it was not until the late 1940s that Vollmer's plan was dusted off for the LAPD. By then, "professionalization" had become the catchphrase of modern policing—though its purpose had strayed far from Vollmer's vision of insulating the police from politics. When the LAPD finally did insulate its ranks from politics, it did so as much to avoid accountability to ordinary citizens as to escape corrupt politicians. And it didn't even have to give up strong-arm tactics. The situation with the lie detector is parallel: touted as a substitute for arbitrary and violent police interrogations, it could readily be used to extend those techniques. And young Leonarde Keeler showed the way.

CHAPTER 7

# "Subjective and Objective, Sir"

*[A]nd even psychology, which he had eagerly awaited, proved a dull subject full of muscular reactions and biological phrases rather than the study of personality and influence. That was a noon class, and it always sent him dozing. Having found that "subjective and objective, sir," answered most of the questions, he used the phrase on all occasions, and it became a class joke when, on a query being levelled at him, he was nudged awake by Ferrenby or Sloane to gasp it out.*

—F. SCOTT FITZGERALD, *THIS SIDE OF PARADISE,* 1920

SOON AFTER VOLLMER QUIT LOS ANGELES, SO DID KEELER. Charles didn't think it safe for his son to stay on without the Chief's protection. To ease the transfer north—to Stanford this time, ostensibly as a premed—Charles sent the dean a six-page hagiography of his son: from his childhood in the company of John Muir to his forays into forensic psychology, plus everything in between: "He is a practical beekeeper and can handle a hive in all conditions." Charles even contacted Walter R. Miles, a professor in the psychology department at Stanford, to ask if he would supervise Leonarde's work on the lie detector.

Keeler soon became a famous figure on campus. As always, his physical grace and personal charm won him admirers. His lie detector gave him an aura of mystery: an undergrad who had tangled with real-life criminals! With his father unable to cover school fees, he hit on a novel approach to tuition: in an abandoned windmill behind his shack in the hills he set up a dairy farm for snakes. Once a week he fed his seventy-five caged rattlers hamburger meat through a bicycle pump, and every other week he gently pried apart their jaws and "milked" their glands for venom, which he sold to a drug company.

Keeler spent no more time in the classroom than he had before. From the moment of his arrival on campus he was at odds with "the billiard ball tops." The profs were dull; his classes were useless. "If study interferes with the college," proclaimed one famous campus novel of 1924, "—out with the study." Instead, Keeler took flying lessons at the local airfield, and cruised the perimeter of the bay in his spanking new Ford: speeding, drinking, and nearly landing in jail after almost losing his life in a triple 360 on wet pavement with a friend—a lovely actress—vomiting behind the wheel. It was enough to make a young man ask: "And what are the symptoms of love-sickness?" Some young women nicknamed him "Rattlesnake"—with affection, it seems. Others called him a peach, "sweetest when stewed."

What remained of his time he devoted to the lie detector. Professor Miles, initially skeptical, was won over by Keeler's charm and provided him with lab space, equipment, and a lab tech. A notebook from this period shows Keeler refining the instrument, developing new interrogation techniques, and experimenting on "normal" and "deviant" subjects. Most of his effort was directed toward securing a patent for his machine. He poured his mother's inheritance into building a test model and hired a law firm in San Francisco to submit his application. Intermittently he updated Larson.

Vollmer had mapped out a division of labor for his two disciples: Keeler was to pursue the mechanical end of lie detection, with an eye on commercial payoff, while Larson validated the technique scientifically. But this division, which suited their personal ambitions, also threatened to drive them apart. While Larson pursued "open science," Keeler sought "proprietary know-how." The marriage of these two strategies has often been credited for America's technological and scientific prowess. But for practitioners they are often in tension, each racked by internal contradictions. And these contradictions ultimately set Larson and Keeler at odds.

The strategy of open science assumes that objective knowledge is produced when the scientist's "disinterestedness" is guaranteed by norms of behavior that spurn venality in favor of the free dissemination of discoveries. The assumption—which emerged fitfully over the course of modern science—was that meritocratic institutions would reward worthy contributions with the resources to continue research. Under such a system, reputation is a scientist's most prized possession. But why would any society sponsor such knowledge? Princely states or private universities might do it to enhance their prestige, but this hardly accounts for the ratio of money spent in the United States on physics and ballet. In fact, the difference is largely due to the additional claim (often advanced by scientists themselves) that open science produced a body of public knowledge that others found useful, if not immediately, then over the long haul. Such considerations have long induced scientists to point their research in directions that serve their sponsors' interests.

The strategy of proprietary know-how, on the other hand, takes social utility as its starting point, and aims to extract profits from knowledge through products or services. To do so, however, often means keeping the knowledge private so as not to dilute its market value or give an advantage to competitors. One way to do this is to keep the knowledge secret in the manner of medieval guilds, the Coca-Cola Corporation, or the Manhattan Project. The problem here is that it is not easy to keep a secret, especially while convincing others that the know-how can be applied. For its part, society worries that valuable secrets will die with their creators, never generating new knowledge. That is why modern societies have created patent systems, which offer a time-limited monopoly over the use of knowledge, but also require inventors to publish. For the inventor, the challenge then becomes one of timing: deciding when to keep the information secret and when to apply for a patent. The problem here is that the profits generated by proprietary knowledge give outsiders reason to doubt the "disinterestedness" of the knowledge-peddler.

These two paths to creating knowledge rarely exist in these ideal forms, of course. It is their uneasy hybridization which holds tenuous sway in our modern republic. Many investigators in pursuit of open science wish to apply their work to solve social problems—if only to confirm their possession of the truth. Many cultivators of proprietary know-how care passionately about their reputation—if only to better market their discoveries or expertise. Open science and proprietary know-how are mutually dependent, even as they subvert one another.

Under Vollmer's prodding, Keeler and Larson both expressed a desire to collaborate, but their promises were fraught with tension. Larson urged Keeler to create a standardized instrument and join him in setting up research protocols. Cooperation was even the best way to get the lie detector accepted by the courts. As he advised his young disciple, the best way to satisfy the demands of the Frye rule was to "let [the courts] come to us, and they will some [*sic*] of these days if enough of us cooperate." Until that time, practitioners had to trust one another, share their work, and criticize each other frankly. Yet Larson often had to pester Vollmer for updates on Keeler's progress.

But for Keeler, success meant seeing his lie detector widely employed—and counting the remuneration. During his six-year struggle to secure a patent, he oscillated between providing Larson with vague reports on his progress (under the impetus of Vollmer's scolding), and jealously guarding his methods. "Yes, I am ashamed of myself," he confessed to the Chief. "I know mighty well that there is no one that can help me more than Larson in this work." But working on the lie detector seemed to breed mistrust.

From Chicago, Larson periodically confronted Keeler by mail. "You might not be glad to receive this communication as it might look as though I were calling you on the carpet." He heard from Vollmer that Keeler wanted "to get everything protected before writing." This silence, Larson deduced, must have been caused by some senior person trying to capitalize on Keeler's work. Larson assured his young friend that he had no such designs, "as from our past relations you surely know that I would take nothing from you, but on the other hand, [I am] trying to stimulate you and to give you every sort of lead possible." Larson had already declined to patent his own device, he pointed out. Instead he had put his trust in Keeler's aptitude for mechanics and experimental work, with Keeler to reap whatever financial rewards ensued. "I then could devote my time to clinical experimentation."

As if on cue, Keeler pinned the blame on Miles. First Edwards the physicist, then Sloan the reporter-machinist, and now Miles the psychologist had each wanted to take control of the machine. The generous psychology professor had been playing a double game: supplying Keeler with lab space and equipment only to gain ownership of the lie detector. He had become reluctant to let Keeler demonstrate the machine publicly. And it was Miles who had dissuaded Keeler from sharing his progress with Larson—or so Keeler wrote. "After all your kindnesses, help and advice, you were the incentive which started me in this work, and you certainly deserve to know what I was doing. I have made it clear to Miles that whatever I am doing under him or in his department that I should be free to disclose any developments in my field to you."

Then, to top it all off, Miles had showed the device to a representative from Tycos Instruments. Professor and undergraduate were now traveling east to confer with the manufacturers. On their way they would pass through Chicago. This would be Keeler and Larson's first meeting since 1923.

That brief stop in Chicago in April 1927 had a lasting impact on Keeler's life. In the morning Keeler met with Herman Adler, director of the Institute of Juvenile Research, who offered him a job—though two years would pass before Keeler could accept. In the afternoon, Keeler secretly met with Larson to show him the drawings of the machine—against Miles's explicit instructions—and to solicit his views on how to manage ownership of the apparatus.

He returned to Stanford to finalize his patent application and begin discussions with a manufacturer in Oakland. Securing that patent—the foundation of Keeler's mystique—would prove frustrating. The problem was that the machine was hardly new.

After reviewing his initial application of 1925 for "Recording Arterial Blood Pressure," the U.S. Patent Office rejected every claim but a minor one about the design of the tambours. Even his own attorneys doubted that Keeler had done more than combine familiar off-the-shelf instruments. Nor was the Patent Office persuaded on the basis of the "great scientific interest" in his device. So Keeler gradually limited his claims to the redesign of the tambour (courtesy of Edwards and Sloan), now robustly fashioned like an accordion of alternating rubber and metal. It still did not give an absolute reading of blood pressure—none of the automatic machines did that—but it highlighted variations in an amalgamated mix of systolic pressure and pulse. On this basis, he was awarded a patent in January 1931 for an "Apparatus for Recording Arterial Blood Pressure."

By then, Keeler had been negotiating for several years with the Western Electro-Mechanical Company of Oakland. Keeler's father, who had complained about his son's obsession with the lie detector, was incapable of remaining unenthusiastic about anything for long. When Leonarde brought him in on the venture as a fifty-fifty partner he suddenly became its biggest promoter, quibbling with the manufacturer over pricing, packaging, brochures, royalties, and even technical matters. Always one to think in universal terms, he wanted to pursue patents in Canada, England, Germany, France, and Italy.

The sticking point was sales. The manufacturer wanted to maximize the multiple of sales and price, and looked forward to marketing the machine to hospitals, police departments, and psychopathic clinics (which is to say, for the sick, the bad, and the mad). Charles wanted to go farther, dispatching sales agents across the nation to sell the machine to the personnel departments of factories, banks, trusts, and insurance companies, wherever people needed to vet the honesty of their colleagues. He even thought the machine could be used to weed out the fakes among spiritualistic mediums, thereby putting psychical research on a scientific footing. More prosaically, vast sales would provide a steady income while his son pursued more lofty avenues of research into the subtle relationship of mind and body.

But Leonarde took a different tack. He insisted that his contract give him a veto over every sale. Keeler realized that the instrument could not guarantee reliable results on its own, and that the reputation of his machine (and hence its long-term sales prospects) might be damaged if he "turn[ed] out machines promiscuously to untrained individuals." Keeler's strategy was therefore based not on selling the machine itself but on selling his personal know-how.

The manufacturer complained that lie detection would never be accepted widely until Keeler sold a standardized instrument—but standardization was difficult to achieve in small production runs. They had their own reputation to consider. Keeler sympathized, but insisted on the restrictions.

There was one final crucial matter to resolve: what would the device be called? Although newspapermen had named it the "lie detector," experts detested that name because the machine did not detect lies as such (though this is in practice exactly what they used it to do). Keeler gathered suggestions for an alternative. Charles Keeler preferred "Emotograph" as both simple and expressive. For a long time, Leonarde stuck with "Respondograph" because it seemed to convey the ambiguity he wanted. In the end he settled on the Keeler Polygraph: "polygraph" was the familiar term for any device that recorded multiple bodily measures. In retrospect Larson came to see this decision as the moment of the fall, the fateful moment when the quest to explore the human psyche in its normal and pathological variants devolved into a "commercial and purely mechanical" venture. As Larson would note, the very familiarity of the description, "many writing instrument," was a sleight of hand "with no specific connotation," leaving only Keeler's personal mastery in view.

Yet every invention worthy of the name must claim an inventor, by which choice it reveals the sort of invention it is. By now the lie detector had accumulated a host of "inventors." First had come Münsterberg with his word-association tests and theories of human automatons. Then had come Marston with his blood pressure cuff and undergraduate games. Then Larson had joined their ranks, publishing papers on how a physiological apparatus could be applied to real-life crimes. And now Keeler had put in his claim, with his patented device. In keeping with its origins in open science, Larson's contraption had been open for inspection. In keeping with its proprietary purposes, Keeler packed the same workings inside an elegant walnut "box" which was simultaneously more portable and more mysterious, and which extruded a thin scroll of lightly ruled paper on which two delicate pens left an inky red judgment, a judgment whose meaning only the box's priestly operator could divine. The oracle of the Keeler Polygraph was Leonarde Keeler.

Keeler honed that personal mastery during his years at Stanford when he developed what one might call the "software" of lie detection: interrogation rituals that allowed the operator to extract maximum information from the subject, ideally a confession. While the subject was mentally preoccupied with what the machine recorded—a classic misdirection, familiar in stagecraft—the "software" allowed the operator to work his will. This software was a key component of Keeler's know-how; it was what made Keeler himself a lie detector, and it constituted his true innovation.

Keeler began by calibrating his instrument against the one Larson had left behind in Berkeley. He tested the device first on himself, only to discover a cardiovascular abnormality, a legacy of his old streptococcus infection that caused his heart to skip a beat every time he was under stress. So he switched to testing the machine on friends and roommates, administering doses of sodium nitrite and amyl nitrite (supplied by his uncle, a physician) and tracking the acceleration of their pulse until he had to dose them with coffee to keep them from fainting. He tested students after depriving them of sleep for seventy-two hours, and took recordings of mental patients at the Palo Alto Veterans Hospital, where he discovered that he could identify paranoid, schizophrenic, and catatonic patients from the pattern of their blood pressure.

Keeler organized these interrogation rituals around a well-conceived geometry. The room was to be quiet and unadorned. In later years, it would be equipped with a one-way mirror. In a pretest interview, the interrogator discussed the phrasing of the questions he would ask, giving the subject time to resolve doubts about their meaning (and reveal additional facts about the case). The interrogator then took an initial reading to set the polygraph's arm cuff at a pressure most people could tolerate for about fifteen to twenty minutes. He then hooked up the seated subject to the machine, took a position from which the subject could not see the machine, and began asking alternating relevant and irrelevant questions at one-minute intervals.

But that was not the end of the interrogation. Instead, the subject was shown the chart and offered a chance to explain any anomalous reactions, which were then reformulated or repeated in a second run on the machine, the whole series being repeated until the record showed no anomalous reaction. This ritual was designed to focus the subject's mind on the matter at hand, make her conscious of the stakes of deception, and put her in the position of having to explain herself. Under those conditions, Keeler discovered, an astounding number of people would tell him everything he wanted to know.

Soon after he arrived in Stanford he added a new twist: the "card test," one of Keeler's signature gambits. Immediately after hooking a subject up to the machine, Keeler presented her with a deck of eight ordinary playing cards, asked her to select one, and then instructed her to deny that each card was hers as he flipped them over one by one—even when it was her chosen card. Then, by examining the record of her physiological reactions, Keeler identified the card. The trick had several purposes. It obliged the subject to tell a lie, providing Keeler with a record he could compare with possible lies down the road. More important, it convinced the subject that he could catch her telling a lie, heightening her fear of being caught and hence the likelihood of her being caught. Keeler boasted that he had been able to catch all but two out of sixty students who played the card test.

But just to make sure, Keeler often used his skills as an amateur magician to stack the deck.

Not everyone fell for his ruse. A young Stanford undergraduate named Katherine (Kay) Applegate, a fellow psych major, first snagged Keeler's attention when she beat the card trick. Miss Applegate was hardly the innocent victim John Larson had discovered in Margaret Taylor. Somehow she concealed her true feelings from Keeler. According to Applegate family lore, Kay had only pretended to look at her card in the first place. She beat the machine by refusing to play by its rules and she soon returned to tweak the lie detector and its owner.

It began as a college prank. Late one night she sneaked into the psychology lab with a girlfriend and inserted a mannequin into the lab's instrument for testing muscle reflexes. When the self-appointed guardians of the mental sciences threatened to sniff out the guilty party by running every major on the lie detector, the women struck again. As she gleefully informed her parents: "We bought some men's underwear, dyed it purple and decorated it appropriately, and went down at midnight and put [the dressed mannequin] in the lie machine. It created a sensation and we are strongly suspected but so far have maintained our innocence quite beautifully."

Keeler was captivated. For an English composition class, he wrote up a version of the sorority sneak-thief case, melding Larson's famous coup at College Hall with his own experiences investigating a sorority in Los Angeles. The resulting soft-core tale differed from Larson's experience in one crucial respect, however. In Keeler's version, the man with the lie detector does not marry the innocent victim. He marries the duplicitous perpetrator—albeit after she reforms.

Keeler and Applegate, competitive pranksters, had much in common. Kay wrote to her mother about her new friend Nard Keeler, "the lie detector man." He had taken her riding in his roadster and out for a milk shake. He had shown her his snake farm and his photographs of the high Sierra. But Mommie mustn't get the wrong idea. This new type of male-female friendship took some explaining to the older generation. "Nard Keeler is the pleasantest thing I know of but that is as far as it goes." It would be several years before they became romantically involved.

Kay was a striking young woman, a blue-eyed, fair-haired girl from rural Washington state, strong-willed and independent, with a sharp tongue and a dash of self-mockery. "Friday night and no news except that I am 5' 6 $^{3}/_{10}$" tall and weigh 129 $^{3}/_{4}$ lb. with no clothes and am in excellent shape." She was as comfortable on a horse as behind the wheel of a car. Her Applegate forebears had been among the first white settlers in the Pacific northwest, and while she was self-conscious about her lack of urban polish, she was proud of it too. She could be difficult, with quick changes of mood, slow speech, and a sarcastic turn of phrase. At times she was beset by a painful social shyness, punctuated by bursts of risk-taking.

Leonarde's younger sister—who grew fond of Kay, but was always protective of her beloved brother—described her as possessed of "an aloof dignity." She kept her longish nose unpowdered, her red cheeks unrouged, her lips colored with just a dash of lipstick. She wore glasses, though none of her photographs show this. She was tough-minded and considerably smarter than Leonarde: "startlingly brilliant," he later boasted to his father. "I only wish I had the keen mentality she possesses."

All her life she showed remarkable bravery: as a rodeo rider, as a criminologist, and as a pilot with the Women's Auxiliary Service in World War II. Her courage included facing down social disapproval. The way to conquer fear was to always tell the truth. She was either all for something, or all against it. After only two months at Stanford she had fallen for a law student. "He is over six feet tall and weighs just the right amount—being beautifully athletic, [and] the best looking man I've seen on the campus." Six months later she informed her parents to "get it through your beans: it is ALL OVER!" She announced that she was too selfish to ever get married.

Thereafter, she set out to enjoy herself. After graduating in 1927, she and a girlfriend stowed away on a steamer bound for Oahu, where she flipped pancakes in a restaurant window, worked as a "forelady" in a pineapple cannery, chauffeured a tourist bus, and even learned to surf. Most exhilarating of all, she piloted a World War I "Jennie," her first taste of flying, the great love of her life.

---

Keeler did not graduate with Katherine or the rest of the class of 1927. Having earned C's and D's in almost every course except experimental psychology, having failed psychological statistics (a required course for majors), having racked up more "incompletes" than any other Stanford student within memory, he stayed on. When he finally left in 1929, he still had no diploma—to his father's eternal shame. It was all Professor Miles's fault, agreed father and son; the psychology professor was exacting his revenge for being cut out of the lie detector.

By then Leonarde had a more important reward: a job at Chicago's Institute for Juvenile Research, where he would be replacing John Larson. Moreover, he at last had a working model of his machine, patent pending. Vollmer wrote to Larson: "Keeler has built a perfectly splendid machine which is now ready for the market. . . . [I]t is a fool-proof portable instrument which gives a quantitative and qualitative curve. . . . It is a very interesting and simple device and will undoubtedly do the job." Keeler promised Larson dibs on the first machine: "You're the only other man I want experimenting with it for the present." It was time to take the machine where the action was.

PART 2

# *If the Truth Came to Chicago*

*Law should be impartial and just. In Chicago it is neither.*

—WILLIAM T. STEAD, *IF CHRIST CAME TO CHICAGO!*

CHAPTER 8

# The City of Clinical Material

*He that has eyes to see and ears to hear may convince himself that no mortal can keep a secret. If his lips are silent, he chatters with his fingertips; betrayal oozes out of him at every pore.*

—SIGMUND FREUD, "FRAGMENT OF AN ANALYSIS OF A CASE OF HYSTERIA," 1901

BOARDING THE "EL" THAT FIRST SUMMER IN CHICAGO OF 1924, John Larson was relieved of $189 by a dip mob. That was nearly a month's pay. He had been squeezing onto the train at the La Salle Street platform—a bulky suitcase containing the lie detector in one hand; a volume of Herbert Spencer, patron saint of social Darwinism, in the other—when two youths pushed past him, and as he turned to apologize for blocking their way, the "dip" lifted his wallet from his inside coat pocket and skittered out the closing doors before Larson realized he'd been played. Too late: the park district cop on the platform waved the train on. Later, at the detective bureau, the desk officer dissuaded Larson from signing a complaint unless he was sure he recognized the culprits among the gallery of rogues. And since Larson, as usual, wasn't *sure* . . .

Welcome to Chicago, "[it's] getting to be a great town," as Charles Drouet said to Sister Carrie as their train pulled into the station. Or in the words of Sidney Blotzman, the real-life juvenile miscreant of *The Natural History of a Delinquent Career,* reminiscing about his first "el" trip to the Loop: "Nothing at home nor at school equaled this joyous occasion. . . . Wild bedlam reigned everywhere and system was unknown. . . . To mix with the crowd, and to struggle from one interesting thing to another, always seeking something more interesting and receiving many pleasant surprises, was one thing that I never grew tired of, no matter how many hours or how many days I spent at it. When I finally arrived home my lying lessons came in handy and proved very useful in pulling the wool over my mother's eyes."

This was how the system worked. The dip mobs recruited young kids from the slums. The delinquents worked in gangs. In return for protection, the mobs cut the cops in on the take. Those who were caught bought their way out higher up the chain.

Describing his rude welcome to Chicago in a letter to Vollmer, Larson mocked himself as a sap, but also expressed his fury. He vowed to look out for the gang on his own, both in the Loop and at the Joliet prison where he was beginning to run deception tests on the inmates. Thanks to the instrument he soon identified the pickpockets as confederates of "Immune Eddy," a racketeer famous for getting out of prison as fast as the authorities could toss him in. Larson began to troll downtown in hope of "playing the loop for these same babies, and if they ever trip us again, if I found anything doing, I think that there won't be any necessity for complaint, for they will be taught a lesson through a few broken limbs."

Larson had always been impulsive, and Chicago brought violent passions to the fore. Larson had begun to study criminal paranoia—"the disease is ushered in by delusions of persecution"—and the city was a web of conspiracies. He was attending medical school and working part-time at the Institute for Juvenile Research, where the intrigues made his head spin. The director there, the psychiatrist Herman Adler, had been angling to get his mitts on a lie machine since first hearing of the device. Adler had even come out to California one summer, working alongside Larson analyzing suspects. Now he had brought Larson to Chicago, paying him to construct a quasi-portable lie detector and giving him a two-room laboratory. The setup seemed ideal, except that nothing in Chicago was as it seemed.

In the 1920s, Chicago was where the action was if you were criminally inclined—or, for that matter, a criminologist. The city was the murder capital of the world: boomtown of gangsters and bootleggers, habitat of rackets and racketeers. Right across the back alley from Larson's office was one of Al Capone's "sugar buildings," where Eliot Ness, the untouchable himself, one day dragooned Larson into helping out on a raid. This sort of depravity was exactly why Larson had come. Chicago, Larson wrote to

Vollmer, was a "city fronting a lake of drinking water drained by a sewage canal." But the city, he acknowledged, had one "redeeming feature": "its vast clinical wealth of disease—especially syphilis and crime and everything that is wrong with the present day civilization."

The challenge of Chicago was that crime was not just on the "outside" in the Loop, the tenements, and the vice districts; crime was also on the "inside," among the supposed enforcers of the law. Larson wrote to Vollmer that Chicago "has any [city] in the country backed off the map for dirty, dumb police work and crooked judicial administration." Everyone was on the take: the thuggish police, the duplicitous prosecutors, the bought judges, the feudal wardens, and the corrupt parole boards—all the way up to the crooked pols who appointed these underlings on instruction from the crime bosses. Fueling the regime was street-level enforcement by the city police. "[T]he cops use every antediluvian method that there is here," Larson wrote to Vollmer. Their method of eliciting evidence was with "rubber hose, black jack, and boot, and I have seen some first hand examples."

This made Chicago the premier test case for Vollmer's reform crusade, and Larson had been compiling a catalog of police abuses—until Adler ordered him not to meddle in police activities. He published the list of abuses anyway, contrasting Chicago's horrors with Vollmer's remarkable achievements in Berkeley, just to prove that an honest police force was "not Utopian." "I sure would like to work with you in this city," Larson wrote to the Chief, "but you would probably be bumped off inside five minutes, as [the] politics is too strong." When a city engineer arranged for Larson to demonstrate his lie detector to the mayor, the politico listened with apparent boredom until the engineer explained how the machine might be used to test city workers about their complicity in graft. That woke the mayor up. He ordered Larson out of his office and into the office of the police chief, who likewise refused to sit for a demonstration and instead grilled Larson and his assistants about their qualifications.

Larson would spend a total of nine years in Chicago. From 1923 to 1927 he divided his time between medical school and research at the Institute for Juvenile Research. Then, after a two-year fellowship in psychiatry at Johns Hopkins and a year at a psychiatric hospital in Iowa, he returned for another five years at the Institute for Juvenile Research, from 1930 to 1935. During those years, Larson made himself into a Diogenes by the lake;

his goal was to transform the "lie detector" into a new kind of diagnostic instrument that could track criminality to its psychophysiological source. Hardly two years passed before he was accusing colleagues of lying, accused of being a liar, and asked to sign statements he considered lies.

Chicago was not only world-famous for crime but also world-famous for what it was doing to address the causes of crime. In the early decades of the twentieth century a coterie of well-to-do businessmen and their well-meaning wives had joined arms with rising professionals in medicine, law, and the social sciences to reform the justice system. These progressives rejected the Victorian assumption that a criminal was a moral failure to be punished by unyielding law. Instead, they converged on the notion that crime was to be treated like a disease. The question was: what kind of disease?

Some of these progressives—the "nurture" progressives—turned to the new social sciences to suggest that criminality was a product of an unhealthy environment. Moral reformers like Jane Addams reached out to impoverished neighborhoods with welfare projects like the Hull-House settlement. Sociologists at the University of Chicago marked out city neighborhoods like laboratories in the hope of studying crime in its ecological niche. Finally, the city's psychiatric and legal elite banded together to find new ways to treat delinquents whose lives had been deformed by violent surroundings. Meanwhile, another group of progressives—the "nature" progressives—focused on the laws of heredity, and blamed the growing crime rate on racial degeneracy. These eugenicists proposed laws to sterilize the unfit and curtail immigration. Both nature and nurture progressives agreed that criminality was, in some sense, driven by forces beyond individual control, and that this disease was, in the final analysis, an affliction of the mind that began young and ultimately warped the delinquent.

These progressives, most of them affiliated with the city's South Side, founded institutions to match their vision. In 1899, under the impetus of women moral crusaders, the state of Illinois launched the world's first juvenile court in Chicago. In 1909 additional philanthropic efforts created the Institute for Juvenile Research to supply the court with social-psychological evaluations of young offenders. In 1910 John Wigmore of Northwestern

University Law School founded the nation's first criminological research center, the American Institute for Criminal Law and Criminology, and its flagship journal. And Chicago itself established the nation's first and largest municipal court to deal with tens of thousands of annual quasi-crimes, plus a rival psychopathological laboratory of a eugenic bent to analyze offenders for deviance and feeblemindedness.

Yet thanks in part to Prohibition, by the time Larson arrived in Chicago the crime rate had soared to fearsome heights, and a new generation of hardheaded civic leaders were demanding that criminals be confronted with "swift implacable justice." In 1919 a group of businessmen founded the Chicago Crime Commission, which took the view that crime was a well-managed business conducted by depraved but rational men, and could be combated only by an equally well-managed "war on crime" fought with scientific means. These new crime-fighters, associated with the city's North Side, sneered at the South Side progressives as effeminate do-gooders. Said the head of the commission, "The mawkish sympathy of good, but softheaded women with the most degraded and persistent criminals of the male sex is one of the signs of unhealthy public sentiment." The concurrent rise of J. Edgar Hoover's FBI was fueled by a similar assertion of manly resolve and scientific incorruptibility.

The one thing on which all crime-fighters agreed was that science had the cure: whether it was the social science of the nurture progressives, the biomedical science of the nature progressives, or the managerial police science of the Crime Commission. The great virtue of the lie detector was that it seemed—at least initially—to finesse all these contradictions. Suspects on the lie detector were acknowledged to be willful, rational beings who knew right from wrong and could freely affirm or deny guilt. At the same time, the success of the lie detector depended on subjects' inability to control their bodily emotional reactions, whether those reactions were socially conditioned or innate. It was while working at the Institute that Larson first explained how this equivocation was captured by the instrument's two measures. A rise in a subject's blood pressure—presumably caused by an irrepressible fear of being caught—signified an involuntary reaction of the autonomic nervous system, or that self which dwells in the heart. A hitch in the breathing rate—presumably caused by a conscious preparation to tell an untruth—signified an action of the central nervous system, or that self which resides in the brain. Willful or conditioned, free or determined, brain or heart, the divided subject was caught either way she turned; damned if she had done the deed, and double-damned if she denied it. Unless, of course, the subject was self-deceiving. In that case, Larson noted, the lie detector became a tool for separating malingerers from psychotics, making the device an aid in psychiatric diagnosis, the core mission of the Institute for Juvenile Research. It was a machine as marvelous and equivocal as the human machine it mimicked.

No one believes in the myth of honest children any more. On the contrary, we assume that kids lie readily. But we still wonder why. Does lying come naturally, or is it learned? Are children malign little devils or fabulists who delight in make-believe? Even as the psychologists of the early twentieth century shucked off the moralistic tone of their Victorian fathers—accepting deception as a natural part of human development—they still considered it a stage to be overcome on the road to adulthood. G. Stanley Hall, the great psychologist of adolescence, suggested that truthfulness "comes hard and *late.*" For Hall, this process of personal maturation recapitulated the emotionally fraught development of his "precocious" homeland as it lurched from savagery to industrial civilization—making the individual passage to adulthood all the more difficult as society became more complex and demanding. Many young people risked falling into a life of crime. Indeed, an apparent epidemic of juvenile delinquency in the 1920s prompted a nationwide obsession with childhood deceit. The portrait of American childhood was disheartening. Studies seemed to show that as many as half of all schoolchildren had cheated in class at one time or other, with dishonesty rates rising as one descended the scale of IQ and social standing. This was read as a sign of America's growing pains, a disjuncture between primitive social conditions and industrial modernity. And no city in America had experienced sharper growing pains than Chicago, the self-proclaimed adolescent metropolis.

William Healy, first director of the Institute for Juvenile Research, conceived of lying as a gateway vice: bad enough in itself, worse for what it led to—and for what it obscured. Healy was a disciple of both William James and Hugo Münsterberg, and he was in close contact with August Vollmer in Berkeley. Healy found lying "excessive and notorious" in only 15 percent of male juvenile offenders and 26 percent of females. Yet in many cases deception obscured the deeper sources of crime. According to Healy, dishonesty was a product of poverty, aberrant home conditions, gang pressures, and "clandestine sexual habits." These social conditions laid down patterns in the child's emotional life, which could be treated only by honest engagement between the child, the child's parents or guardians, and (if need be) professional psychiatrists and social workers. Reduce childhood deception, Healy argued, and you just might straighten out any criminal tendencies before the youth grew crooked.

When Herman Adler replaced Healy in 1916, he expanded this mission. At the time the institute was placed under the Illinois Bureau of Public Welfare's Criminology Division. In addition to the 1,000 children evaluated for the juvenile court each year, the institute became responsible for psychiatric evaluations of all state prisoners and parole applicants. Access to these populations made the institute an ideal testing ground for the lie detector.

As fate would have it, the most notorious juvenile crime in American history hit the front pages just as John Larson arrived in Chicago. Overnight the nation became consumed with why Nathan Leopold and Richard Loeb, two precocious friends from privileged families in Chicago, had capriciously but deliberately murdered a fourteen-year-old boy they hardly knew. Larson, watching the affair, bemoaned the farcical light the trial shone on the so-called "impartial administration of justice"; the "hysterical" newspaper accounts; and especially the blatherings of the self-important psychiatrists of every stripe.

According to the state's official psychiatrists, Leopold and Loeb were fantasists engaged in games of dominance and deception. The young genius, Nathan Leopold, they reported, had often fantasized that he was a willing slave to a king whose life he had saved, and whom he deceived. Handsome, suave Richard Loeb had long fancied himself a master criminal; as befitted a skillful and unrepentant liar, he had a reputation for being frank and truthful. Even when his examiners knew he was lying, they admitted that they could not detect any telltale sign. Healy, back in town to testify for the defense, carried out a battery of psychiatric tests. His testimony hinged on the divergence between the young men's superior intellect and their warped emotional life, a divergence so great it resembled a split personality. Yet as he admitted on cross-examination, as both were consummate liars how could he be sure?

Larson was sure he could do better—get "good stuff on them"—if he could only run them on his machine. Not only did it prey on the disjuncture between the intellect and emotion; it was ideally suited to distinguish psychiatric illnesses from fakery. Yet not until Leopold and Loeb were saved from the electric chair did the Institute for Juvenile Research assume control over America's most notorious killers. In the Joliet prison, Larson subjected them to his own tests, including a session with Loeb on the lie detector. As was his wont, he even tried to befriend the pair, seeking over the years to solve the riddle that had eluded so many rival experts. This attempt would prove his undoing.

In the meantime, as the institute's staff swelled to 100—thanks to philanthropy spurred by the fear of another Leopold and Loeb—Larson had to defend his approach against rival methods. All happy families may be alike, but do-gooders don't always get along. The lie detector may have initially seemed to finesse divergences within the reform movement, but it also threatened to expose fissures. Sociological investigators urged delinquents to tell "their own story." Those in the Healy-Adler "orthopsychiatric" camp approached their patients from a medical-psychiatric point of view. Still others, like John Larson, looked to physiology for evidence of criminal pathology. The institute tried to weave these strands together by having each patient examined by a team of professionals, including a social worker, a physician, a psychologist, and a recreation worker. All these examinations culminated in a series of in-depth psychiatric interviews. But friction developed beneath the surface.

Even among the physiologists, there were differences. Larson's colleague Chester Darrow tested his subjects in a "Photo-Polygraph," an experimental cocoon in which pubescent boys and girls were subjected to various stimuli while festooned with gauges. Boys were read phrases like "masturbation pleasant," then smacked on the head. Girls were read words like "menstruation," followed by a slap. Darrow envisaged his apparatus not so much as a lie detector but rather as a way to explore the general relations between emotions and reflex reactions, showing for instance that forms of stress which provoked the flight-fight response also produced sweaty palms, which presumably had once been useful for gaining traction. Darrow helped found the science of psychophysiology.

Larson, by contrast, focused on individual cases and cures. Between 1923 and 1927, he questioned over 200 children. Some were frightened by the test and cried, but when told they would come to no harm, all submitted, except one five-year-old girl. About half had been referred for making false accusations (usually of a sexual nature) against adults, for faking psychiatric conditions, or for confessing to crimes they had not committed. The other half were suspected of lying to a nurse or had refused to speak. Larson extracted confessions from over half the children—without, he noted, directly accusing them of lying. He then returned them to their caseworkers with their resistance removed. For instance, he got a boy who lied repeatedly about sex to confess to masturbation, whereupon the boy showed no further reaction on the instrument. Once this boy was freed from the lie, psychotherapy could begin. Only 8 percent had their stories confirmed by the test. This left a large group of children, nearly 40 percent, whose tests Larson marked "disturbed" and sent back with a notice that the polygraph test should not be used as evidence in court. Yet even in these cases Larson thought he had identified the complexes inhibiting therapy.

Instead of winning him their gratitude, however, Larson's work threatened to expose his colleagues as dupes. And to be honest—always Larson's policy—he let them know it. Thanks to the lie detector, he boasted, he was "showing up some of the methods of these pseudo scientists and psychologists and psychiatrists who think themselves so almighty." The sociologists had no way to vet the tall tales the children fed them (and it was no good claiming, as they sometimes did, that the patients' braggadocio was the phenomenon they were studying). Larson had particular contempt for the female social workers, who acted, he said, as if they "run the roost." He didn't let them sit in on his cases. "One woman, especially has been pestering me to get in, and just because she [is] a pest I didn't let her in." The women couldn't understand how he could get confessions in ten minutes on cases that had baffled them for ages. "It seems these women have an idea that I use some brutal or unscientific method of getting the confession and they want to get the goods on me."

He didn't think much more of the psychiatrists. It didn't help that they were mostly Jews. Larson considered Adler, with whom he was cowriting a paper on self-deception, a self-promoting fraud and "colossal joke." He told Vollmer that he already considered himself a better psychiatrist "than this baby here," thanks to his police experience—and his machine. In one quick session on the machine, Larson solved the staff's learned controversy over the hallucinations of "Tony" (age fifteen). Larson got the boy to admit he was making the whole thing up: "Dose guys were kidding me and I went one better, gees doc. . . . I made suckers out of dose boids didn't I?"

These resentments boiled to the surface when Adler, under pressure to cut the budget, fired Larson's assistant and insisted that Larson quit medical school to work full-time at the institute. He also wanted to know why Larson had not upgraded the lie detector, or spent more time interrogating prisoners at Joliet. That did it. Larson exploded. He had not come to the institute to work on machinery; he was counting on Keeler for that.

As for Joliet, Adler himself had blocked Larson from telling the truth about the nefarious goings-on there. "Any honorable work there was a farce. Either the convict was innocent and framed in by officials whom Adler said we could not and dare not expose, or he was guilty and was buying his way out through channels we must ignore." In a series of lengthy midnight letters to Vollmer, typed at a frantic clip, Larson laid out a plot worthy of 1920s Chicago in the raw. The story revolved around Walter Stevens, the "immune gunman" himself, a onetime labor slugger, now top triggerman for Johnny Torrio, who once boasted, "I own the police." Stevens had been arrested 300 times yet had never paid a fine or done time, lodging comfortably in a policeman's apartment while the cops searched for him, until at long last he was brought to trial for the murder of a cop in Aurora. A dozen big-city pols, including the chief clerk of the criminal court of Cook County, confirmed Stevens's alibi, repayment for services rendered during the governor's recent trial for embezzlement, during which Stevens had destroyed evidence and intimidated witnesses. But the trial was in Kane County and the jury voted to convict. Even then it took public pressure and a Supreme Court ruling to dissuade the governor from granting him clemency. "Nobody knows just what the mysterious influence is that protects [Stevens] and imperils the public," said the *Chicago Tribune*, "but most folks express it in the word 'politics.'"

In prison, the dapper Stevens had sat for a session on Larson's lie detector, boasting afterward of his complicity in multiple murders, his pull with the authorities, and his confidence that the parole board would set him free after the primary elections. Larson had the temerity to write all this down. Moreover, he got his hands on a carbon copy of a letter from Stevens—smuggled out of Statesville by a trustee who had risked his life—indicating that Walter Jenkins, director of Public Welfare (and hence Adler's boss), was holding out for $2,400 to secure Stevens's release by the parole board. Not long after the primaries, lo and behold, Stevens was paroled and promptly disappeared.

According to Larson, Adler "nearly had a hemorrhage" when he saw the complaint. Adler demanded that he drop the case, as it was "loaded at both ends," and forswear his accusations. Larson refused. How could he admit to being "wrong and a liar in everything," when he was the one who was telling the truth? How, Larson wanted to know, could Adler square this sordid tale with his claim that the Institute for Juvenile Research merely provided scientific advice to the parole board? His only other option was resignation. Their exchanges became shouting matches. Adler accused Larson of using "caveman tactics"; Larson accused Adler of breaking his promises, of becoming hysterical.

Sensing conspiracies everywhere made Larson paranoid. Whenever he spoke with Adler on the phone, he made sure his wife secretly listened in. He took out $5,000 in life insurance. On several occasions he asked Vollmer to "ditch" his letters "for paranoid reasons." Winding through the maze of corruption, he sometimes wondered if he himself was lost. He admitted that his letters were "scatter-brained." He was so apoplectic that he sometimes felt like exploding. To bear the burden of the truth was agonizing.

From California, Vollmer tried to calm his disciple. The "strong-arm methods" of police work were not suitable for his colleagues. "Your reactions of necessity will be the full-blooded American reaction, but be patient and your time will come." He warned Larson not to run afoul of men who could louse up his chance for a medical degree, and advised him to drop the Stevens scandal, bide his time, and then, once he got his M.D., show them all up. In the meantime he needed the institute's access to subjects. With one quick phone call, Vollmer persuaded Adler to take Larson back at the institute.

In his letters to Vollmer, Larson had vilified Adler as a liar, loyal only to his own interests; no wonder, Larson said, no one trusted the psychiatrist. Larson, by contrast, considered himself loyal to his ideals, uncorrupted by material gain or personal advantage, and unswervingly honest. If anything, he was proud of his "natural inclination . . . to tell people what I think of them." These were the qualities of a trustworthy man. "I at least have open enemies," he typed in a letter to Vollmer, adding by hand, "if any."

In 1927, John A. Larson, M.D., Ph.D., left the cesspool of corrupt Chicago to take a fellowship in psychiatry at the Johns Hopkins Medical School. Adolf Meyer, head of the Phipps Clinic there, was America's most influential psychiatrist, a Swiss-trained neurologist who had brought modern clinical methods to the New World, only to discover, under the influence of William James and John Dewey, the possibility of a "truly American advance" toward "a new conception of man." Again, it was Vollmer's connections that sprang the fellowship for Larson: Meyer had first been introduced to Larson in Vollmer's office. Meyer called his approach "psychobiology," and he explained its principles in a lecture before Larson and the Institute for Juvenile Research in 1926.

Psychobiology sought to unite all contending schools under one pacific faith; it combined the Freudians' emphasis on dynamic psychology, the behaviorists' focus on public actions, the psychiatrists' mania for classification, the neurologists' focus on brain function, the physiologists' researches into the bodily correlates of the emotions, and even the sociologists' understanding of the environment. What organized this contradictory mélange was an intense focus on the individual patient, the "person," as Meyer disarmingly put it, with each individual to be considered "an experiment in nature." The task of the psychobiologist was to help the patient adjust to the conditions of life, much as animals either adjusted to their circumstances or faced the Darwinian consequences. Meyer's method was both simple and humane. But to its critics, psychobiology was both simplistic and moralistic.

Take the problem of eliciting the facts of a life history. Meyer had a naive faith that the psychiatrist could elicit the most relevant facts from each patient, whereas Freud and his successors were convinced that only roundabout maneuvers, like the analysis of dreams or free associations, could coax patients to reveal traumas hidden even from themselves. For Freud, the manner in which self-deluding patients represented their lives was as apt a key to their psyche as what had "actually happened." Because patients' lies were as revelatory as the truth, a lie detector had little value. For his part, Meyer was confident that any fact worth recording was the sort of public, objective fact the physician and patient could readily agree on. To aid in this, the psychiatrist might well use intelligence tests, memory tests, and even physiological tests like a lie detector. Hence, Meyer was intrigued by Larson's technique, even as he insisted that it was only a tool of diagnosis.

From a different angle, others accused Meyer of moralizing. Like the "personality studies" he inspired, psychobiology was meant to coax human beings into conformity with the social norm, honing the personality into a tool for advancement. In this, psychobiology suited the managerial cadres who were transforming American life and for whom "people skills" were the means of social and professional success. Here too the lie detector intersected with Meyer's program. Even if it was not used to detect lies as such, by measuring emotional deviations it might be used to coax individuals back to the straight and narrow. Emotional self-management, meaning control over bodily functions and their outward display, was the new ideal of American maturity.

Larson spent two intense years under Meyer's influence and was tempted to continue to pursue his research, but there were obstacles. In the first place, Meyer quickly lost patience with Larson's fixation on the lie detector, and insisted that he stop conducting "autopsies on old records." Moreover, the Depression had given Larson several households to support, including those of his in-laws and his own separated parents, and no fellowship could cover his needs. Nor could he return to the Institute for Juvenile Research in Chicago as long as Adler was in charge there. This intellectual and financial conundrum brought out Marge Larson's loyalty. In a "private note" to Meyer, she solicited a post for herself as an office nurse so as to free her husband for research. She explained that "something in his personality seems to draw him irrestably [*sic*] toward unsolved problems—and I can't bear to have him turn from his natural inclinations because of financial pressure."

In the end, Larson took a well-paying job at a psychopathic hospital affiliated with the University of Iowa. Three months into his stay he was already on the outs with his colleagues and angling to get to Chicago. Big changes were afoot in the capital of crime. Not only had Keeler persevered in Larson's place at the Institute for Juvenile Research, but August Vollmer himself had been offered a position as a professor of police administration at the University of Chicago. More, Volmer's allies were fanning rumors that he might be appointed the city's police chief. With the team from Berkeley reassembled, perhaps Chicago was finally due for reform, ready or not. Then, as luck would have it, Adler quit his job as director of the institute, and Vollmer persuaded the new director to hire Larson. And more: Vollmer had also arranged for Keeler and Larson to team up on a definitive study of lie detection.

CHAPTER 9

# Machine v. Machine

*There ain't a police force in the country could do its job with a law book. You got information and I want it. You could of said no and I could of not believed you. But you didn't even say no.*

—RAYMOND CHANDLER, *THE LONG GOODBYE,* 1953

LIKE A REFORMERS' TAG TEAM, VOLLMER'S BOYS WERE determined to clean up Chicago. The racketeers and naysayers may have temporarily run John Larson out of town, but Keeler had arrived to take his place, and he had brought with him a new lie box, much improved from Larson's crude device. Also, Keeler was a smoother operator, more politically savvy and with better connections. He knew how to handle people. He was determined to make his mark. And he would do so, though his ascent proved precarious. Even as Keeler promoted his machine as an antidote to police violence and political corruption, he had to sell it to the police as a technique that served their interests.

Keeler shared Larson's commitment to police reform. On his first day in Chicago, Keeler confided his plans to his private journal. Herman Adler had given him a tour of the Joliet penitentiary, where he would live while he conducted his research. "The prisons are well in the grasp of politicians, most of whom are ignorant and irresponsible," he wrote. Chicago would remain corrupt "until the whole system is conducted by non-political scientists." The warden—a former fish and game warden who had helped Governor Len Small out of his embezzlement scrape—was an "old double-saggy-chinned, fish-eyed, cigar-sucking fool [who] can hardly sign his name." The trick was to convince the warden and other officials that reform was their idea. As for the convicts, they would have to be duped too. Ruled by favoritism, they would suspect any reform as "an underhanded method for furthering their suppression." Stealth and secrecy would have to light the way to the dawn of science.

In the meantime Keeler dined at the officers' table, where he was waited on by a laughing "smoke," an African-American stickup man; got his hair cut at the barbershop, where the trusties had all tested negative for syphilis; and received his morning shave from "a nice kindly fellow who raped his daughter."

Such was life under the old regime. In the coming era, even the convicts would enjoy "freedom of personality," by which Keeler meant that each "will be treated according to his ability to respond properly to his social environment." The previous warden had already set up a "progressive merit system" to track each prisoner's rehabilitation. Keeler now wanted to weight these merits and demerits scientifically: subtracting, say, seventy-five points for planning an escape, ten points for sex offenses against others, and 2.5 points for sex offenses against oneself; and adding points for productive labor, cooperation with the authorities, and so forth. By correlating this behavioral scorecard with statistical data and the findings of the lie detector, the prison's mental health officer—aka the "bug doc"—would advise the parole board on whom to release. Over time, he hoped similar protocols would diffuse into the ranks of the police, probation officers, and social workers, and into every corner of American life.

In 1929 Chicago was in the process of violent transformation. In the six years between Larson's arrival and Keeler's, the consolidation of organized crime begun under Johnny Torrio had reached maximum ferocity under Al Capone's syndicate, to the point where the monopolization of crime was giving birth to its logical counterpart: a consolidated political machine. Into these "corrupt cesspools of humanity," the lie detector shone the bright light of truth. Three days after his arrival in Chicago Keeler wrote to his father, "absolutely on the Q.T.":

> But all this is about to change. I am the first shot from the gun of destruction of political graft—and construction of an orderly scientific management. More and more the administration of this penitentiary

will be from this office. Already things are popping, but we must go slow and let the politicians still believe they are the bosses.

It was machine versus machine now. Against a patronage machine which thrived on the exchange of favors and boodle, Keeler, Larson, and Vollmer imagined a justice machine that ran to the logic of fair play. Theirs was a characteristically American solution to the problem of justice, one analogous to the contemporary push for intelligence testing and Taylorist industrial management. Like them it was proposed by a coterie of applied psychologists, and like them it was subject to severe criticisms.

Consider intelligence testing. Intended by the French psychologist Alfred Binet to diagnose gross mental deficiencies on an individual basis, the IQ test was refashioned during World War I by American psychologists such as Robert Yerkes of Yale and Lewis Terman of Stanford, who made it into an instrument to assign millions of middling Americans to a fixed rank on a continuous scale. They designed the grading to require so little judgment that a machine could do it, and one soon did. But the test was not just intended to cope with the scale of a mass society; it was also meant to introduce a semblance of scientific evenhandedness to the contentious new competition for spots in the education meritocracy and civil service. Indeed, the egalitarian appeal of the IQ test and its successor, the SAT, lies in the way they ostensibly treated everyone alike. In practice, of course, the content of these tests could (and did) favor some social groups over others. Nor did these paper tests necessarily predict those professional abilities that went into being an inspiring military leader, a resourceful police officer, or an attentive physician. Some Americans even challenged the notion that narrow criteria of this sort could ever define a citizen's worth. As Walter Lippmann wrote in 1923: "I hate the impudence of a claim that in fifty minutes you can judge and classify a human being's predestined fitness for life. I hate the pretentiousness of that claim. I hate the abuse of scientific method which it involves." Yet the IQ test triumphed in America as nowhere else in the world.

Similar logic drove the contemporary movement, led by industrial engineer Frederick Taylor, to introduce scientific management to American business. By calculating the "one best way" to perform each worker's task, Taylorist time-and-motion studies appeared to provide a scientific solution to the bitter dispute between labor and capital—seeming to ease the workers' burden, while standardizing a more profitable degree of exertion. In their wake, a new breed of industrial psychologists, led by Hugo Münsterberg, extended Taylorism to the realm of behavior, treating workers' attitudes as a "human factor" to be engineered like their bodies. As critics have complained, both then and since, the new industrial psychology enforced a new code of emotional self-management in the workplace, discouraging even legitimate displays of passions, like anger at shoddy conditions of labor. Aspects of Taylorism also appealed to Europeans (notably Lenin), but only in America was applied psychology touted as a solution to industrial conflict.

The lie detector belonged to this same American strain of the Enlightenment project to replace personal discretion with objective measures, and political conflict with science. It proposed to do this by redefining its object of inquiry—the lie—in narrow yes-or-no terms, much as IQ redefined intelligence as a single factor "g," or Taylorism redefined the worker's task as a series of bodily motions. Just as Americans dreamed of resolving disputes over merit and industrial relations with machinelike objectivity, Vollmer and his disciples hoped that the lie detector would enable them to administer justice with machinelike fairness. Indeed, some in the African-American community welcomed the device as an antidote to prejudice in the criminal justice system. This explains the main appeal of the lie detector in the United States: the charade that it is the polygraph machine and not the examiner which assesses the subject's veracity.

Such a project, of course, had little appeal for politicians who made patronage appointments or police officers who selectively enforced their authority on the streets. No wonder, then, that most old-style officials resisted the lie detector—at least initially—especially as they were among the first Americans subjected to the test. But like intelligence testing and Taylorism, the lie detector offered only the veneer of science. Behind the public facade, the polygraph, depending on how it was operated, did not necessarily restrict the discretion of examiners. Indeed, as Keeler conceived it, the lie detector might even enhance the power of the police, by becoming a psychological third degree. And it was here that Larson and Keeler would part company.

---

Within a few weeks of his arrival at Joliet, Keeler was running prisoners on his new machine as fast as the prison psychologist would clear them. And unlike John Larson, he was having success. All he had to do, he discovered, was put himself on the same social level as the inmate, and the prisoner would loosen up and confess. At that point "the man's conflicts, his secretiveness and his dishonesty become a thing of the past." Not that Keeler necessarily thought of the inmates as his social equals: "Damn rotten bunch of money grubbers," he wrote to his aunt.

Success bred success. Just as Keeler had predicted, his tests of prisoners netted him the chance to test law enforcement officers. The new warden, a former military man, authorized Keeler to refuse employment to any guard at Joliet who failed to meet his standard for honesty. At the tender age of twenty-six Leonarde Keeler had become the solution to that ancient conundrum *Quis custodiet ipsos custodes*—Who will guard the guardians? Henceforth, he wrote to his father, he would see that the prison employed only a "higher type of man."

Keeler's father responded warily. He warned Leonarde against making the machine the sole determinant of a person's fate. He urged him to think of the device as a "truth finder" and use it to attack the problem of "the human personality in relation to society and to life." He imagined Leonarde working in conjunction with the Cosmic Society to bring harmony to the relations of science and religion, mind and body, mind and spirit, and perhaps even confirm the presence of psychical powers as yet unacknowledged by official science. He wanted his son to conceive of the lie detector as a first step in tracking "the soul to its lair." In his own half-hearted way, Keeler was trying to do just that, with his own soul as the hare.

Several times a week for an hour before work, Keeler lay on a couch, like a psychiatric patient, providing "a rescitation [*sic*] on my life and habits" while a device akin to his own lie detector went "swinging its telltale record" and a waxed cylinder recorded every word. His analyst was Harold D. Lasswell, a charismatic professor at the University of Chicago, who wanted to categorize the psychopathology of politicians and nations, and hoped the lie detector would calibrate his analysis for diverse individuals and cultures. His methods horrified orthodox psychiatrists, and Keeler dismissed the sessions as "a lot of nonsense," though he did admit to

Vollmer that Lasswell might well "get my wrath up before we're through." The topic often turned to Keeler's relationship with his father. Lasswell seems to have written Keeler up as "subject A," a young man who pretended to cooperate, while his polygraph record indicated considerable inner tension. Subject A was secretly so wary of suggestion and "anxious to maintain his independence" that he postponed interviews and was discontinued after only minimal insight into his "aggressive motivations." Lasswell was especially disappointed that he tried to "discipline himself into emotional passivity." Obviously, if anyone was going to do the interrogating, it would be Keeler himself.

Then, in 1930, Vollmer himself arrived in Chicago. Only the year before he had surveyed the terrain while writing the chapter on police corruption for the *Illinois Crime Survey.* Now he was to teach police science at the University of Chicago, while reformers maneuvered to have him appointed police chief if the coroner, Herman Bundesen, won the Democratic primary for mayor. No such luck. Instead, for the next two years, Vollmer achieved his greatest national influence as a force behind the National Commission on Law Observance and Enforcement, known as the Wickersham Commission. The most shocking revelations came in the volume devoted to police lawlessness. This volume walked readers through a national horror show of police abuses that reached its maximum intensity in Chicago.

In theory, persons arrested in Illinois had a right to be promptly heard by a judge; even the threat of violence was forbidden; and convictions based on coerced confessions were to be overturned. In practice, the police secured confessions by alternating extreme heat and cold; depriving suspects of sleep; and taking them on trips to a "goldfish" room where they were lashed in the stomach with a rubber hose, half-drowned, and kicked in the shins, and had the Chicago telephone directory slammed against their heads until they saw goldfish. As the report noted, the Chicago phone book "is a heavy one."

The members of the International Association of Chiefs of Police reacted with fury to these slurs on their conduct. They called the report "the greatest blow to police work in the last half century," and denied the existence of the so-called third degree. They also insisted that the police couldn't do their job without it. Besides, only guilty suspects were beaten up. Moreover, the rough justice of the station house deterred crime and secured the confessions that prevented criminals from slipping through the judicial system. One police official in Chicago warned that without the third degree 95 percent of his department's work would be nullified.

The goal of Vollmer and his disciples was to persuade their colleagues that these methods were counterproductive: that they often produced false testimony, led juries to acquit, or obliged appellate courts to overturn verdicts. In the mid-1930s, the U.S. Supreme Court finally began overturning state court convictions based on coerced confessions, announcing: "The rack and torture chamber may not be substituted for the witness stand." Meanwhile critics were urging a more radical course; they wanted the courts to reclaim from the police their monopoly on interrogations, a monopoly the courts still held in Britain.

It was on just this point that Vollmer and the Wickersham Commission offered the police an out. If the police valued their prerogative to interrogate, the report noted, they would need to drop the third degree and adopt new scientific methods such as the lie detector. Persuading the cops would not be easy, of course. Offered a chance to adopt the lie detector, one police official in Chicago had held out a clenched fist and answered, "Here's the best lie detector."

Keeler's challenge was to convince the police that his technique could get the sort of outcome they wanted—ideally a confession—while assuring the public that the instrument would constrain police violence and illegality. Two cases from this period showed how he tried to navigate between these contrary tugs.

In Seattle, a five-time loser, Decasto Earl Mayer, was suspected of murdering a young naval officer named James Bassett and stealing his blue roadster. Bassett had been missing for a week when Mayer was picked up in Oakland with his stepmother in Bassett's car. A year later, in 1929, with Mayer unwilling to help locate the cadaver, the prosecutors consulted August Vollmer, who persuaded them to ask Larson if he would be willing to try the experiment. Larson generously referred them to Keeler.

Mayer had already been thoroughly worked over—including being forcibly injected by Dr. House with his truth serum—so Keeler devised a guilty-knowledge test. For seven days, eight hours a day, Keeler interrogated Mayer on the machine. He began with general questions: Is Bassett's body in the lake? Is Bassett's body in a well? When Mayer seemed to react to questions about cemeteries, Keeler asked Mayer whether he had placed the body in a grave. Had he placed it under a concrete slab? By Sunday Keeler had zoomed in on a cemetery near Mayer's old house and was making the prisoner go through the graves one by one, when Mayer ripped free of his restraints and did what Helen Graham had only threatened to do: he "smashed the machine." Two deputies had to drag him to his cell.

Keeler repaired the instrument and resumed the interrogation that evening with the prisoner wearing a metal Oregon boot that tightened each time his leg moved. The next morning, according to the prosecutor, Mayer privately offered to confess, saying: "I know what that machine is. I know it's recording the truth. I can't beat it." Keeler (listening through the door) wired the good news back to Chicago:

> TOP BLEW OFF KETTLE THIS MORNING STOP PLENTY OF EXCITEMENT STOP WE ACTUALLY OBTAINED CONFESSION AND SOON WILL HAVE BODY STOP.

But no sooner had the story gone out on the wires than Mayer denied his confession and asked to be returned to the machine, announcing, with apparent fatalism, that "the machine will tell." Yet he still refused to look at cemetery maps. When two deputies tried to pry his eyes open, he rolled his pupils up into his skull. At that point he had a seizure and had to be sedated.

The next morning Mayer's bleary mug shot stared up from newspapers across the nation, gaunt as a skeleton, his eyes smudged with black despair, alongside a "setup" photo picturing Keeler bent over the lie detector. Later that day Mayer's lawyers accused the prosecutors of torturing the detainee. Superior Judge Malcolm Douglas agreed. He called Mayer's treatment "more in keeping with [the] days of . . . the Spanish Inquisition than of this present enlightened civilization." He banned any further use of the lie detector against the detainee's wishes. For several days, investigators dug up graves, without success.

The Mayer case seemed to belie the promise of the lie detector. *The Nation* called it torture. The *Harvard Law Review* was appalled. The muckraking book *Our Lawless Police* wondered how an interrogation begun with the ultramodern lie detector had degenerated into medieval barbarism. Even the Wickersham Report called it "one of the outstanding third-degree cases in the United States." Larson was ashamed and apoplectic.

Keeler saw things differently. In a thinly disguised account, his allies in Chicago scored the case as an exemplary "win" for the lie detector, even though they had to "adjust" the facts to reach this conclusion. In their version, the interrogation took place in a "laboratory," the test quickly identified the correct cemetery, and its results led the prosecutors directly to a disturbed grave where the victim's body was discovered. All this was a complete fabrication.

Seven years later, Mayer's stepmother confessed to a cop disguised as a priest. She said the corpse had been buried in dispersed sites, none of which were cemeteries. Finally able to charge Mayer with murder, the prosecutors were robbed of victory when he hanged himself in his cell, his nose and throat filled with wadded-up paper. The Mayer case, like the case of the Ku Klux Klan in Santa Barbara, showed how easily the lie detector could become an adjunct to the third degree, rather than its antidote.

Soon after, Keeler took on a case in Chicago that seemed its antithesis: four cops accused of killing a trick canary! Trivial as it seemed, the case taught a potent lesson. For years Probate Judge Henry Horner had tried to stop the police from robbing estates that they were meant to be guarding. Only a few years earlier he had charged fifteen policemen with stealing a priceless violin from a dead man's shop. This time he turned for advice to the visiting professor August Vollmer, who recommended the services of Leonarde Keeler. The newspapers had a field day with the judge's efforts to get to the bottom of this comical crime. But Horner had a serious lesson in mind; he wanted to show the police that they would be bound by the law.

These were the facts. On October 6, 1929, Anna Gustafson, the impoverished proprietor of a tearoom in Chicago, committed suicide by taking gas. Her body was found by her lodger, who also discovered her most prized possession, a trick canary named Nimble, still singing in its cage despite the gas. The police came, removed Gustafson's body, and stood guard over the premises. The next day, when the lodger came by to retrieve his possessions, he noticed that the bird (estimated worth: $100) was missing. Three days later another friend of Gustafson's came by and found a dead bird atop the sewing table, but doubted that it was Nimble. She also noticed that some silverware and clothes were missing.

Gustafson's friends accused the police of taking these items and substituting a dead bird for Nimble. The police denied these accusations and suggested that the bird must have died of the gas. The coroner's office conducted an avian autopsy, which revealed that the bird had died of "fractures of the cervical vertebrae caused by wringing of the neck." A few days later the bird's body was exhumed so an ornithologist could affirm that it was not a member of a singing species.

As Gustafson had died intestate, jurisdiction over her estate fell to Judge Henry Horner, a prominent jurist soon to be elected governor of Illinois. Despite Horner's stern cross-examination, all four police guards denied any hand in Nimble's fate. Indeed, one officer, William Tobin, had the effrontery to claim that while he was on duty Nimble had flown out the window; that five days later the same bird had flown back in the same window; and that the bird had then promptly died.

On February 13, in Horner's chambers, Keeler and Vollmer examined all four officers on Keeler's apparatus (AC current having been installed for the purpose). Keeler concluded that Tobin reacted strongly to questions about the theft and about the complicity of another officer, named Gall. Gall's reaction likewise indicated participation. A third officer gave a clear record, except when asked about Tobin and Gall; he was allowed to go. The fourth officer, named Dompke, gave a clear test except when asked whether he had taken anything from the house. After a second test Dompke confessed that he had taken a lamp shade, worth sixty cents. He was told to go home. As neither Tobin nor Gall would confess, Horner fined the officers, pending appeal. The civil service commission threw up its hands: "This looks as if there's been a fix. Case under advisement." But Horner would not back down, despite being threatened by phone. He kept bringing the officers in for testing—with the newspapers covering each twist—until Tobin privately promised to bring in the guilty party. Later that day, he brought in Gall, who confessed but said that Tobin helped him.

A year of interrogation and humiliation had solved the "canary murder case." Not even the lawman's code of silence had withstood the lie detector's ability to focus the public's attention; the device was like a magnifying lens that brought the conscience of the many against the one. "Another notch in our gun," gloated Keeler. "Judge Horner doesn't believe there'll be any more stealing by police officers." Trivial on the surface, the case was a telling victory against police corruption. Much like sorority sisters, police officers are insiders with their own codes and secrets. Cops, moreover, were used to operating in a legal gray zone, pocketing the "takings" they considered a perk of the job. What others called corruption, they considered part of the customary practices of their trade. But thanks to Horner's backing, Keeler's lie detector had drawn a sharp line between lawful and dishonest behavior.

Whenever they could, Vollmer's disciples put cops on the lie detector. When one of Vollmer's college cops was appointed police chief of Evanston, a suburb of Chicago, he immediately invited Larson to subject officers to tests. By 1933 Keeler was using the technique on the chief's successor in Evanston, accused of graft, selling liquor, tolerating roadhouse gambling, and threatening opponents. The department in Berkeley had added an obligatory session on the lie detector for all recruits. And in Wichita, Kansas, another disciple of Vollmer's ordered lie detector tests for police applicants; he discovered that eight out of twenty-two had lied on their forms or exhibited psychopathic reactions.

To maximize the publicity in Chicago, Judge Horner arranged a command performance, letting himself be tested on Keeler's machine in the presence of a reporter for the *Chicago Tribune*; the police commissioner; the coroner, Bundesen; and a businessman, Burt Massee, who was a sponsor of Northwestern University's Scientific Crime Detection Lab, and whose latest hire was none other than Leonarde Keeler.

Four months later, on August 8, 1930, Judge Horner administered a truth test in a short ritual of his own, uniting Leonarde with Katherine Applegate in holy matrimony. When Horner asked, "Do you take this woman . . . ?" Keeler replied, "I do"—though Leonarde's family and friends seemed to think (as his father put it): "She cared more for you than you did for her." Sensing that something was afoot, Charles had warned his son that Katherine Applegate was "on your trail," and advised him not to be swayed by "the romantic and sentimental associations of the moment." But Leonarde had assured him that their compatibility had been gauged by the most up-to-date psychological tests. He had even persuaded Adler to give Katherine a job at the institute working alongside him.

Once the deed was done, however, Charles quickly sent a conciliatory note to Kay: "[W]e know that the girl he loves is fine and true." He also gave the couple an original Keith landscape as a wedding present. Nard's sister Eloise sent the newlyweds a toy handgun, which came with directions for managing "wayward husbands." The toy proved more dangerous than it looked, however, raising a welter of pimples on the bridegroom's arm. Kay promised never to use it again.

What difference would it have made, one wonders, if instead of simply plighting their troth before a judge, bride and groom were instead hooked up to a lie detector while they took their vows? The proposal was not far-fetched. Within a year, Keeler and his associates at the Northwestern University crime lab had arranged an honest wedding for Harriet Berger and Vaclav Rund. The *Chicago Herald-American* announced it on the front page, "SO THIS IS LOVE!" with a graphical record of the ceremony: the groom's blood pressure starting high (prenuptial doubts?), then declining steadily during the ceremony; the bride's blood pressure starting low (feeling faint?), before a gradual ascent that leaped upward when the groom was asked "Do you take this woman . . . ?" And when the judge pronounced them husband and wife, the two lines crossed.

More hard-core methods were coming. A year later, the cardiologist Ernst Boas in New York devised a tachycardiogram (with a 100-foot lead) to record the heart rates of 100 volunteers performing mundane tasks, including a poker game; and a husband and wife engaged in coitus, during which the woman experienced four orgasms in twenty-five minutes, each driving her heart rate to 145 beats per minute, the last coinciding with her partner's, whose heart rate likewise peaked at 145. Though the test enabled Boas to draw a medical lesson about the risk of sudden death during coitus, he acknowledged that the relationship between heart action and human emotions was "infinitely more baffling." This did not stop his colleague from trying to market the device to the FBI as a lie detector. And by the middle of the 1930s Freud's heretical student, Wilhelm Reich, went them one better, measuring the galvanic responses on the nipple, tongue, clitoris, and glans of the penis. By 1939 Reich had arrived in America, a magnet for machine-body therapeutics, where he constructed large crates lined with alternating metal and organic materials—his so-called orgone boxes—to channel bioenergy and cure mind-body imbalances, including cancer and sexual dysfunction. He reported excellent results, but then, so do practitioners of lie detection.

Despite Charles's concerns, Nard and Kay seemed ideally suited: a handsome couple with common interests and passions. Just before their wedding they took a "pre-moon" in the Canadian woods, a two-week canoe expedition during which they sighted moose, bear, and deer, and caught all the fish they could eat. But as much as he derided the filth and corruption of the city, Keeler was always eager to return. Though still subject to bouts of illness, he'd put on weight, and everyone agreed he looked great. Kay too seemed content in the city, though she complained about the grime. The city fulfilled her endless search for novelty. She took up fencing. Unable to afford fine clothes, she rented a sewing machine and learned to make tailored suits. Meanwhile, she had been conducting her own experiments at the institute on the relationship between emotion and galvanic skin resistance. It wasn't long before this crime-fighting husband and wife were making regular appearances in the Chicago papers.

Soon after their marriage, the Keelers quit their jobs at the institute in favor of full-time positions with Northwestern University's Scientific Criminal Detection Laboratory, the nation's first forensic lab. For a decade Dean John Henry Wigmore, America's foremost expert on the law of evidence, had been urging his university to create a criminal lab on the model of Europe's new police labs in Paris, Berlin, Vienna, Rome, and Lyon. It took the Saint Valentine's Day massacre to realize his ambition. When five members of Capone's last rival gang were gunned down on February 14, 1929, along with two bystanders, the city's newspapers discovered evidence that pointed to police involvement. The killers had arrived in a police-style touring car, wearing police uniforms, and firing police-style Thompson submachine guns. To investigate these ties, the coroner's office impaneled a special grand jury, whose wealthy foreman, Burt Massee, solicited the advice of Wigmore. Wigmore advised him to call on the new forensic sciences: specifically the testimony of Colonel Calvin Goddard, a former army physician and amateur gun collector who had developed a ballistic analysis that could match each bullet to its individual gun.

Goddard tested every Thompson machine gun owned by the Chicago police and testified that none had fired the bullets found at the scene. Then, when Fred Burke of the Detroit Purple Gang killed a cop in Michigan, Goddard found a match. Science had absolved the police.

Massee, impressed, wanted to bring Goddard to Chicago permanently. But as a member of Chicago's Crime Commission, Massee knew the byways of official corruption. He insisted on creating a lab "outside the pale of politics and divorced from all petty and narrow influences." Wigmore persuaded him to finance the lab as a private corporation affiliated with Northwestern University. The nonprofit corporate structure would assure outsiders that the lab's experts were not beholden to financial interests. A modest fee for private cases would make the lab partially self-sustaining, with the staff allowed to consult on their own cases so long as they split any fees with the university. And the lab would supply its services free to government authorities, as well as train police officers, prosecutors, and other forensic experts. Goddard was appointed director and dispatched to Europe to study the world's most advanced police labs.

Wigmore also wanted to recruit the lie detector for this effort. Twenty years earlier, in the days of Hugo Münsterberg, Wigmore had mocked the psychologist's attempt to replace the judgment of the jury with a battery of unproved tests. But Wigmore had not ruled out hope that science might one day help the law distinguish truth from falsehood. During the Great War he had supported Marston's studies. He had closely followed the "promising" work of John Larson. And in 1923, in the second edition of his monumental *Treatise on the Anglo-American System of Evidence,* he had asserted, "If there is ever devised a psychological test for the valuation of witnesses, the law will run to meet it." Indeed, Wigmore had come to believe that science was on the verge of making such a test practicable. Guilty knowledge left a trace as surely as a fingerprint. "As an axe leaves its mark in the speechless tree," he wrote, "so an evil deed leaves its mark in the evil doer's consciousness." Might Keeler's new machine be the first to detect that mark? Soon after Keeler arrived in town, Wigmore invited him to demonstrate the new machine to his class at the law school. Against all expectation, Wigmore was impressed. "My former skeptical attitude has been very much altered since I listened to your statement and demonstration."

Goddard had already hired specialists to help with ballistics, toxicology, the microscopic analysis of dust, and crime-scene photography. Now he added Leonarde Keeler, initially half-time, then full-time. The lab also found a job for Katherine Keeler as Goddard's personal assistant, though she soon acquired her own forensic expertise.

Keeler proved a godsend. The stock market crash ended Massee's attempt to raise an endowment, and Keeler's fees from lie detection became the lab's main source of revenue and public fame. It was not long before the lab needed additional hands. One day an old pal from Berkeley called in on his way to New York. Charlie Wilson never got back on the train. He had practical skills in electronics and soon became an expert polygrapher. Affable but hot-tempered, he played bad cop to Keeler's good cop. Once at a state fair, told that he couldn't take target practice because he had had too much to drink, he pulled out his service revolver and began blasting away. Charlie and his wife, Jane, soon became Nard and Kay's closest friends, and Jane volunteered around the lab as well. The university also hired a young professor of criminal law named Fred Inbau, whom Keeler trained on the lie detector.

Everyone in the lab was devoted to publicizing its work. This was not just good for business; it was the key to enhancing the lie detector's aura of effectiveness, and hence its actual effectiveness. Keeler demonstrated his machine to dozens of business groups, ladies' clubs, criminologists, scientists, and journalists. He scored a press notice whenever he solved a sensational case. In addition, the lab published the *American Journal of Police Science,* the first journal in the United States devoted to the science of criminal investigation. The inaugural issue contained Keeler's first write-up of his lie detector. Also, the lab began a series of biannual monthlong courses, training fifty police officers at a time in the collection of evidence and the use of the lie detector. One student that first year was the FBI agent Charles Appel, who returned to Washington, D.C., to organize the FBI's crime lab along similar lines.

The most influential alumnus of the course, however, was neither a cop nor a prosecutor, but a cartoonist. Chester Gould, a graduate of Northwestern University, signed up, he said, "to further my knowledge of criminology." Gould was the creator of "Dick Tracy," the *Chicago Tribune*'s premier crime-fighter, and the first to bring the violence of the pulps to the comics. Central to Tracy's success was his use of science to defeat crime and the political corruption that fed it. Tracy himself could be read as a composite portrait of the members of the Northwestern University crime lab, with a passing resemblance to Colonel Goddard. And Gould put his own forensic lessons to good use, having Tracy tap telephones, identify bullets, test dust particles, distinguish animal from human blood, and put dozens of suspects on the lie detector.

"Dick Tracy" was fiction, that form of lie in which the forces of law and order triumph over crime on a serial basis. Chicago was real, that form of truth in which human criminality works its daily will. Though Keeler had begun to accumulate an impressive scorecard of cases, he had yet to win the approval of the Chicago police. He regularly shared a whiskey in Chicago's finest men's club with patrons such as Governor Horner, Dean Wigmore, Massee, and Vollmer. He dashed around town in the crime lab's Nash, which had gold shields in front and at the rear, plus a huge silver siren up top; so that, in the words of his glamorous crime-fighting wife, "We take our place in traffic now." In 1933, at age twenty-nine, he was honored for making a "most outstanding" contribution to civic life in Chicago. Yet the Chicago police still wouldn't let him near their suspects. Asked to comment on Keeler's award, the city's chief of detectives laughed derisively: "I don't know anything about the 'lie detector.'"

## CHAPTER 10

# Testing, Testing

*She frowned. "But, Mr. Beaumont, why should people think it unless there's some sort of evidence, or something that can be made to look like evidence?"*

*He looked curiously and amusedly at her.*

—DASHIELL HAMMETT, *THE GLASS KEY*, 1931

UNDER VOLLMER'S GUIDANCE, LARSON AND KEELER RESUMED cooperation. Larson promised to list Keeler as coauthor of his book on the scientific validity of lie detection. Keeler offered to contribute case results and help Larson find a publisher. In the short run, the two were to work with a committee of Chicago's leading psychologists and criminologists—chaired by Vollmer—to resolve what Larson called [*sic* throughout], "the problems of standardization, scientific controlled and objective evaluation of validity."

But proximity can sharpen differences, and overlapping goals can make divergent strategies divisive. Working side by side, Keeler and Larson discovered they didn't trust each other. The Keelers had settled on the North Side to be near the crime lab. The Larsons took up residence on the South Side, with access to the Institute for Juvenile Research and the Joliet penitentiary. They might as well have been living in two different Chicagos. Keeler had signed on to the North Side strategy of manly managerial crime-fighting, Larson to the South Side strategy of therapeutic reform. The outcome would be two distinct lie detectors.

---

Their collaboration began with a moral victory. In April 1930, two collection agents from Chicago, doing business with a bank in Black Creek, Wisconsin (population 500), returned that afternoon—according to the deputy bank teller and some other victims—to rob the bank of $733. It looked like an open-and-shut case. But the men had a good alibi and a good lawyer, who sent them to be tested by Keeler. When Keeler affirmed their innocence, the defense lawyer asked him to testify.

Without a college degree, Keeler would have been an easy mark on the stand. So he immediately got on the horn and "shouted loudly for John L. with his experience and many degrees." And Larson agreed to help, despite his previous opposition to using the lie detector in the courtroom. In later years he ascribed this lapse to his need to gather scientific information for his book; at the time, it sounded more like a matter of friendship.

On the eve of the trial's final day, the jury was dismissed so that Keeler and Larson might explain their findings. When the chief prosecutor responded to Keeler's proposition with a sneer and lawyerly contumely, Keeler challenged him to a card test and identified his card on the spot as the king of clubs. This coup was to no avail. The judge refused to let the lie detector usurp the jury's role. And with so many eyewitnesses, a conviction seemed assured.

Yet the next day, in a reversal worthy of Perry Mason, the defense attorney's daughter, Appleton's sole female lawyer, reported that she had just obtained a confession to the Black Creek robbery from two men being held in Minneapolis. The trial was halted, and the accused men were set free.

Keeler and Larson spent a celebratory weekend in Chicago trying to persuade the Institute for Juvenile Research and the Northwestern crime lab to hire both of them half-time. Unfortunately, as Keeler informed Vollmer, "doggone John, the poor fish" had blown it. He had "arrived [for his interview] in one of his hyper states of mind, without a damn for his appearance, talked around in circles instead of the questions at hand and had his feet so far off the ground that he made a poor impression on the [Northwestern] group. In consequence he has been ruled off."

Instead, Larson resumed work under the old arrangement: half-time in medicine and half-time as an assistant state criminologist at the Institute for Juvenile Research, conducting lie detector research for Vollmer's committee. For six months all seemed to be going well. Then Vollmer abruptly resigned his position in Chicago to return to Berkeley (for reasons never clearly identified), and everything went sour. In short order, Larson and Keeler went from being collaborators to being bitter rivals.

Though Vollmer's committee drew on the best experts in Chicago, none met with Larson's approval. L. L. Thurstone, a psychologist at the University of Chicago, insisted on the statistical analysis of artificial tests in which undergraduates told lies about meaningless games. The two other professors—Harold Lasswell, Keeler's erstwhile analyst; and Chester Darrow, the physiologist—knew nothing about criminal matters.

Initially Larson exempted Leonarde Keeler from his criticisms. Still, as the senior scientist, he also felt entitled to lay down standard protocols so that their case reports could be compared. He insisted that Keeler have a physician present at every test; a girl had fainted during a recent exam, and someday a subject might suffer a heart attack or even die. Larson wrote a preamble for Keeler so that subjects would be advised that they might opt out if they objected to the test. He laid down the proper sequence of relevant and irrelevant questions, their timing, and their maximum duration, warning Keeler not to exceed these limits. He gave precise instructions on how high to inflate the blood pressure cuff, both to standardize their procedures and to ensure that the cuff did not cause discomfort. He insisted that Keeler stick to specific crimes and not go on "fishing expeditions" to elicit guilty reactions. Finally, he demanded that subjects who confessed be asked to sign a statement affirming no "rough means" had been used.

Larson was soon dishing out secondhand criticisms of their protégé to Vollmer. Apparently Keeler ignored his instructions. Worse, Larson had heard that Keeler was publicly touting the machine as "infallible," an exaggeration which tarnished the field's reputation among serious scientists. Also, Larson's assistant at the institute had told him that Keeler cheated on the card test by relying on clues from the audience and marked decks, just like a showman. Larson even wondered whether Keeler was too restless for research—not, he hastened to add, that he agreed with the psychiatrist in Berkeley who had once diagnosed Keeler as having a psychopathic personality. Finally, Larson had heard rumors that Keeler was turning the Northwestern University crime lab into a commercial operation. Larson urged Vollmer to rein in "our young enthusiast."

There had always been a paradox at the heart of Larson's commitment to open science. Larson recognized that Keeler had produced a polygraph machine without which researchers like himself could not hope to create a science of lie detection, and he advised all his scientific colleagues that the Keeler Polygraph was the "best modification available at the present time." But now that Keeler had actually received his patent, Larson reported that physiologists were criticizing Keeler, saying, "This sort of thing is not done in research circles." Suddenly, Larson was priding himself on not having taken the "unethical" route of patenting his own earlier device—despite the urging of friends. Unlike Keeler, he noted, he had always published his results in journals of criminology and psychology, as it was priority that mattered in science. He wrote to his former disciple, "You have never recognized the difference between *meum* and *teum* [mine and thine] as far as priority and scientific research is concerned." He feared that Keeler would sell machines "to every Tom, Dick, and Harry," allowing poorly trained operators to ruin the reputation of the new science.

Larson's strict protocols and lofty principles did not sit well with Keeler's sense of his prerogatives as inventor, or with the sort of interrogation Keeler wished to run. Yet there was a paradox at the heart of Keeler's ambitions too. Keeler needed to produce reliable machines, accepted as standard in the field. But he realized that the best way for him to make money from the device was, as he put it, to "control the instrument and lease his services." Thus, when Walgreens wanted to buy several machines to set up its own in-house security team, he refused, and offered to consult for the company instead. As he confided to an associate, he made only $125 from each machine sold, and each sale created a competitor. That is why Keeler retained a veto right on every sale. So he was not being disingenuous when he assured Larson that he was doing all he could to prevent the "prostitution and promiscuous use" of his polygraph. But this short-term sacrifice of sales was worthwhile only if he cultivated his public renown.

Keeler and his colleagues at Northwestern University certainly had a knack for inserting themselves into crime stories. The machine didn't always break the case, but it always got a headline. When the attractive stepdaughter of a merchant in Rock Island, Illinois, was found dead, trussed up and half naked in an icy Illinois river, Keeler's machine fingered the neighbors' son, who later confessed and was convicted of murder. When a paunchy assistant embalmer from Rockford claimed that two highway bandits shot his wife on her fifty-third birthday, Keeler put him on the machine and soon had him confessing in tearful sobs. When a prominent woman physician of sixty-two was suspected of murdering her daughter-in-law on the operating table in the basement of her gloomy Chicago mansion, a nonstop lie detection interrogation produced two confessions: the first by the loyal son seeking to shift blame away from his mother, the second by the steely mother seeking to shield her beloved son. The mother was imprisoned on manslaughter charges.

The risk of this strategy was that it stoked a demand Keeler could not satisfy, tempting rivals to enter the market. His patent offered little protection. Moreover, his rivals had pedigrees of their own. Chester Darrow's machine by the Stoelting Company was already on the market. Captain C. D. Lee, one of Vollmer's oldest aides, sold a Berkeley Psychograph, which was almost identical to Keeler's device except that its tambours were still all-rubber. As Lee explained, Keeler's metal ones were prone to fracture and required frequent calibration because they needed long, fragile pen arms to amplify their slight movement.

Under this pressure, Keeler gradually switched strategies. Through most of the 1930s he sold his machine only to police officers trained in his courses at Northwestern University. Then in the late 1930s he allowed Walgreens and a few others to purchase machines. But as long as he remained at the crime lab, Keeler preferred to trade on his personal renown as an interrogator.

More divisive than their fights over the hardware of the polygraph were Keeler and Larson's disputes over the "software" of interrogation. Larson had found that as many as half of the subjects reacted with guilt when asked a broad question about complicity in crime. Larson's obstacle was Keeler's opportunity. One month after the crash of 1929, still working at Joliet, Keeler confided to Vollmer an idea he had been "tossing about among the fleeting clouds." Many department stores, he noticed, were losing money to their employees' "crooked twists." What if he subjected each employee to a lie detector test every six months, weeding out the "lifters" and putting "the fear of the Lord" in the rest? Marshall Field lost $50,000 a year to pilfering by employees; if he reduced the theft by, say, 75 percent, the retailer could pay him $12,000 and still come out ahead. With American businesses as a whole losing nearly $337 million a year to employees' pilfering, there was a huge market for testing honesty.

One early case showed how easily it could be done. When the state's attorney called in Keeler to test a bank teller for embezzlement, the polygraph test showed no evidence of guilt; but when Keeler ran other employees, three tellers confessed to pocketing the bank's money and the vice president showed signs of guilt. Within a year Keeler was offering businesses in Chicago his services as an enforcer of honesty. In 1931 he signed an agreement with the Chicago representative of Lloyds of London, who promised banks a 10 percent reduction in their insurance premium if they let Keeler test their employees periodically. What he found shocked managers. In bank after bank, about 10 to 25 percent of employees confessed to thievery. Most of the thefts involved a furtive dip into petty cash, losses that would not show up in the account books.

The managers' reflex was to fire these employees, but Keeler urged that they be quietly retained and retested regularly. There was no need to make a fuss while the public was skittish about bank closures. Besides, he promised, the confessed tellers would henceforth be the bank's most honest employees. Indeed, of the 1,000 tellers tested during his first two years at Northwestern University, only three resumed pilfering—or admitted as much on Keeler's return. The lie detector become a psychological deterrent as much as a catcher of thieves. This strategy also brought Keeler back in for another round of remunerative testing. And because he split his fees fifty-fifty with the crime lab, his screening venture soon became its financial mainstay.

As for employees who protested, Keeler advised managers to inform them that the insurance company had made the test a requirement for bonding. "Under the circumstances employees have very little or no resentment to the test." Many, he said, actually felt better after owning up to their misdeeds. In reality, though, many tellers were fired or resigned after failing tests; of the twenty employees at Lake View Trust, three were discharged for taking previously unnoticed sums, and two resigned rather than accept a transfer.

Agnes de Mille, Keeler's old college friend, had a rare peek at how Keeler pulled this off. While visiting Chicago with her dance troupe, she watched with bank managers through a one-way mirror as Keeler examined an applicant for a teller's job. The young man appeared nervous. At one point, when Keeler pressed him for information about his past, he asked to be released and began to disentangle himself from the apparatus. "At this point," de Mille recalled, "Leonarde got rough. I have never heard him talk to anyone with such abruptness. He told the boy, in effect, that unless he took the test, he could not have the job." Cowed, the young man answered in a subdued tone. Suddenly, he admitted having served time for theft. Inside the observation room, de Mille heard the bankers gasp. But Keeler did not register disapproval. He completed the interrogation, unhooked the subject, and handed him a cigarette lit in his own mouth. Then came the payoff. The trouble was his mother-in-law; she had ruined his marriage. For a half hour, the young man poured out his soul. Keeler listened sympathetically, shook the young man's hand, and sent him out the door. Then he turned to the bankers and told them the young man was a safe bet. "He's perfectly honest as far as I can tell," he said, "but he did serve a jail term."

De Mille was shocked, though less by the young man's fate than by her sweet Nard's ruthlessness. Still, she could not help but be impressed—as one choreographer to another—with Keeler's skill at striking a powerful emotional chord in the body of his singular audience—and then releasing the tension. As a further demonstration, he strapped one of de Mille's male dancers into his machine and announced, "Now I'm going to ask you an embarrassing question"—presumably about his sexuality. The needles jerked so dramatically that Keeler didn't even ask the question. The nervous laughter was a tribute to the test's insidious power.

Keeler exploited this same technique to extract confessions in criminal investigations, at a clip of 75 percent among "guilty" suspects. In his courses at Northwestern, Keeler showed the police and prosecutors how easily it could be done. First, maneuver subjects into taking the test by telling them that only guilty people feared it; this was not difficult with suspects aware that their chance for release depended on cooperating. Then convince the subject that the lie detector worked, and repeat the questions until a full confession was forthcoming. It wasn't science, but it did the trick.

After all, the courts may have ruled out polygraph results at trial, but they still accepted confessions obtained during an exam—even when the confession was oral, even when there was only one witness, and even when the operator falsely informed the subject he had failed the test. Such shenanigans had long been a legitimate tool of police interrogation. Keeler's colleague at Northwestern, Fred Inbau, wrote the book on how to pull it off, in terms any police officer could understand. His slender volume *Lie Detection and Criminal Investigation* became the bible of police interrogators, teaching officers the byways of psychological coercion, tricks deemed so likely to induce false confessions that they earned Inbau a sharp rebuke from Chief Justice Earl Warren in the *Miranda* decision of 1966.

How successful were Keeler's methods? In 1933, while demonstrating his so-called jury box in front of 200 members of the Kansas Bar Association, he was on the verge of getting a young man to confess to a car heist—until a reporter from the *Topeka State Journal*, onstage a few feet away, hissed loudly, "Don't be a darn fool." More to the point, Keeler's technique could be used by others.

At the end of the 1930s Keeler privately surveyed police units using his polygraph technique. The thirteen respondents were East Cleveland, Toledo, Indianapolis, Kansas City, Buffalo, Honolulu, Madison, St. Louis, Cincinnati, and Wichita, plus the state police of Michigan, Indiana, and North Dakota. By their reckoning, of the total of nearly 9,000 subjects tested, about 97 percent had "voluntarily" agreed to take the exam, with only 1 percent refusing, and 2 percent confessing before even being strapped down. Of the one-third of subjects labeled "deceptive," an impressive average of 60 percent were persuaded to confess. Of the steadfast remainder, roughly half were later convicted and half had their cases dismissed. Contrariwise, of the two-thirds found to be not deceptive, only 0.3 percent went on to be found guilty at trial.

This survey, long buried in Keeler's private files, may offer the best picture we will ever have of how the police deploy the polygraph when they think no outsider is watching. Not that we ought to take these statistics at face value. If anything, they show how the lie detector had become a self-fulfilling prophecy, a way for police and prosecutors to triage suspects: quickly resolving cases through confessions, marking out suspects to prosecute aggressively, and setting aside those to consider for release. The director of police in Kansas City admitted that "if he had to choose between the polygraph and the rest of the detective department, he would choose the polygraph." An officer with the Michigan state police estimated that his polygraph alone, by disposing of cases, had saved $25,000 in court costs in 1938.

Of course, this efficiency came at a price. An examiner in Madison, Wisconsin, boasted that on being threatened with the machine four subjects had confessed to a crime—or five, he noted in the margin, if you counted the one who committed suicide. The police counted no more than 1 percent of these confessions as false. Americans like to think that we have left the age of inquisitorial justice behind, but the confession, for all its troubling ambiguities, remains our "queen of proof."

But the one dramatic exception among Keeler's respondents belies the polygraphers' confidence in the worth of these confessions. The operator with the Indiana state police achieved a confession rate of only 6 percent, one-tenth the rate of the others. Why the huge difference? The operator in Indiana was the only one trained not by Leonarde Keeler but by John Larson. Because suspects often made faulty admissions, the operator wrote to Keeler, "we rely on the Polygraph simply as an aid in investigation and not a means of bluffing an individual into a confession."

A noble sentiment—but Keeler's bluffing better suited the sort of justice then becoming the American norm. The sad truth was that an ever-expanding caseload had long pressured both police and prosecutors to resolve cases short of trial—while judges winked. And it was not until the end of Prohibition that the distinctly American practice of "copping a plea" suddenly became scandalous. Data released at the time indicated that in Cook County in 1926, only 4 percent of those charged in preliminary felony hearings faced a jury trial, and 76 percent of those indicted for a felony had pleaded guilty to a lesser charge. Today, the subterranean process of plea bargaining still resolves some 90 percent of criminal convictions and is the predominant method of determining what counts as the "truth" in American criminal law. Such bargains, as one critic noted at the time, constitute false confessions or "legal fictions . . . , psychologically more akin to a game of poker than a process of justice." In this process, the lie detector plays a crucial role in determining which suspects prosecutors will pursue and which will be given less scrutiny. So the device remains central to the process of American justice even though it is banned from the courtroom. In the game of justice poker, the polygraph bluff plays a significant role.

This bluffing explains a central paradox of lie detection. Despite tremendous advances in physiology, the hardware of the polygraph has hardly changed since the 1920s. At unguarded moments, polygraphers have admitted as much. "The lie detector has not been improved since John Larson invented the thing in Berkeley," noted one expert in the 1950s. "It has been beautified. It has a better looking box and costs a lot more. Otherwise it is the same machine." Given the nature of the ruse, the interior workings of the machinery are almost beside the point.

Indeed, police examiners regularly wring confessions by putting suspects on sham devices with contemporary flash. In the 1930s a high school principal in Newark, New Jersey, got students to confess to a mock lie box; and a policeman in New York City used a towel and a ticking alarm clock to get a youth to admit filching $10 from his parents. For those interested in something with more action, *Popular Science* laid out "Two Simple Ways to Make a Lie Detector," with advice on how to amaze your friends with the card trick. In the 1980s cops were extracting confessions by putting a suspect's hand on a photocopy machine filled with paper printed with the word "LIE!" In our era of cognitive science, cops have taken to placing the suspect's head in a colander with wires attached. Naiveté of this sort among criminals may elicit a chuckle—it got a big laugh in the U.S. Supreme Court—but the joke is double-edged. Though some suspects may be duped into confessing to a mechanical placebo, cops and prosecutors are also locked in a self-fulfilling prophecy: deciding the fate of suspects on the basis of a doubtful test.

This is a general problem. Psychological studies suggest that when subjects are convinced that an apparatus can measure their true feelings—even when it is an empty box—they admit having socially unacceptable opinions. It is the strategy recommended in every interrogator's handbook; to get a subject to confess, lower the stakes by convincing him that you already know what he has done and that you will not personally judge him. This is why the Roman Catholic Church grants confessors and penitents privacy and a degree of anonymity; and besides, parishioners are told, God already knows and forgives all your sins. Likewise, in confessing to the lie detector, the suspect is led to believe he is confessing not to a person, but to science, which knows and understands all. The judge and jury, of course, may not be so accommodating.

This is not to say that all—or even most—Americans have ever put full faith in the lie detector. Despite the best efforts of Keeler and his confederates, a whiff of hokum has always trailed after the device since its early days in Berkeley. But whatever a person's degree of skepticism, there always remains a residual skepticism about skepticism—the sort of self-doubt that P. T. Barnum knew how to exploit so well. There is always a lingering suspicion that the damn machine just might possibly work. Wasn't it plausible, after all, that guilty knowledge made cheeks blush and hearts pound? Who wouldn't feel anxious while being strapped into the chair? And then, when the operator guessed your card the first time out. . . .

The one major technical innovation in the polygraph since the 1930s actually confirms the power of this ruse. In the 1990s new computer algorithms were developed that could analyze the subject's physiological responses with mechanical neutrality. But because the algorithms might preclude operators from accusing subjects of lying (whatever the machine said), the nation's top examiners at the Department of Defense Polygraph Institute report that most operators usually turn the computer off.

In sum, Keeler and his followers operated his lie detector according to the same logic as judicial torture. This explains why the police, despite their initial resistance, ultimately welcomed the technique. So not only did Keeler pack the lie detector into a "box" that almost anyone, even a minimally trained police officer, could operate; but because of the way he conceived of its operation, it actually enhanced the discretionary power of the examiner, who was less interested in the polygraph record per se than in using it to screen suspects, intimidate detainees, and extract confessions. In this way, Keeler spurred the replacement of the violent third-degree methods of the old-style municipal police with the psychological third-degree methods of the modern professional police.

From Larson's vantage point, however, Keeler's project was a betrayal. One year after their triumphant return from Appleton, when Keeler pushed again to introduce his lie detector into court, their split became public. In

1931, a case confirmed everything Larson despised about Keeler's methods even as it conjured up demons that Keeler found hard to contain.

The case actually turned on an exoneration. In March 1931 Virgil Kirkland, a high school football star of Gary, Indiana, was tried a second time for the rape and murder of his "sweetheart," Arlene Draves. In a written confession, Kirkland had admitted that after getting Draves drunk at a party on wine laced with formaldehyde, he and four friends had driven her around town for hours and repeatedly "ravished" her in the backseat before they realized that she might be dead and deposited her at the house of a local physician. Kirkland's defense was that she had died from a drunken fall, not a blow from Kirkland's fist.

For months the case made headlines in Chicago, spiced by tidbits like the report of the medical expert who exhumed the victim's body to perform a second autopsy and concluded (somehow) that Draves had not been a virgin at the time of her rape. When the first jury returned a verdict of first-degree murder, Judge Grant Crumpacker offered Kirkland a second trial, as he personally did not believe a blow from a fist constituted premeditated murder. On the first day of the new trial the defense attorney tried a new strategy: he announced in the presence of the jury that Kirkland had taken a lie detector test which had proved him "one hundred percent innocent."

This strategy was the handiwork of Dr. Orlando Scott, a surgeon in Chicago who had become a courtroom alienist and had twice been cleared of perjury charges for filing false medical reports. It was Scott who had led the exhumation and autopsy. It was Scott who arranged to have Keeler test Kirkland at midnight in his jail cell and saw to it that the results were reported in the local newspaper as the jury was being seated. And it was Scott who orchestrated the testimony of the members of Northwestern University's crime lab in support of the lie detector, with Keeler expressing his willingness to test Kirkland in open court if that would help.

It was no use. The prosecutor, John Underwood, denounced "any fantastic device which presumes to supplant the greatest lie detector in the world, twelve good men and true." And Judge Crumpacker conceded that he could not let his court be "used as a testing room for an unproven device." "Such a step," he added, "would subject me to ridicule and probably result in a mistrial." (He had no problem, though, with his son's serving on the defense team.) When the second jury exonerated Kirkland of the major charges, the shockingly light verdict caused pandemonium. Dr. Scott was picked up that evening for drunkenness and tossed into the cell adjoining Kirkland's.

In his campaign to get the machine into court, Keeler had again tried to enlist the testimony of John Larson. This time, Larson not only complained to Vollmer but wrote directly to Judge Crumpacker, then denounced the charade in a police magazine. As Keeler sheepishly acknowledged, "I'm sure he was quite upset by the whole affair."

Worse, while Larson was criticizing the lab for grandstanding, Orlando Scott was out-grandstanding the lab. It's hard to say which vexed Keeler more. Soon after Kirkland's trial, Dr. Scott entered the field of lie detection himself. In short order he was promoting his own 100-percent-effective "Thought-Wave Detector," which tapped, he said, the electrical currents of the brain. In an interrogation room resembling the laboratory setup in the movie version of *Frankenstein* made in 1931, the subject was seated in a giant chair, a metallic halo was fastened around the skull, and electrodes were attached to the base of the skull and both hands. Then Scott would flick the switch and commence the interrogation while a giant needle swung from "True" to "False" on a dial. Not even psychopaths could escape detection, Scott explained; he just cranked up the juice when a dullard was in the chair.

In fact, his instrument measured not "brain waves" but galvanic skin resistance, a proxy for sweat. Yet with this unholy contraption, Scott got results. And thanks to his status as a physician, he was able to pioneer the introduction of lie detector evidence in civil courts. Scott's machine confirmed that a sixteen-year-old was telling the truth when he said he had killed a twelve-year-old with a stone by accident; the award was reduced to $500. Scott's lie detector affirmed a wife's claim that her husband had fathered her child; the divorce court judge ordered the man to pay the expenses of the delivery. Scott's lie detector belied the alibi of a seventeen-year-old boy accused of the statutory rape of a fourteen-year-old girl who then had a baby; the boy was ordered to pay child support. A federal compensation examiner praised the device for exposing false injury claims. As Larson ruefully acknowledged, "Despite his mumbo-jumbo and perhaps because of it, Scott was remarkably successful in getting confessions."

No wonder Scott infuriated Keeler: the alienist outdid him at his own game. Whereas Keeler restricted sales of his machine, Scott refused to sell his machine at all lest it prove too potent in the wrong hands. Whereas Keeler's patent gave him a limited monopoly but obliged him to publish his design, Scott showily kept his superpowerful circuitry top secret. Whereas Keeler had provided his services to a limited number of private enterprises, Scott proudly advertised his National Detection of Deception Laboratories with the motto, "Diogenes searched for them—We find them." Soon Keeler was pleading with Scott to stop introducing his contraption into the courtroom and was having Northwestern University threaten to sue Scott if he didn't stop advertising his one-time affiliation with the school.

Scott's success confirmed everything Larson despised about Keeler. It's almost an American axiom: No matter how much you pander to the public, there is always someone willing to take your idea farther down-market than you would dare. Conversely, no matter how much you pride yourself on your intellectual integrity, there is always someone more exigent. Even as Keeler was being chastised by Larson for being inadequately scientific, he had to contend with rivals like Scott for the popular stage. And even as Larson was chastising Keeler for failing to make the scientific grade, he was himself "in the doghouse" with the medical and psychological authorities.

Larson was working days running cases for the Illinois state criminologist, and nights on his big book and report for Vollmer. The more he studied the technique—and the more he saw of Keeler's methods—the more he was convinced that he himself had been right all along: the lie detector's results ought never to be determinative. What was new was that Larson had found a framework to express this. The lie detector had to be understood in the context of the medical clinic.

Laboratory psychologists like Marston who boasted about achieving a success rate of 90 to 100 percent were conducting meaningless games. And crime-busters like Keeler who boasted about getting 75 percent of the "guilty" to confess did not know the actual number of guilty individuals and had no way to estimate the number of false confessions. These statistical claims dressed the lie detector in scientific clothes, while evading deep problems involved in the interpretation of emotions. Without a theory to explain how deception produced bodily responses distinct from the anxiety of innocent subjects, there was no way to draw inverse inferences from a record alone. Hence it was no surprise that the field's two most experienced workers—Larson and Keeler—had disagreed on the interpretation in several prominent cases. On the contrary, the lie detectors' overconfidence was a strike against them. Larson reminded Vollmer that "all scientists become suspicious of the technique and method of investigation if [it] shows up 100% when dealing with such factors as human emotions."

This was a matter not just of science but of justice. The criminal justice system, like medicine, demanded that subjects be treated individually. Even knowing that 90 percent of patients with typhoid fever died, a physician was obliged to treat each patient as one who might survive. Similarly, even if there was only a 10 percent error rate in distinguishing the innocent from the guilty, the lie detector ought not to assign guilt. His training with Adolf Meyer had taught Larson to treat each person as a unique experiment in nature. "We have made mistakes and will doubtless continue to make mistakes," he reminded Vollmer. In that light, it would be a miscarriage of justice to convey these mistakes to a jury.

This did not mean that Larson considered the polygraph useless. After all, no disease had an absolutely certain diagnosis, yet physicians had a profound understanding of many illnesses. The machine could help diagnose mental illness by "probing for complexes and in the removal of resistances." If this coincidentally solved a crime (by getting corroborating information), that was a worthy result. But the device ought not be called a "lie detector," and Larson was adamant that exams be conducted only by a fully trained psychiatric expert, working in conjunction with experts in psychology, criminology, social work, and police procedure.

Larson's scientific modesty disappointed Vollmer. Of course Larson was free to write up his research as he saw fit. But Vollmer warned that these conclusions might lead readers to believe the lie detector had no value. "My own opinion is that you cannot throw in the waste basket all of your labors." Perhaps, Vollmer suggested, Larson's modesty was itself a form of overreaching. How could Larson be so sure that his failure to isolate an identifiable deception response meant there was no such thing? Perhaps his technique was faulty, or his criteria of discrimination had been poorly chosen.

This time Larson stood firm. "I flatly disagree with you," he wrote back, "but any disagreement is due to the difference in approach to the problem and to the training involved." Larson had found a new Chief—that was how Larson now referred to Meyer—and for a scientist, nothing matters more than the respect of the respected. Larson made his choice. And he published.

The book had a difficult gestation. In 1926, when Larson first offered to make Keeler his coauthor, it comprised 600 pages. By the time he returned to Chicago in 1930, it had swollen to 1,000, of which Keeler had supplied five. After sitting on the manuscript for a year, Keeler mailed it back: "Golly, I'm sorry I haven't had time to wade through the stuff, but . . . my head buzzes when I think of the things I really should do but don't." As a consequence, Larson told Vollmer, he could no longer in good conscience include Keeler as his coauthor. He recalled phoning Keeler one day for help running a test; Kay Keeler had phoned back to tell him they were off for an outing to the Indiana dunes. "While this work was being done," Larson told Vollmer, "I didn't get any outings." Thus have ants always resented grasshoppers.

To Vollmer, however, Larson's repudiation was a "breach of faith." He reminded Larson that in the Berkeley police school "a [man's] word was as binding as a National Surety Bond." But Keeler himself—back at Stanford for the summer to finish his degree—generously waived any claim to the book. As a compromise Larson offered to list Keeler as a "contributor." And so Keeler appeared.

Perhaps because of these conflicts *Lying and Its Detection*, published by the University of Chicago Press in 1932, turned out to be a strange book. Larson himself referred to it as a "source book," meaning a compilation of work in the field, rather than his own views. The book opened with no fewer than three prefaces: one by Vollmer, one by the physiologist who directed Larson's dissertation, and one by a leading psychiatrist. Adolf Meyer had refused to contribute; he felt that this parade of authorities "savored of cheap publicity." The prefaces were followed by three major sections of extracts from commentators on deception, printed in minuscule type. First came the philosophers and psychologists on the prevalence of lying in different sorts of persons: children, women, psychopaths, and "normal" men. Then came the legal authorities on the history of exacting truth: by trial by ordeal, by torture, by "third degree" methods, and by adversarial justice. Third came a tour of recent physiological research on deception. Only in the fourth section did Larson present case studies from his twelve years of research, grouped as penitentiary cases, confession cases, "insider" cases (including the College Hall case), and murder cases. Amid these he slipped in oblique slurs against Keeler's handling of various cases, such as the Ku Klux Klan case, the Mayer case, and the Kirkland case.

The book offered no insight into the mind-body relationship that presumably governed the action of the lie detector. It did not explain why the technique worked or failed. It gave no statistical assessment. As Larson was wont to say, "Each case is a clinical entity within itself." By implication, however, the book did make two contradictory claims. First, it suggested that the so-called lie detector was a diagnostic tool in the hands of a well-trained criminologist. Indeed, the book's historical story line implied that scientific progress had finally put the interrogation of suspects on a humane footing, protecting the innocent and insulating the police from politics. But then, in a contradictory vein, the book suggested that the so-called lie detector would probably never determine a person's guilt or innocence, except as an adjunct to police interrogation, principally by extracting confessions.

Even reviewers who praised the book as "ultra-conservative" for its refusal to perpetuate popular myths about the lie detector noted that Larson provided no alternative scientific account. Others pronounced themselves baffled and disappointed. What sort of science offers a parade of undigested authorities and refuses to make generalizations? For his part, Keeler did not fail to notice the slurs. Larson, he wrote to Vollmer, "published every slanderous thing he could think of about me. . . . I feel that I can hardly trust him in the future."

It was time for Vollmer to choose among his disciples, and he did not hesitate: between Larson's hyperactive scrupulousness and Keeler's cool effectiveness, there really wasn't much choice. He wrote back to Keeler to console his young disciple. Larson, Vollmer said, had acted ignobly; indeed, the psychiatrist might not even be right in the head. "It is my opinion that Larson may be slipping slightly. . . . It is possible that the poor chap is overworked." For the future, he suggested that Keeler ignore Larson's barbs and treat him with indifferent courtesy. And Keeler agreed to do so, despite Larson's provocations.

> Our good friend John Larson is still up to his old tricks. Gosh, I've done my best to be friendly, to give him all the credit due him for his good work, and to cooperate with him whenever possible. He always seems so friendly in my presence, but behind my back that's a different story. To individuals and in public talks and articles, he slams and pokes and tells some of the darndest lies you ever heard.

Keeler thought Larson's accusations of scientific duplicity were themselves two-faced. How hypocritical of Larson to accuse him of peddling polygraphs to incompetent examiners and simultaneously gripe that he was holding back sales. And Keeler had heard rumors that Larson was under the "delusion" that Keeler was trying to get him fired from his job as Illinois state criminologist, and in "retaliation" was seeking to get Keeler fired from Northwestern.

Keeler admitted to Vollmer that his patience was running out, but that he would continue to "try to speak kindly of John and . . . ignore his foolishness." He even professed a willingness to offer Larson a half share in his patents if that "could save him from a complete mental breakdown"—though Keeler doubted it would help. In any case, Keeler never made the offer. But he did make one last effort to reach out.

As for the Chief, he practiced what he preached. He wrote a brief note acknowledging receipt of Larson's book, then dropped all contact with Larson for twenty years.

Keeler likewise blandly thanked Larson for sending him a copy. After all, it's easy to be gracious when you've won. Larson's reply feigned oblivion: "I am glad to hear from you as I feel we can do rather good work if we keep together." As it happened, the two men made one final, halfhearted effort to collaborate. Chicago was preparing for a spectacular world's fair, a potential showcase for the lie detector.

As soon as Keeler got wind of the fair he stumped for a crime lab exhibit, with the lie detector as its centerpiece. The organizers of the fair were happy to oblige him. The idea of hosting a world's fair in Chicago had originated with businessmen close to Mayor Cermak who wanted to clean up the city's reputation. Around the world, wherever Chicagoans traveled, they heard,

"Bang, bang, Al Capone," accompanied by either the old-style index finger or the new-style double-fisted rat-a-tat. Calvin Goddard promised the organizers that the exhibit by the Northwestern University lab would forever dissociate the words "Chicago" and "crime," so "inseparably linked" in the public mind. The exhibit would also dovetail with the grand theme of the fair, which was called "Century of Progress"—the promise of science.

To those Depression-era Americans who doubted that technological progress was their ally, the organizers offered this stern rejoinder, emblazoned over the fairground entrance: "Science finds, Industry applies, Man conforms." Privately they agreed that the fair of 1933 would offer an "unconscious schooling" in the lesson that science was the cure for social disorder. At a minimum, this meant the fairgrounds themselves had to be well-policed. To this end, the organizers hired one of Vollmer's college cops to re-create in rough corrupt Chicago an oasis of Berkeley-like lawfulness.

Yet the exhibit suffered from the same paradoxes as the lie detector. Unable to find donors to mount a free exhibit, the lab decided to charge an admission fee. This, in turn, meant luring an audience with "features which might be termed sensational." When Larson, who had contributed his cardio-pneumo-psychogram to the display, saw the "half-witted" exhibit, he called it an advertisement for Keeler's lie detector. In the end, the exhibit split in two. On the eve of the show, Colonel Goddard was asked to resign as director of the crime lab, in part to relieve Northwestern University of his hefty salary, in part because of rumors of his "overt activities" in answering the "feminine call." Goddard retaliated by mounting a rival crime-fighting exhibit, a "bally-hoo side-show" featuring a display of horror crimes, with instruments of torture and an electric chair, plus a personal appearance by Dorothy Pollock, "Chicago's Most Beautiful Murderess."

Keeler considered the legitimate exhibit a success. Each day that summer 10,000 people passed by the second floor of the building where he and the other lab regulars answered questions about crime-scene methods and the lie detector in particular. He hoped the exhibit would foster "a more sympathetic understanding between the public and the police." It also served a more practical purpose. When a lockbox was stolen from the Michigan state exhibit, the fairground police used the shoeprint and fingerprinting kit featured in the exhibit to recover the money. No one ever caught the culprits who twice burglarized Goddard's gun display.

A year after the fair was over, Larson wrote a short note to Keeler asking him to return the borrowed lie detector. By return mail Keeler replied that he was "snowed under" with work, and asked Larson to drop by the crime lab and pick it up himself, or if not, then Keeler could bring it around later. Only twenty blocks separated the South Side Institute from the North Side crime lab, but neither man made the first step. Twenty years later, Larson was still bitter that Keeler had never returned his machine. As the contemporary social critic Philip Wylie wrote of those times, "People . . . made the farcical assumption that because their machines were efficient and honest, they too partook of those qualities."

CHAPTER 11

# Traces

*"I was fascinated by him," Dorothy said, meaning me, "a real live detective, and used to follow him around making him tell me about his experiences. He told me awful lies, but I believed every word."*

—DASHIELL HAMMETT, *THE THIN MAN*, 1933

IN THEIR APARTMENT EIGHTEEN STORIES ABOVE THE GOLD Coast, Nard Keeler played Nick Charles to Kay's Nora. Like the dashing detective and saucy heroine of Dashiell Hammett's *The Thin Man*, the handsome young marrieds joshed, drank, and solved crimes together, while disdaining the ugly thugs, crooked cops, and nosy newshounds who disrupted their cocktail-party repartee in the witty heights of the vertical city. He played the sardonic charmer who could handle the heavies. She played the sexy wife who could handle a gun. They had no kids, but plenty of adventures—with Chief, their clever German shepherd (named in honor of August Vollmer) in the role of Asta, the Charleses' terrier. The Keelers even knew how to leaven murder mystery with screwball comedy. A year after the release of *Bringing Up Baby* (1938), Mr. K. brought home from the Brookfield Zoo a six-month-old baby jaguar, which promptly slashed Mrs. K.'s stockings, ripped her curtains, and caused domestic mayhem—another choice item for the press.

Beneath their windows, the Depression ground grimly along. But who said marriage couldn't be sexy? Nard admired Kay in her modern finery: her pert hats, homemade suits, and "fantastic pajamas." The consensus among Nard's family and friends in those early years (especially among the women) was that Kay loved Nard more than Nard loved Kay. She certainly didn't like his trips away from home. "If I had a hundred a day," she wrote to her parents, "I'd pay him the same to stick around Chicago." But he didn't like her to leave town either. "I miss the grand girl," he wrote to Vollmer. Kay took a sardonic pride in her husband's charm. When a glamorous actress raved about Nard after one of his public demonstrations, Kay remarked wryly: "He certainly has a gift of attraction for men and women alike." Nard admired Kay's brains, her looks, her pluck. Their marriage was founded on playful competition and mock jealousy. Writing to her father-in-law, the prudish poet, Kay teased:

> Nard is down in his lab hunting for a confession in a bank embezzlement case which he has been working on for several hours. I hope he is getting something, as he is on the last male employee, and if he isn't guilty Nard will have to start on the women.

They made a smart team, and people constantly dropped by their apartment: friends, family, family friends, friends of friends, colleagues from the lab, business acquaintances, drinking companions. The Keelers entertained so often that it was a year into their marriage before Nard could say they were finally alone together. They had returned to California for the summer so Nard could finish his degree at Stanford. No sooner had he wondered "how bride and groom will get along without chaperones" than old friends of Kay's showed up bearing chocolate milk shakes. After Leonarde scraped through the exams, they returned to Chicago, the city they had made their own.

After that, Nard and Kay managed to get away periodically. In the summer of 1932, they left the wilting heat of civilization for a two-week canoe trip in the cool woods of Wisconsin, just the two of them and a nine-pound tent. They didn't reach the halfway hotel until three in the morning; yet even at that late hour—while Nard (with a stomachache) undressed for bed—Kay set the scene for her mother: the two of them hurtling north through the darkness at fifty-five miles per hour as their radio played snatches of news, music, and police dispatches about a "man in a hotel causing trouble . . . , [a]nd beside me was the man who made it impossible to lie with any feeling of security. Very ultra moderne indeed, Madame!"

To his credit, Leonarde did everything he could to help Kay enter the lists of scientific crime-fighters. Soon after their marriage she set out to become the lab's document examiner and handwriting expert. At Vollmer's criminology class at the University of Chicago she studied with Albert S. Osborn, America's foremost expert; and she completed her apprenticeship under Herbert J. Walter, the mild-mannered man who had testified in the income tax case against Al Capone. In 1931 twenty-five-year-old Katherine Keeler became America's first female practitioner of the world's original forensic science.

As the first courtroom witnesses to claim the illustrious title "expert"—a claim dating back to the Renaissance—handwriting analysts served as the archetype for all subsequent forensic scientists, including the lie detectors. All forensic scientists make a fundamental assumption that each person's bodily actions leave an idiosyncratic trace. For instance, according to handwriting experts, the way each writer forms characters is a distinct expression of his or her own personal character (a term derived from the Greek word for a pointed stick that made a mark). Over time, this bodily habit—whether innate or acquired—becomes so deeply engrained that it cannot be entirely disguised or imitated. Indeed, early modern handwriting experts advised that this sort of personal consistency was the best surety against having one's handwriting forged. Paradoxically, this meant that to identify a person's signature—a supreme act of the conscious will—experts relied on the person's unconscious habits, in much the way the polygraph operators implied that unconscious physiological reactions affirmed or belied a person's willful statements.

As Katherine liked to explain to jurors, it was just a matter of reading the clues, something that ordinary men and women did all the time. "Can you tell the tracks of a running deer from those of a walking deer . . . ? Can you tell machine sewing from hand sewing? It takes no Sherlock Holmes to answer these questions affirmatively." Of course, each individual's handwriting differed each time the person set pen to paper, depending on age, circumstances, health, and mood. This is why handwriting analysts compiled a set of comparative documents from the writer's past, much as polygraph operators took a "normal" for each subject by asking irrelevant questions to compare with (possibly) deceptive answers. Properly read, such traces were more reliable indicators than whimsical memory or the obfuscations of a criminal mind. The problem is that any two things are bound to differ in some respects while being similar in others, making it all too easy for experts to emphasize just those similarities and differences that made their case according to their personal predilections. Indeed, all too often handwriting analysts disagreed with one another. This is not to say that the art of comparison was entirely without value; the trick was to bolster the plausibility of the identification by amassing corroborating inferences: not just in the handwriting, but in the ink, the paper, and the spelling. This was rarely an option available to the lie detectors, for whom the only external corroboration was a confession, the very reaction they sought to elicit.

As Kay was the first to admit, her work attracted extra attention from the press (and hence from clients), thanks to the contrast between her comely looks, her arcane skill, and the foul deeds at issue. "Mrs. Keeler, [a] tall girlish blonde with blue eyes, appears entirely too young and pretty to be associated with the grim business of fighting crime." She had mixed feelings about this attention. She shaved two years off her age for all official business. She was ecstatic when she first appeared on the front page of the *Chicago Tribune,* after testifying against an extortionist who had sent death threats to four society brides—though this was the sort of case, she noted with exasperation, that would have put Nard on the front page for a week. (Also, her photo, on the back page, was perfectly rotten because she had forgotten to turn up the collar of her new coat.)

But she soon found herself featured in the papers "to a nauseatingly disproportionate degree." The Sunday supplements thrilled to the idea of an attractive "girl" sleuth. A typical photo spread—"How Science Wars on Crime"—showed Katherine, in a velvet dress and marcelled hair, in the lab preparing slides for a microanalysis of fibers, photographing a dubious signature, and pointing a test gun straight at the reader.

The one time that her sex actually helped her do her job was the most dangerous. During the Depression, Sears, Roebuck began losing $1 million a year because of forged checks from "Bloody" Breathitt County, Kentucky. One brazen schemer signed his name "E. Normous Wealth." Private investigators were warned off with shotguns. But Kay laid out an ingenious plan to get comparative signatures. In June 1933, she and her best friend, Jane Wilson, posed as doctoral students doing eugenical research on "pure blooded" Americans, and rode horses into the Kentucky hills. Whenever they found a family with new Sears equipment, "I'd look at some small boy and say, 'Jane—see he has the perfect type of eyes!' Jane would paw over the glass eyes in the box, pick out one, and agree that the boy had an unusual example of type Y-62 eyes." Then they would ask their subjects to sign their names to certify the research. One day, a ten-year-old boy accosted them with a rifle, "You'uns better git out of these mountains—quick." They did. A year later, the Post Office Inspection Service had used their identifications to convict 150 persons. Leonarde was immensely proud of his wife's pluck.

It took equal courage to battle the electoral fraud that was the sustaining crime of the Chicago political machine. In the primaries of the 1920s—marred by kidnappings, shootings, and murder—ballots were weighed, not counted. But after the Democrats consolidated control in the 1930s, the vote padding took place peaceably, behind closed doors. The *Illinois Crime Survey* listed twenty-five ways to rig votes, each with dozens of variations. Scams multiplied as fast as officials could write rules to prevent them—and no wonder, since the officials benefited from the scams. Kay's investigations of election fraud in Chicago convinced Leonarde that "[t]he ballot, the one weapon of the honest citizen, is a meaningless scrap of paper."

When the human witnesses were in cahoots, an election judge had to rely on traces left behind unconsciously. In 1934, in one recount in Cook County, in which 29 percent of ballots were found to have been falsified, Kay's testimony proved that hundreds of votes on tally sheets had been tacked on after the fact. Instead of being irregularly marked, as would have been the case had the tallier shifted from candidate to candidate as their names were called out, the marks were all of the same length and angle, as if ticked off in rapid succession. Moreover, the marks showed a tiny characteristic uptick at the bottom of each line as the tallier's pen anticipated the downward stroke of the next line. This minute trace was enough to convict. In another case, she relied on faint indentations, visible only in sunlight, made by pen pressure on one ballot atop another. By 1940 she had assisted in 100 convictions involving fraudulent ballots, part of the same scientific battle her husband fought against Chicago's political machine.

By the mid-1930s, Leonarde and Katherine Keeler were the crime lab's mainstays, with Nard serving as unofficial director. Despite many promises to shift from crime-fighting to scientific research, he preferred running cases. And no case better dramatized the crime lab's capabilities—and allegiance—than the Valier bombing. Described as the nation's first case constructed wholly around forensic evidence, it tested the skills of every member of Keeler's staff, beginning with Nard's facility with his lie detector. Prosecutors were skeptical that this sort of scientific evidence could persuade a lay jury anywhere, let alone in rural Illinois. But as the local paper reported, the culprits had not grasped "the extent that modern science has progressed in pointing the finger of guilt at those who think that punishment may be evaded when no person actually sees a crime committed." From fragmentary traces found amid a chaotic wreckage, the lab recreated a coherent chain of events. It was as if science could run the film of time backward and piece Humpty Dumpty back together again.

The Valier mine was located in coal-rich "Little Egypt" at the southern tip of the Illinois dagger. When it was founded by the Chicago, Burlington and Quincy Railroad in 1917, it was the largest and most modern coal mine in the world. Twenty years later it still employed 500 miners, all members of the Union of Mine Workers (UMW), accused by the breakaway Progressive Mine Workers of colluding with mine owners. During the Depression their conflict had escalated into something like a civil war. A nearby march by 12,000 members of the Progressive Mine Workers was violently dispersed by club-wielding sheriffs and deputized UMW members in the Battle of Mulkeytown. Keeler tallied some 300 violent incidents between 1933 and 1935, including at least ten murders and 100 bombings—all unsolved.

Then at two o'clock in the morning on August 26, 1935, a huge explosion destroyed the Valier engine house. An hour later four deputies affiliated with the UMW roused Mitch McDonald and Robbie Robertson from bed. Only that afternoon the two had attended a memorial for a Progressive leader killed by a member of the UMW during a strike in 1933 that had cost them their jobs at the Valier mine and led them to join the Progressive Union. Though the deputies found nothing incriminating and didn't have a warrant, they hauled McDonald and Robertson off to jail. That morning, the mine manager called Governor Horner; within twenty-four hours, Leonarde Keeler and Charlie Wilson were on-site seeking clues.

Amid the physical wreckage of the engine house Keeler picked out a half-shattered alarm clock, which he assumed was the bomb's timer because of its copper leads and adhesive tape. And amid the human wreckage of this labor struggle he picked out evidence of guilty knowledge: a physiological reaction from McDonald and Robertson after an eighteen-hour interrogation on the lie detector that was so intense that Robertson had ended it by smashing the machine with his fist. Keeler's goal was to connect these two sorts of evidence.

Though he failed to get a confession from either man, Keeler extracted an admission almost as telling. Robertson admitted he had tape "like" the tape on the alarm clock in a first-aid kit in his home, adding, "You can go out to my house and see if you want to." McDonald similarly admitted owning fishing line and copper wire "like" those found with the alarm clock. These items, considered innocuous by both the police and the suspects alike—Who didn't have a first-aid kit? Who didn't own fishing line?—were taken to Chicago to be analyzed in the Northwestern University lab. At the trial seven months later the staff returned to declare a match. Charlie Wilson matched the microscopic scratchings on the household copper wire with those on the alarm clock. Katherine Keeler matched the jagged cuts in the tape from the first-aid kit with the tape on the alarm clock, confirming that the thread count and weave were identical. Ed O'Neil matched the fibers in the fishing line with those on the bomb mechanism. And fragment by fragment the crime lab pieced the past together.

The defense struggled in vain to impugn this scientific testimony, arguing that any piece of wire or tape from the same manufacturer would have been similar. But without its own experts—or access to the physical evidence—the defense could not cast doubt on the specificity of the similarities. Besides, the defense attorney was in a drunken stupor throughout the trial. Ostensibly, all this evidence was meant to corroborate the accusations of the defendant's roommate, Peter Benetti, a twenty-four-year-old ex-convict who had been granted immunity in exchange for his testimony. In practice, the honest witnesses were shreds of tape and broken wire. "Oh, Pete Benetti can lie," summarized the prosecutor, "Bobbie Robertson can too, and so can Mitch McDonald, but those can't lie. . . . They speak for themselves and say, 'I Mitch, and I, Bobbie, and I, Pete—we made that thing.'"

*Res ipsa loquitur,* as the law says: the thing speaks for itself—thereby eliding the seven months that Nard and his fellow experts had labored to coax those accusatory cries from the torn fabric of the first-aid tape, severed fishing line, and mangled copper wire, not to mention the years of training that had gone into learning how to make these wounds in the world sing in unison. No, whatever else the forensic sciences may accomplish, they do not make things speak for themselves. Yet amid the silence of the defense, the expert testimony spoke loud and clear.

The jury of local farmers took an hour and a half to reach a verdict of guilty, and Judge Hill sentenced the men to twenty-five years. Thanks to science, proclaimed the local paper, the "reign of terror" was over. As for the team from Northwestern, their victory paid dividends, starting with an "anonymous" gift of optical equipment and $1,000 from the mine manager, who took Nard, Kay, Charlie, and Jane in his personal railroad car to Cheyenne, where Kay rode in the Frontier Days rodeo.

Then, a year later, Keeler achieved the breakthrough so far denied him: he formally presented results from his lie detector to a jury. Yet the case also set a limit on the machine's acceptability. Ever since seeing Keeler demonstrate his technique, Judge Clayton F. Van Pelt of Portage, Wisconsin had been on the lookout for a trial that might showcase the lie detector. Tony Grignano and Cecil Loniello were charged with fleeing a pharmacy holdup in which a cop had been killed. They had much to fear from a jury, and the prosecution must have known that two of its own witnesses were perjurers. Hence, Van Pelt was able to secure an agreement in advance from both the prosecution and the defense that they would allow Keeler to testify to the jury, no matter how his test came out. On February 7, 1935, the prosecution called Keeler to the stand, and for three hours he explained that the two young men had been untruthful, that his technique was 75 percent effective, and that his results ought not to be the sole basis for conviction. According to the judge's private survey, the jurors found the lie detector offered "corroborative evidence in connection with other facts proved," and they voted to convict.

Modest on the witness stand, Keeler was less bashful afterward in the press. "It means that the findings of the lie detector are as acceptable in court as fingerprint testimony." This statement was a whopper, but then Keeler was always willing to read similarities and differences to his advantage. The case did, however, set a legal precedent: prior stipulation remains the sole basis for accepting the polygraph tests in most criminal courts.

The *Frye* ruling of 1923 had declared that before scientific evidence could be heard in court it had to be based on science that was "sufficiently established to have gained general acceptance in the particular field in which it belongs." This begged the question of who constituted the relevant experts. Polygraph examiners claimed this mantle for the science of lie detection, and argued that the polygraph had proved itself 90 to 100 percent reliable in hundreds of laboratory studies. But the courts instead looked to academic psychologists, who pronounced themselves skeptical of its merits in surveys in 1926 and again in 1952.

During this same period, however, the judiciary invoked the same Frye rule to admit many other forensic sciences treated with considerable skepticism outside the immediate circle of practitioners: handwriting analysis, ballistic identification, and forensic psychology, to name a few. The lie detector alone has been banned. As several judges have hinted, the courts rejected the lie detector not for its failings but for its power—what one court called its "aura of near infallibility, akin to the ancient oracle of Delphi." The concern dates back to the *Frye* case itself. Because lie detector evidence goes to the heart of the defendant's guilt or innocence, jurists fear that polygraph experts—were they to be believed—would unduly influence, or even supplant, the jury.

Keeler hoped as much. Not only did he aspire to sway juries with polygraph evidence; he hoped in time to see the jury system abolished. Lay jurors were too easily swayed by pretrial publicity, rhetorical flash, and emotional appeals. As jurors were also incapable of evaluating sophisticated psychological tests, he agreed that they ought not hear polygraph evidence either. Instead, he advocated trying criminal cases before expert criminologists wielding a polygraph, with a judge to rule on legal technicalities. Keeler looked forward to a justice system run with the efficiency, precision, and impersonality of a machine, regardless of whether this ran roughshod over constitutional guarantees or how many lawyers were sidelined.

Perhaps it is not so surprising, then, that the judiciary kept the polygraph out of their criminal courts—while, of course, allowing it to play a role in the invisible 90 percent of criminal cases where it functioned as just another chip in a game of plea bargaining. In any case, though, this formal repudiation was in large part Keeler's own doing. The main obstacle to credible tests is the large number of "incompetent" examiners who use the test to bamboozle defendants into confessing. But it was Keeler who pioneered the quick training of operators and cultivated a vast marketplace for the kind of expertise that thrives on enhancing the examiner's discretion. That is, Keeler's style of lie detection succeeded at its principal task—extracting confessions and intimidating subjects—only if the operators consistently refused to be bound by even the most basic norms and standards. If polygraphers have thrived, it is because they are consummate antiprofessionals.

As for the general public's credence in the lie detector—the reason judges have been so anxious to shield juries from polygraph evidence—it was Keeler and his successors who cultivated the public myth of the lie detector's effectiveness, not only to increase the demand for their services, but to make the lie detector that much more potent. Indeed, the lie detector is a placebo science in that it works to the extent the popular culture has been convinced it works—even though it works best when its operators lie.

By the time Keeler decided that his mania for publicity was counterproductive, it was too late. Thus, when his father suggested that he use his lie detector to solve the kidnapping and murder of the Lindbergh baby—the number one crime saga of the 1930s—Keeler wrote back in indignation, even though he was angling to do just that, as were all his competitors. William Marston, Orlando Scott, and even John Larson wanted to test Bruno Hauptmann, the enigmatic immigrant who had been convicted of the crime and was petitioning for a pardon from the governor of New Jersey. Publicly, Keeler announced, "I would like to be the one making the examination." Behind the scenes, he even asked his friend Governor Horner to put in a good word for him. Yet when Keeler's father drafted a letter to Governor Horner suggesting just this course of action—plus a proposal that the Keeler Polygraph be given before all state executions—years of accumulated resentment poured forth.

Leonarde sternly rebuked his father. It was time for Charles to let him stand on his own merits. "All scientists"—among whom Keeler numbered himself—"fear press reports." Too much coverage was ruining his reputation. "So PLEASE never—never—under any circumstances write to anyone about me or concerning my work. . . . In other words, father, you've been the grandest father in the world—but your role as a protecting go-between for your children is a thing of the past."

But when it came to indignation, no one could top Charles Keeler. Leonarde's rebuke had hurt him "as deeply as a son could wound a father. . . . *Please* never-never under any circumstances write to your mother or me another such letter unless you wish knowingly and deliberately to bring on an attack that will prove fatal." And he elaborated in a sixty-page letter. Alas, Leonarde's criminological work had made his son "harder and less charitable," even as it crowned him with material success. On further reflection, however, Charles supposed he had only himself to blame if his son had turned against him; he had wanted too much for his children, given them too many spiritual gifts. For the first time he told his son of the nightly words he had recited by his bedside to encourage independence of thought; plainly, this hypnotism had backfired. "I was playing God," he now admitted, "and I left out the side that has so often caused me to suffer at your hands, turned you away from me without meaning to or realizing it."

By return mail Leonarde graciously begged forgiveness. He blamed his emotional outburst, "which I should have controlled," on his overactive sense of shame. The truth, he admitted, was that he admired his father for not giving two cents what other people thought, whereas Leonarde still feared what people thought of him, not just in his personal life, but in his public life as well.

> I wanted publicity in the past because I thought it would help us in our work. I've wanted to be friendly with everyone, including newspaper men. But now I have an abhorrence for publicity. I fear it because it always brings criticism from the more worthwhile people. I'll be happy if I never see my name in the press again.

Somewhat mollified, his father advised him to rid himself of "doubts and suspicions and fears" and regard his fellow man—and his father—with "free and easy confidence."

If only it were that easy. But a career of mistrust had made Leonarde wary. One year later, he did just as his father suggested, and the result was the most damaging scandal of his career.

Keeler's friend Governor Horner was in a political bind. Joseph Rappaport, a resident of Chicago, had been sentenced to die for the murder of a police informant preparing to testify against him about a heroin sale. The mother of the dead man claimed to have witnessed the killing, at South Lawndale and Nineteenth Street, although the police had seen her arrive after they did. But the defense witnesses proved even less credible. One denied, in the morning, that the killer resembled Rappaport, then returned in the afternoon to admit having invented the tale at the behest of Rappaport's sister. In June 1936, the Illinois supreme court upheld the murder conviction and set a date six months hence for the execution.

From death row Rappaport's family solicited testimonials from prominent Orthodox and Reform rabbis to put pressure on the Jewish governor. Though personally opposed to the death penalty, Horner had sworn to carry out the law. Yet he granted Rappaport five more stays, including one for the Jewish holiday of Purim. Then, on the day before the execution, Rappaport's sister accosted the governor at a train station in Chicago. According to the Yiddish papers, her pleas brought tears to the governor's eyes. In a dramatic flourish, she even invoked an ancient curse—presumably in Yiddish—warning of what would befall the governor should he fail to offer clemency. Cornered, Horner conceded that he "might" consider another postponement if Rappaport passed a lie detector test.

So at nine o'clock at night on March 1, 1937, four hours before the scheduled execution, Keeler sat Rappaport down in his jail cell on death row for a session on the polygraph. Two hours later he phoned the governor to say that Rappaport's denial was a lie. Two hours after that, Rappaport sat down in the electric chair.

The story was wired around the world, and Rappaport's record was unfurled across front pages like "the mute lines of doom." Some people called the test a primitive ordeal; others considered it scientific proof of the killer's guilt. The British press was horrified that a man's fate should be decided mechanically. Keeler's colleagues were flabbergasted. How could an operator take a "normal" reading when the subject knew that if he failed he'd be dead within two hours? Larson wrote a commentary in the *Chicago Daily News* that was remarkably restrained, considering the provocation. Privately, however, he excoriated Keeler.

For his part, Horner was defiant. "I can't afford to be a sentimentalist in this job," he said. Yet for weeks after the execution, according to his biographer, the governor's pulse rate was high and he suffered from cardiac arrhythmia. Around the same time Keeler also began experiencing severe heart palpitations and dizzy spells. On doctor's orders Kay had been feeding him a slab of liver each morning, beefsteak at lunch, and two pieces of lamb plus liver soup for dinner, with generous portions of oatmeal, fruit, vegetables, rice, and potatoes. Unfortunately, he washed it all down with alcohol. In addition, he smoked heavily. Kay suffered from fainting spells too, and other unspecified ailments. Since the mid-1930s the couple had been making periodic visits to the Mayo Clinic, ostensibly to test Keeler's apparatus on mental patients, but in fact to receive treatment: Leonarde for his heart condition and Katherine for infertility. Leonarde assured his father he was not disappointed that she had failed to conceive. "Whatever happens I know we'll always be mighty happy. She is a wonderful helpmate."

Then, only three weeks before Rappaport's execution, Keeler had blacked out while he was boarding a train home from Boston. "My sympathetic nervous system got tangled up, giving me a rare thrill that I hope comes only once in a lifetime." The attacks had since resumed, and his physician urged a temporary leave from the stressful business of inducing stress in others.

Rappaport's aggrieved sister did what she could to exacerbate these woes. She had fixated on the lie detector as the cause of her brother's death, and she began sending Keeler bitter letters laced with curses, prophesying that he would be punished for his sin, that he would be paralyzed, that he would be struck dumb. Keeler contemplated taking legal action against her, even as he continued, against the advice of his friends, to read the letters. His sister thought that the letters directly affected his health. A year later, while he was traveling in Cleveland, his left arm and leg were briefly paralyzed.

Another calamity followed. When Charles Keeler read in the *Berkeley Gazette* about Rappaport's execution, he was deeply distressed. The next day, while minding his eldest daughter's twin girls, he tripped over a small dog and broke his femur. For several months he convalesced in bed, and then, on the verge of recovery, he died of a heart attack on July 31, 1937. The idealist, dreamer, and poet of utopian Berkeley was dead at sixty-five.

The newspapers, which had once hung on Charles Keeler's every word, hardly noticed his passing. His poetry was outmoded; his Cosmic Religion had flopped; his inquiries into the realm of the spirits had proved fruitless. Yet until the day of his death he never gave up on his ambition to find a purposeful intelligence in nature, or on his plea for mutual understanding, or on his hope for his own Leonardo. "I've been trying to think of something new for you to invent, but have been too weak to think to much purpose of late. But you don't want to just have the Keeler Polygraph as your contribution, however important that may be."

Nard, who had just returned from heart treatment at the Mayo Clinic, rushed back to Berkeley to meet with his sisters. While he tried to put a brave face on his loss, he feared that his own heart had the same weakness. Because of the cardiac arrhythmia he had discovered as an undergrad while testing himself on his polygraph, he could not buy life insurance. The doctors at the Mayo Clinic urged him to get "his nervous system in shape," and Kay understood this to mean that his illness, like his father's, was "most of it . . . mental." In his grief, Keeler turned to Vollmer. "I suppose it sounds foolish for a guy my age to say it—but you know when a father goes, one looks to someone else to take his place—and of all the people in the world, Chief, you're it."

He needed the support. That summer, while the Keelers were cruising in the Caribbean on a much needed vacation, Leon Green, the dean of the law school at Northwestern University, entered into secret negotiations to sell the crime lab to the city of Chicago for $25,000, with all its equipment and personnel intact. From the university's point of view, the sale represented a success: having nurtured a forensic lab with a reputation for scientific neutrality, Northwestern could now pass it on to the city that needed it most. For its part, the city had long coveted such a lab; but before signing the deal the police had just one nonnegotiable condition: under no circumstances would they employ Leonarde or Katherine Keeler.

No explanation for this condition was ever given. Perhaps it was enough that the Keelers were friends of Governor Horner and hence enemies of the Chicago political machine. The full explanation was probably more general and more personal. The Chicago police had never liked the lab's getting credit for solving cases, and Keeler's mania for publicity had won him a long list of enemies in law enforcement. The university accepted the deal.

To be cut in this way was humiliating. Because of this, and other betrayals, Keeler considered Dean Green a two-faced liar, and as much as told him so. The bitterness reached such intensity that the morning after the sale, Katherine Keeler absconded with the lab's photographic equipment and Nard's lie detector records, which she announced they would keep "until Mr. Keeler decided what to do with them." To Fred Inbau, the lab's newly appointed director, this was simply theft. It made no difference that Kay had used the equipment for seven years, or that the records represented Nard's lifework—the sort of customary claim that Keeler prosecuted mercilessly among bank tellers, wayward cops, and retail clerks. Only when Inbau threatened to inform the city police did Kay return the equipment. But Leonarde kept the polygraph charts for the book he always said he was planning to write.

In truth, Kay considered the demise of the lab a blessing. For years she had envied Nard the fact that the lie detector was his own invention, whereas she had never done anything original. Within a month she had leased an office in the Continental Bank Building downtown in the Loop and opened the nation's first all-female detective agency. Joining Katherine were her friend Jane Wilson, whom Kay had trained in photography; Edna Howie, an attractive, poised young woman from Wichita, whom Kay was training in handwriting analysis; and Viola Stevens, the crime lab's former secretary. Nearly 200 people, including John Henry Wigmore, attended her grand opening.

Initially Nard and Kay had talked of opening a crime lab together—as a real-life Nick and Nora—but Kay didn't think it wise. In the first place, lie detection and handwriting analysis couldn't readily share the same office space. Then there was Kay's desire for personal recognition. "As long as I am with Nard people have a confused impression that I am a wifely assistant. Now it is clear to everyone I have my own business well defined along certain lines." Finally, Nard had a tendency to procrastinate, and she didn't like to dawdle. Besides, they would avoid conflict by working apart. "We get along better now that our businesses are separate."

Kay was also convinced that the demise of the crime lab would do Nard good. In 1939, he opened his own firm devoted to lie detection cases.

CHAPTER 12

# A Science of the Singular

*It was well said of a certain German book, that* "er lasst sich nicht lesen"*—it does not permit itself to be read. There are some secrets which do not permit themselves to be told.*

—EDGAR ALLAN POE, "THE MAN OF THE CROWD," 1850

THE PSYCHIATRIC PROFESSION HAS DEVELOPED A KNACK for "making people up," a phrase coined by the philosopher Ian Hacking to describe how the process of classifying people can gradually transform their sense of self until they exhibit the symptoms that confirm their diagnosis. In the middle decades of the twentieth century, under the amplification of the new culture of mass testing, this looping process likewise transformed the great mass of "normal" Americans. Hence, when a polygraph operator obtained the "normal" for a particular citizen—calibrating the device for that fraught state of elevated blood pressure, accelerated pulse, and rapid breathing known as "test anxiety"—he was simply confirming conditions of life under the testing regime. Add to this those gambits that increased the subject's apprehension—such ruses as the card trick (known today as the "stimulation test")—plus the foreknowledge that any misstep might indicate complicity in wrongdoing, and "normal" refers to an individual afraid of being accused of failing a very serious exam. Welcome to *Amerika,* true-false edition.

John Larson wanted to transform the so-called lie detector from a method of inducing anxiety into a technique to relieve it, from a pseudoscientific oracle into a diagnosis of the self. He was determined to rein in the mechanical monster he had unleashed. He had energy and persistence and an endless supply of criminal minds to unlock. His young wife had put it best: "Something in his personality seems to draw him irrestably [*sic*] toward unsolved problems." It was for this reason, as much as any other, that he was run out of Chicago.

Sometime between 1932 and 1934 Larson shaved off his Hitler-like toothbrush moustache. He was living on the South Side, working part-time at a municipal neurological clinic, squeezing in two days a week of research at the Institute for Juvenile Research, and running deception tests on inmates at Joliet. Though he had stopped writing scientific papers, he fired back a rejoinder whenever Keeler appeared on the front pages with more egregious abuse. In January 1934, the nation's leading police journal published an interview with Larson headed "Has the 'Lie Detector' Failed?" This was a rhetorical question. In the interview, Larson claimed credit for the machine and disavowed any responsibility for its abuse. He singled out Keeler as his "first pupil," then blamed Keeler for training unethical interrogators. He insisted that the instrument was fit only for psychiatric diagnosis, then offered his services free to the Chicago police to test their criminal suspects.

At the core of Larson's ambition for the lie detector was its ability to distinguish between genuine delusion and faked belief. The technique tested for duplicity, that doubleness of self occasioned by a split between the body's emotional response and the mind's willful stratagems. It would not flag either a pathological liar or a person of sincere delusions. Even Keeler often pointed out that the lie detector "does not apply to small children, morons, unethical savages, or insane persons, but for normal civilized persons it is a very reliable guide." In other words, the polygraph hunted for a conscience—civilization's most elevated product—in the very people accused of having violated society's moral code. At the limit, it could test for the subject's belief in his own self, a matter of some uncertainty in the case of Allen R. Hammel.

On December 15, 1933, Hammel, a guard for Brinks Express, fled the Chicago Loop with $39,000. But the man arrested four months later with $935 in his pocket and Hammel's security badge in his apartment claimed to be not Hammel but "Burt Armstrong"—and persisted in this claim, despite being identified as Hammel by former colleagues, by a tattoo on his arm, and by his young children and distraught wife. "Lady," he told her, "I never saw you in my life." To complicate matters, a woman in Ohio identified Hammel as her missing husband and the father of her children.

The prosecutors in Chicago thought Hammel's claim of being "Armstrong" was bunk. But some psychiatrists speculated that Hammel suffered from disassociation, his amnesia being the result of a dual personality. His brother acknowledged that Hammel's skull had once been fractured in a motorcycle accident. A family physician also testified that Hammel had long suffered from epilepsy, although another physician called it "pseudo-epilepsy" and denounced him as a faker. This was just the sort of mystery Larson's lie detector was meant to resolve: the relationship between surface and depth, authentic and inauthentic self, conscious will and irresistible reaction.

But the prosecutors who asked Larson to examine Hammel were disappointed by his findings. At no time did Hammel register anything other than equanimity. "His answers all agree with his contention that he is Armstrong," Larson announced; he "believes what he is saying." No matter: the prosecutors easily persuaded the jury to convict Hammel, despite his dramatic seizure on the witness stand.

For the next year, with Hammel in Joliet, Larson treated the inmate with hypnosis and monitored Hammel's mental progress with his instrument. Sometimes Hammel's elaborate stories about being "Armstrong" rang "true as truth"; a week later, the prison psychiatrists would conclude that he was playing them for suckers. After three years in prison, Hammel was declared insane, and he died a year later. Yet even then Larson wouldn't relinquish the mystery. He tried to get Hammel's autopsy report, or better yet, access to his dissected brain. Obsessed with individual cases, Larson could never come to any final conclusions. Yet his job obliged him to do just that, and failure had consequences, for both his subjects and himself.

In his capacity as assistant state criminologist Larson was expected primarily to predict human behavior, specifically to advise the parole board—at a rate of 100 to 150 inmates a month—as to which ones posed a risk to society. But predicting human behavior is never an easy job, and it is especially hard when the stakes are high and time is short. The Illinois plan for variable sentencing had been a pillar of reformed justice in Chicago, letting prison administrators tailor reward and punishment to each case, control the prison population, and manage the social risk posed by recidivists. Unfortunately, this discretion also gave new scope to political corruption. So the reformers in Chicago had again tried to swap scientific methods for insiders' pull and set objective criteria for release. The question was which science and which criteria.

Like his fellow psychiatrists, Larson championed individualized assessments. The rival approach, advocated by his colleagues at the Institute for Juvenile Research, such as Ernest Burgess (also of the University of Chicago), was to predict success statistically, correlating recidivism rates with factors such as age, crime, job skills, and alcoholism, along with behavior as assessed according to the prison merit system. This effort formed the core of the new quantitative sociology. One intriguing variant let prisoners predict whether their fellow inmates would go straight, and then tried to systematize those hunches. The prisoners secretly selected to predict the behavior of their fellow inmates were none other than Leopold and Loeb. Indeed, Nathan Leopold, under a pseudonym, had published papers on statistical prediction of success on parole.

Joliet's most famous prisoners were still America's most famous prisoners, and Larson still wanted to use the two young men to validate his approach to psychiatric lie detection. By comparing their polygraph records against those of unconfessed killers, he hoped to identify the signs of unassuaged guilt. Leopold, the younger, more brilliant murderer, refused to sit for the exam. Having his blood pressure taken, he said, made him "neurotic and overactive." He had always presented himself as a Nietzschean superman whose mental abilities transcended others'. But he realized that he was hardly likely to get parole on those grounds. So Leopold published papers suggesting which criteria might be used to set the optimal number of factors that would predict success on parole: too few, and prediction suffered; too many, and each prisoner became a category of one. This helps explain why Leopold never had much respect for Larson or for Larson's lie detector. He considered the individualist approach insufficiently scientific.

Larson, by contrast, felt that a statistical approach slighted the person's total history. (For the record, Larson judged Leopold one of the "precocious, egocentric, psychosexual deviates" at Joliet.) His individualized lie detection technique would succeed in conjunction with a team comprising a psychiatrist, a psychologist, a social worker, a sociologist, and a statistician. By the late 1930s he had amassed the records of 2,200 prisoners on his lie detector, a quarter of whom he judged mentally disturbed.

Richard Loeb, the slightly older, more affable murderer, happily cooperated with Larson in this project. His polygraph showed "no significant disturbances" when he was discussing his confessed murder. Only when Loeb told a deliberate lie at Larson's request did the apparatus record any disturbance. Over the years, Larson became quite fond of Loeb and thought he understood Loeb well.

It was all the more shocking, then, when in late January 1936 John Larson responded to cries in the prison corridor and became the first official on the scene as Richard Loeb staggered out of a private jailhouse bathroom, his naked body dripping with water and gushing blood. Standing above him was his former cellmate James Day, also wet and naked, but unbloodied, an open razor still in his hand. After dispatching Loeb to the infirmary (where he died), Larson had Day hauled into his office for a severe interrogation.

Who needed a lie detector? *Res ipsa loquitur:* for once the thing really did seem to speak for itself. Or did it? Day claimed he had acted in self-defense. Originally sentenced to one to ten years for stealing $26 at gunpoint at the age of nineteen, he had been denied parole for violent behavior and transferred to Joliet, where he fell into the orbit of "Dickie" Loeb and became Loeb's cellmate. Loeb had shared his special privileges (and possibly more) with Day until two weeks earlier, when Day was transferred out of Loeb's cell as part of a crackdown by a new warden. Loeb, Day said, had then tricked him into entering a private shower, locked the door, and threatened to cut him with a razor if he didn't change his "narrow minded" attitude. Day had gone so far as to disrobe before he got a chance to kick Loeb in the groin, get the razor away from him, and slash blindly while the hot water poured down on them both. There was presumably more to this tale, but suddenly Day clammed up. "Jesus God, I can't tell you all this to put down."

Prison officials agreed; anxious to suppress lurid tales of misrule, favoritism, and "abnormal proposals," they took their own statement from Day the next day, and ordered Larson to stay away from the press, inquest, and trial. Even so, the papers in Chicago, eager to fan the flames of scandal, quoted Larson extensively. Readers learned that Leopold and Loeb enjoyed breakfast rolls each morning in the privacy of Leopold's cell, ran the prison library and school, had access to a sizable bank account and private bathrooms, and abused their position "to reward 'broadminded' convicts with money and favors."

Horner's rivals in the forthcoming primary took aim at the governor. "Can we forget that Loeb and Leopold, two degenerate murderers whose families are lifelong friends of my opponent, have been allowed to direct the politics of the entire penitentiary, teaching the prisoners, getting work assignments for them at will, conducting gambling operations, and being provided private baths?" Behind the scenes, Horner's prison administration set about suppressing the story, while in the *Chicago Tribune* Larson denounced corruption in the prison.

This time, without Vollmer to rescue him from the consequences of his own integrity, Larson was soon fired, nominally for recommending that Day hire Larson's wife's cousin as his attorney. (In fact, Day was found not guilty by reason of self-defense.) Worse accusations against Larson were making the rounds. Police gossip had it that Larson had lost his job for being too close to the prisoners. "Poor old John," Keeler wrote to Vollmer, "finally stumbled and badly stubbed his toe." According to Keeler, Larson had threatened to commit Day to a criminal mental facility "for the rest of his life" if Day didn't hire Larson's wife's cousin as his lawyer. Though Larson quickly found a new post in Detroit at the Psychopathic Clinic attached to the municipal courts, Keeler did not predict success. "I fear," he wrote to Vollmer, "that he will embarrass the authorities there and be on his way again."

In Detroit, Larson nursed his grievances. He was painfully conscious of having surrendered Chicago—world capital of crime and criminology—to Keeler and Keeler's cronies. From outside, Larson tried to rally the opposition. He resumed contact with William Marston in hopes of forming a united front against Keeler. He contacted J. Edgar Hoover at the FBI to denounce Keeler's "false claims" about the lie detector and his training scheme—a "racket" that had ruined the field with "quacks."

His obsession with Keeler made Larson see conspiracies everywhere. As fast as he secured new allies, he turned on them. He publicly announced that he had stopped using the Keeler Polygraph; and he permitted Captain C. D. Lee to advertise his Berkeley Psychograph as the "same as used by Larson," then accused Lee of exploiting his reputation and complained that Lee's machine did not work very well either. When a former assistant at the institute proposed a National Polygraph Operators Society to set professional standards for the field, Larson blasted the effort as Keeler's secret handiwork. The assistant's retort ended up in Hoover's files: "You have a strange way of showing any appreciation for loyalty, by accusing those who would be your friends of ulterior motives and selfishness." Behind the scenes, Keeler—no friend of polygraph standards—had been trying to quash the society.

Larson did take positive steps. He and Marge rented a house on an island that backed directly onto the Detroit canal. Their son Bill was born there in 1937. Moreover, it was in Detroit that Larson finally assembled the "clinical team" method that became his model for testing deception. Patterned on the teams used at the Institute for Juvenile Research, it combined several experts to diagnose psychopathology, expose false complaints, resolve domestic disputes, and identify embezzlers. In 1938 he organized a symposium that brought together physicians and lawyers to contrast his method with Keeler's mania for publicity and exploitative practices.

Yet when a sensational case came along, Larson could no more resist its allure than Keeler, even if it meant sharing the stage with his rival. There was something about a terrorized public that drew lie detectors like moths to a flame. Add to the mix a psychotic serial killer—like Cleveland's "Torso Murderer"—and Larson could not keep away. As he put it to a friend, "You didn't think I would let a case like that slip by." In the end, he proved no more capable of solving the case than Keeler—though he did conclude that the principal suspect, a deranged surgeon from a prominent family in Cleveland, "revealed disturbances indicative of guilt."

In his frustration Larson looked to other bodily tracings for clues about the mental state of subjects. He began by trying to transform Vollmer's successful program for classifying crimes according to the modus operandi of their perpetrators into a scheme for classifying criminal minds by correlating their modus operandi against the lie detector results he had obtained in 1,700 murder cases. "The crime," he suggested, "may be considered as the entering complaint in the case of disease for which the doctor is consulted." Larson's method—a forerunner of today's "psychological profiling," touted by the FBI and featured on crime dramas—did not meet with success. Nor did his effort to interpret another bodily mark that he thought might reveal the workings of the criminal psyche: the tattoo. He could not find any meaningful pattern in the 7,000 tattoos he gathered over the course of a decade: of the 63 percent of subjects he categorized as "feeble-minded, alcoholic, schizophrenic" (itself not a well-defined category), he found that they sported 65 percent of the "mother" tattoos, 66 percent of the heart tattoos, and 74 percent of the crucifix tattoos. Unable to deduce any meaning from those numbers, he revisited a hypothesis of his youth: that fingerprints were not only inherited but expressive of "psychopathology, schizophrenia and other organic factors." Yet he never settled on an interpretive scheme that could explain his 12,000 cases of modus operandi; his 7,000 cases of tattoos; his 2,200 lie detector records; or his thousands of fingerprints samples. Vollmer had once teased him about his "magpie den"; Larson was a collector of singularities, not a systematic thinker.

Where Larson did succeed was in treating patients. This required not so much an orderly mind as an open and generous one. Yet even this success did not bring Larson solace. As his onetime adviser Adolf Meyer noted, there was restlessness in his soul, an unsatisfied curiosity that fed upon his sense that he had left his main work undone. Time and again Larson returned to the theme of how his discovery had been hijacked by charlatans; how the public accounted him its "inventor" despite his insistence that "no one has invented anything"; how his reputation, the thing he cared most about, was based on a misunderstanding about what he had done and why. Larson's obsessive attacks on Keeler were attempts to resurrect his own version of the machine.

Meyer had urged his disciple to treat each patient's life as an experiment in nature. But what does any single life prove? In the early 1940s, Larson gave up his job in Detroit to take a series of positions in various mental institutions across the country and work on a sequel to his book on the lie detector. Writing from one of these far-flung posts, soliciting another job reference from Meyer, Larson explained that there was "a method in this apparent nomadism"—by which he meant both his bodily and his intellectual travels. He was still determined, he assured his mentor, that he would set the record straight, reminding the world of what his invention had been and still could be.

CHAPTER 13

# Fidelity

*Janet Henry laughed. "I didn't make all of it up," she said, "but you needn't ask which part is true. You've accused me of lying and I'll tell you nothing now."*

—DASHIELL HAMMETT, *THE GLASS KEY*, 1931

AS LARSON'S DIAGNOSTIC PROGRAM FALTERED, KEELER'S methods commanded the nation's trust. Gradually in the 1930s, and then with increasing appetite in the years after World War II, the managers of American businesses turned to Leonarde Keeler's mode of lie detection to supervise their workforce. First thousands and then millions of American job applicants and employees were obliged to pass a mechanical fidelity check.

When nineteenth-century Americans engaged in marketplace commerce with strangers, they assessed the reliability of buyers and sellers with the aid of characterological sciences like physiognomy, phrenology, and graphology, reading creditworthiness from a person's look, head, or handwriting. But as these sciences were themselves hardly reliable, Victorians also made sure to build the cost of potential fraud into the price of the transaction. Economists argue that modern corporate capitalism emerged in the early twentieth century in large part as an end run around the "transaction costs" of such marketplace deceptions and uncertainties. Businesses increasingly found it more cost-effective to pay managers to run their own suppliers than buy doubtful materials on the open market. Thus arose the vertically integrated firm. But with expanding scale and accelerating turnover of employees, businesses came to have less and less personal knowledge about their workers' characters. How could modern managers trust their employees any more than their grandparents trusted duplicitous traders?

This is where the lie detector stepped in. Not only did the instrument promise to vet employees for honesty; it acted as a deterrent against future deception as well. What Keeler had done for banks could be adopted by almost any large organization. Some social scientists even welcomed the device as a substitute for the old religious and moral injunctions which had once ensured honesty, but whose force had been weakened by the anonymity of modern urban life. The pioneering sociologist Ernest Burgess frankly called the lie detector "a scientific aid for social control" to deter those in positions of financial trust from "yielding to temptation." It would serve as the scientific conscience of a rootless nation.

Only in America was the lie detector used to interrogate criminals and vet employees. Abroad, it was disparaged as a typical American gimmick. Even those foreign forensic scientists who took an interest in lie detection treated it with skepticism. Keeler sold only one machine outside the United States, to Selfridges in England. Even in Canada the American instrument was spurned by both the police and business.

This is not to suggest that Europeans (or their police) were more intrinsically honest than Americans, or that they lacked scientific know-how, or that they somehow missed out on the great economic transformations of the twentieth century. The difference lies not in human nature, expert capacity, or market forces, but in the way that human institutions have been assembled over time by distinct cultures. The European police forces, for instance, had long served as a direct arm of the central state and were consequently far less subject than the American municipal forces to the tug of local politics, and hence less obliged to assuage public opinion about their neutrality or conceal their harsh interrogations under the veneer of science. Touring the United States soon after World War II, the president of the International Criminal Police Commission was horrified to see an ordinary patrolman interrogating a suspect on the device, which he deemed of dubious scientific worth, and worse, a "violation of the conscience." The commission expressed deep reservations about the machine.

As for the use of polygraphs in the workplace, given the resentment that lie detection aroused among employees even in America, it is not surprising that lie detection made no headway in Europe, where workers have generally had greater success in forming unions and negotiating labor conditions. (Tellingly, the first American states to forbid lie detectors in employment were those with the highest rates of unionization.) Besides, European bureaucracies have their own pseudoscientific means of vetting employees. In France, for instance, job applicants have long been required to supply a handwritten cover letter so that one of the nation's 30,000 graphologists could assess their character. In Britain, employers used class markers like accent to evaluate applicants.

At the deepest level, the difference between American and European attitudes toward the lie detector turned on the proper sort of relations one should have with one's fellow citizens and intimates. In André Maurois's satire of 1937 on American life, translated as *The Thought-Reading Machine*, a French professor of literature visits an American university where a physicist has invented a "psychogram" that can surreptitiously record unspoken thoughts. The first person the professor thinks to turn the device on is his wife—*naturellement*—only to discover to his dismay that she still pines for her ne'er-do-well cousin. When his wife—furious at this "disgusting" violation—surreptitiously turns the device back against him, she discovers his half-baked notion to seduce a divorced student. Yet these revelations do not destroy their marriage: husband and wife agree that these daydreams are less "truthful" than their lifelong fidelity, and they agree never again to peer unbidden into each other's thoughts. So the machine becomes a commercial hit in America, where it is used to unmask campus scandals, steal football plays, apprehend criminals, test the fidelity of lovers, and monitor the thoughts of employees; but in France, after some lamentable incidents—such as when the professor's sister-in-law commits suicide upon learning of her husband's infidelities—the device is roundly ignored. Besides, people there soon learn to mask their inner reveries, much as they have long masked their feelings' outward show. In the end, the professor concludes, the "psychogram" is a curse—not because it drags sinful thoughts into the open but because it suppresses the free range of human feeling. Even the inventor concludes, "Interior language is no more authentic than ordinary speech; the latter is protection against others and the former against ourselves." The "psychogram" has simply extended deception to self-deception. Rather than enable citizens to trust one another, it renders them incapable of trusting themselves.

This was a common refrain in foreign commentary on the lie detector: that social life would grind to a halt if people were actually obliged to tell the truth to one another and to themselves. Even some Americans fretted that the machine would stop landlords from renting, hostesses from giving parties, lovers from wooing, and politicians from running for office. As for family life: "How would you like to have a little Lie Detector in your home? No, thanks? You'd rather have a washing machine? So would we all." We are better people for asking others to believe us, even if we sometimes lie. As Molière instructs us in *The Misanthrope:*

> For if all men were to speak the truth
> And all hearts were honest, fair, and true
> Most of our virtues would serve no use.

In 1939 Keeler set himself up as Keeler, Inc., "personnel consultant," and began to franchise trustworthiness. Whenever possible, he still preferred to work cases himself, though he hired an assistant and trained Jane Wilson—Katherine's friend and the wife of his partner Charlie Wilson—as the nation's first female polygraph operator. But he now began to allow selected graduates of his course to buy his machine and set themselves up as Keeler Polygraphers with exclusive license to some territory. One former cop, Russell Chatham, took title to Indiana; another got Michigan. By contract the buyer could not resell the device or let anyone else operate it, and Keeler could even reclaim the machine if he decided that the buyer was in any way "prostituting the field." Keeler's personal expertise was still his most valuable asset.

For just that reason he switched manufacturers. Over the years Keeler had been bombarded with complaints about the quality of his machine: the tambours leaked; the paper jammed; the pens failed. His manufacturer cannot have been satisfied either; in the span of a decade the firm had sold fewer than fifty machines all told. In 1939 Keeler dropped Oakland's Western Electrico-Mechanical Corp. in favor of Chicago's Associated Research, Inc. At this time, he also added a galvanometer to his apparatus. The price of his machine rose from $550 to $900, but the problems with quality persisted.

This put Keeler at a disadvantage against competitors like Orlando Scott and the Jesuit psychologist Walter G. Summers, who boasted of "100 percent accuracy" with a Psychogalvanometer of his own design. Each interrogator had his shtick. Scott had his medical mumbo jumbo; Summers had his clerical collar, plus two Ph.D.s. When Summers died in 1938 his successor was the son of a founder of the Union of Orthodox Jewish Congregations. All Keeler had was his salesman's self-confidence.

Not until after the war, when his patent expired, did Keeler abandon all restrictions and tell his manufacturing company to "go ahead and sell to anybody." At the time Keeler Inc. was still the only place in the nation to go for training in lie detection: either a two-week orientation course for $30 a week, or the more extensive six-week courses for certificate as a graduate of "Leonarde Keeler, Incorporated"—though Keeler always pointed out that it took at least a year of supervised casework to become a proficient examiner. Among the challenging questions on the final was this one:

> Multiple choice. For practical purposes, an interrogation room for a suspect should contain:
> (a) a large window from which the suspect can see the surrounding territory.
> (b) an imposing array of technical equipment such as "recording machines" etc.
> (c) only four bare walls, chairs and desk visible to subject.
> (d) a complete display of all known instruments of torture.

On graduation, the "boys" were given certificates; a cocktail party with five quarts of bourbon; and a copy of *Gracian's Manual,* containing the wisdom of the Counter-Reformation Jesuit who believed that personality was a mask adopted for strategic purposes. "No. 26: Discover each man's thumbscrew. It is the way to move his will, more skill than force being required to know how to get at the heart of anyone." Afterward the class went to a nudie show at the French Casino.

Keeler's success was publicly recognized (and extended) by celebratory articles in popular magazines like *Forbes* and *Reader's Digest,* which merged his scientific and business successes into a seamless sales pitch. When Keeler was hired by a chain of department stores that was annually losing $1.4 million in pilfering by its 14,000 clerks, he claimed to have discovered that 76 percent of the clerks were stealing; six months later, when he came back for his retest, the cheats had been reduced to 3 percent. The numbers were impressive. Keeler's machine caught 90 percent of the guilty; it worked 99 percent of the time. No proof of the claims was ever forthcoming.

To be sure, there were naysayers. An article in *Esquire* in 1941 was the first to expose the lie detector publicly as "the bunk" and offer advice on how to beat the machine: intensify your emotions during the irrelevant questions, bite your tongue, never believe that the operator is your friend, and above all don't be fooled into thinking that the machine works in the first place. Readers were reminded that no one was obliged to take the test, that the courts did not accept its verdict, and that it often served as "the third degree dressed up in white tie and tails."

Keeler did acknowledge that some employees resented the exams, but he dismissed their worries as the concern of namby-pamby "women's clubs." After all, "the cash register with recording tape was also taken as an insult by indignant employees when it was first introduced." What Keeler refused to acknowledge was that whereas the lie detector strictly separated what belonged to the employee (his labor) and the employer (everything else), many workers did not consider their actions as theft so much as one of the perks of their low-paying jobs.

As before, it was Keeler's competitor, Orlando Scott, who took his methods to their logical limit. At economical rates—$15 a person for high-volume clients—Scott promised to test employees for "integrity, intentions, loyalty, competency, intuitiveness, stability, alertness, efficiency, ambition, vocational stability, sabotage, etc." He even had a jingle to close the sale.

> Little Brainwave Lie Detector's come to our Lab to stay
> It measures ev'rything one does, a pure 'lectrical way,
> It catches ALL folk that chisel, ev'ry place they work
> Sure it ain't healthy anymore, this grown-up childish quirk
> An' so with all one's employees aknowin' HOW it's done,
> They just set 'round a feelin' that thiev'ry's on the run.
> An' our Lie Detector'll git YOU
> Ef YOU
> Don't
> Watch
> Out!

The lie detector did more than intimidate employees; it taught the virtues of "emotional management." The industrial psychologists who ran the famous Hawthorne experiments at General Electric from 1924 to 1933 had concluded that labor productivity depended more on morale than the actual conditions of work—hence the need to reduce workers' "irritability" and rage. Personality testing became a new tool of pacification in the workplace, with personnel departments matching individuals to tasks for which their temperaments were said to be suited. Their triumph was the creation of the "organization man" of mid-century, an employee trained in emotional opacity and able to control fear and resentment behind a mask of "positive thinking." Only those at the very top of the corporate hierarchy were exempt from this regimen; their appetites and aggression were the engines of capitalism. Rank-and-file personnel were to heed the upbeat admonitions of Dale Carnegie, author of *How to Win Friends and Influence People* (1936), who taught his white-collar readers the emotional price to be paid for eliminating transaction costs.

For his part, Keeler always struggled to keep a tight rein over his own emotions. In a "Personal note to self" entitled "Be Friendly," he admonished himself, "Whatever you do in dealing with people, think of <u>their feelings.</u>" Among the pieces of advice he underlined: "<u>Never allow anyone to know you are disturbed or angry.</u>" It was advice he tried to heed. For instance, he always adopted a friendly pose in his dealings with John Larson. He reproached himself for shouting at polygraph subjects who refused to bend to his will, and he tried to maintain an outward calm no matter how high his own blood pressure rose.

In sum, Keeler affected an "American cool," the pose of emotional detachment favored by corporate salesmen, noir detectives, and research scientists—all for similar reasons. Emotion, or at least its public display, was unmanly. It was a sign of clouded judgment and self-interest. To succeed at his job, the salesman, the detective, or the scientist—or someone, like Keeler, who passed himself off as all three—had to appear objective, meaning that he had to see things from the other fellow's point of view, if only to close the sale, solve the crime, or address the problem to the satisfaction of his peers. If he became too involved with any particular sale, case, or problem, it would be that much more difficult to move on to the next. This was bad for business, bad for crime-fighting, and bad for research. Passions like love, jealousy, or vengeance were dangerous distractions in a service economy. Dale Carnegie himself once praised Leonarde Keeler as a fellow laborer in the field of applied psychology.

Of course, self-control, both mental and bodily, is also the first prerequisite for being a good liar. And Keeler always claimed that he could beat his own machine.

The ethos of emotional detachment had another virtue at mid-century; it was considered good for your health. The same psychological stresses that made the lie detector's needles twitch could also inflict bodily harm. The mind-body traffic ran both ways. Walter Cannon—the physiologist who first identified psychosomatic diseases like "soldier's heart"—advised patients to minimize emotional disturbance. Digestive, cardiovascular, and neurological disorders could be exacerbated by bouts of intense rage or grief or passionate love—perhaps even by attempts at their suppression. It was on the latter front that Keeler needed help. His most effective aid in his struggle against emotion was alcohol.

Keeler's principal patron and drinking companion in those prewar years was Gene McDonald, the founder and president of the Zenith Radio Corporation, known as "the Commander" for his expeditions above the Arctic Circle. McDonald was a self-made millionaire and hard-drinking womanizer who entertained revelers aboard the *Mizpah,* his 185-foot motor yacht, outfitted with three ostentatious staterooms, a radio station, and a crew of twenty-seven. The Commander—a titan of capitalism who did not believe in self-control—lived on board with his wife, Inez Riddle, a musically talented beauty queen whom he bullied about her secret doings.

McDonald brought the Keelers along for cruises, including a trip to the Caribbean to help President Trujillo of the Dominican Republic reorganize his police force. Life on board was a continual party. When drunk, McDonald often became violent and fired out of the portholes. Among his gifts to Keeler was a copy of *Liquor, the Servant of Man,* a book defending alcohol as integral to a balanced life: good for the heart, less addictive than smoking, a spur to sexual ardor (though, in excess, a cause of temporary impotence), and an all-around stimulant and social lubricant.

The Commander was an iconoclast, open to new scientific vistas, among them mind reading, spiritualism, and graphology. Like Keeler's father, the Commander hoped Keeler's lie detector would serve as a gateway to the parapsychic realm. In 1938 McDonald promoted a series of Sunday radio programs in which listeners were asked to "read" correctly the order of J. B. Rhine's famous ESP cards as McDonald flipped them over in the studio: wave, square, star, cross, circle. Among the 1 million responses, McDonald had statisticians identify 379 "sensitive" individuals, who were then invited to try their psychic skills at eight o'clock in the evening, Central Time, when he and the Keelers, broadcasting from the *Mizpah* over KFZT, would conduct a series of activities: flip over some more ESP cards; deliberately misset a clock; open a book to a random page; and finally let Mrs. K. do something to her husband aboard ship, with listeners to determine what she had done and whether it was pleasant or nasty. Unfortunately, the listeners failed miserably; that is, they did no better than chance. The majority wrote in to say that Mrs. Keeler had done something pleasant to her husband; some guessed a kiss; others guessed that she shook his hand. In fact, Kay had jabbed a pin into Nard's left thigh.

In formal settings, Kay could be awkward. In action, she was fearless. She did not back down from a challenge: not in the hills of Kentucky, not on the witness stand, not with her husband. Nard had always taken pride in his wife's intelligence, courage, and accomplishments. But her recklessness—and her success—discomforted him. Leonarde Keeler, for all his hard-boiled pose, was, by his own admission, a deeply conventional man. Billed by the newspapers as partners against crime, husband and wife were also rivals. It was in a rueful tone that Keeler acknowledged that her detective agency pulled in more money than his.

The *Chicago Herald and Examiner* ran a flattering twelve-part serial about her, "The Girl on the Case." Though the all-female agency soon reverted to a one-woman operation, Kay pushed ahead. She prized her independence above all. In the view of her onetime secretary, "Kay was temperamental. Friendly one day; the next you couldn't please her. She had a complex personality."

No sooner had her lab begun to prosper than she lost interest. Suddenly she began devoting all her spare time to learning to fly. Nard joked that he would have to take up flying again if he wanted to spend any time with her.

She bought her own honey of a car: a Buick convertible coupé with a green body and khaki top. When she drove through town, heads turned.

Soon everyone was noticing the friction. Keeler's adoring sister certainly noticed it: "You're both too intelligent, interesting and attractive not to iron out your maladjustments. . . . Then if the difficulties are insurmountable, break it up." Even Agnes de Mille, who now thought Nard "incurably" in love with his wife, saw that they were unhappy. "It was the core of Nard's nature to dominate," she recalled, "and Kay refused to be dominated." When one of her document cases netted Kay $2,000 she secretly sent $500 to her parents. Not that Nard would have objected, she hastened to add, but "it does make life simpler not to discuss things with people who may raise a lot of questions and might not agree with what you intend to do anyway."

They fought about politics. Nard, like his father and August Vollmer, had always been a reform Republican who believed in the rule of law. He despised Franklin Roosevelt, and as the election of 1940 approached, he complained that "Rosenfeldt" (as he called the president) was taking America to war in order to usher in totalitarianism at home. But Kay had long ago gravitated toward the Roosevelt camp. At the Democratic Convention in Chicago in 1932, she surprised herself with her passion. She admitted that most of their respectable acquaintances were Republicans. "Maybe that's because most respectable people are conservative." Soon she was an avid New Dealer, thrilled with Roosevelt's reelection in 1936, and distressed when some workingmen celebrating their victory mistook her for a fancy Republican just because she was wearing a fur coat. In 1940 Agnes de Mille overheard Nard explode when Kay told him she was planning to vote for Roosevelt again: "You're voting for that megalomaniac?" De Mille recoiled at the vehemence of his words, but Kay did not blanch. The repartee of Nick and Nora had turned sour.

In the dissolution of a marriage there are events that leave no trace. During the spring of 1940 Nard and Kay's differences reached a breaking point, and the two separated. Nard took a room in the Chicago Towers Club, and Kay moved into a large apartment in a stately three-story building on Dearborn. Kay had regrets—"I've made so many mistakes lately that I am losing confidence in my judgment"—but she was determined to start a new life.

For a short period later that year, they reconciled and Nard moved into the apartment on Dearborn. As late as October 1940, he was still pretending to his family that things were fine between them—though Kay was more often at the airfield than at home. Then, on December 11, while interrogating a suspect, Keeler "cracked up." His systolic blood pressure shot past 200, a physician was called in, and an hour later he was in the hospital, where he stayed for four months. The doctors ordered him to cut out alcohol and cigarettes. They advised him to reduce his workload. His pals Burt Massee and the Commander made daily visits. "Kay drops in when she has time." By the time he was released in the spring of 1941, Kay had taken a room at the Lake Shore Athletic Club. He moved back into the apartment on Dearborn, where he lived for the rest of his life. The time had come to call it quits for good.

Their friends told him they were better off apart. Nard's sister announced that she had always felt they weren't suited to each other. Even Vollmer admitted that he had always considered Kay "not quite normal." The awful truth was that Nard had finally done what a criminologist on a case would have done long ago: he had had his wife tailed by a private eye, who discovered that she had betrayed him, repeatedly, with many lovers, over many years.

It was a new twist on one of the oldest jokes in the world: Did you hear the one about the lie detector's wife?

The detective's report has not survived, nor has any of the correspondence between Nard and Kay. So how do we know, at this historical distance, what happened? Because Keeler reported Kay's betrayal to someone he trusted. "[He told me] the same story, over and over," recalled Agnes de Mille in her unpublished memoirs. "How Kay had betrayed him, how he had put a detective on her and found out that she had betrayed him with lover after lover." And we know that he confronted Kay with her betrayal, because members of the Applegate family recall her telling them that he had once had her followed by a detective.

The world's premier lie detector, skilled at reading bodily responses for signs of deception, had been fooled by the woman whose body he supposedly knew better than any other. As George Orwell once noted, nothing is more painful for a policeman than being laughed at.

Kay initially planned to file for divorce on the grounds of cruelty. This idea horrified Nard, mostly because of the publicity it would entail. Luckily, she had no evidence for her unspecified charge, and he refused to acquiesce to her accusations. In the end, to get a quick divorce, she filed on the grounds of "abandonment." On June 20, 1941, Katherine Keeler resumed her maiden name, Katherine Applegate.

In public Nard affected nonchalance. He informed the press that the couple still planned to work as a sleuthing team when a case required it. "Sure my machine works," he informed one columnist a few years later. "In fact, it works so well that my wife and I made a pact never to test it on each other."

Even with his family, Nard put on a brave face. He called the divorce his "new birthday." "I am now free, white and twenty-one," he boasted to his sister. Kay, he explained, had chucked her criminology business to pursue her dream of flight. "I have a hunch that she is attempting to simulate Amelia Earhart and expect that she will be flying bombers to England. She's just that nuts!"

It was nearly true. Within a year Kay had left town with her lover, René Dussaq, a handsome Latin American adventurer with a pencil moustache and colorful past. Born to a Swiss family in Argentina and raised in Cuba, where his father was an ambassador, he had been an Olympic oarsman, a Golden Gloves boxer, a tango dancer, a Hollywood stuntman, a wing-rider, and a deep-sea diver. He had been involved in several South American revolutions, held degrees in several scientific fields, and was fluent in six languages. Kay had found a man even more adventurous than Nard. By December 1942 she and René were married and living in Washington, D.C., where Dussaq had been recruited into the Office of Strategic Services (the forerunner of the CIA), while Kay took up civilian aviation defense, then joined the Women's Auxiliary Service Pilots (the WASPs).

Leonarde, in the meantime, had returned to the empty apartment on Dearborn to face another betrayal. The previous year, during his troubles with his health, Keeler had hired one of Larson's former assistants to help out on cases. While Keeler was in the hospital, the assistant had copied Keeler's client list and files with the intention of starting his own business. Keeler secretly recorded the assistant's confession to these deeds, then fired him.

Why is anyone unfaithful? Kay's threat to obtain a divorce on grounds of "cruelty" might seem to suggest that Leonarde had himself been unfaithful, or violent, or otherwise abusive. Certainly in later years he had a series of mistresses. And he was known to have an explosive temper. Or perhaps it was inadequacy on his part, caused by alcohol, stress, and high blood pressure. Or perhaps Kay did what she did because she wanted to. What other explanation is there, really? She was always a woman of iron determination. Kay's mother wrote a letter of consolation to her former son-in-law:

> [M]arriage, even where two people really love one another, is a very difficult situation, too often fraught with the desire for complete possession or complete domination. And the human spirit demands freedom and rebels against tightening bonds. Try to look at it all realistically and see how patient you were when the dominating tendency showed itself and how much you took that was competitive and belittling. Where such an attitude develops it just isn't in the cards that a marriage can hold, and love turns to hate which is but the reverse of the shield and a true sign that love is still there.

What reason do people need to do what they want to do? What matters is that Kay's departure confirmed some terrible blackness in Keeler's soul: his view that the world was poisonous, cruel, and traitorous. He had come to the conclusion that the human race could be divided into "those who wished me dead and those who didn't care."

Friends tried to comfort him. "I think so well of you as a man," wrote Agnes de Mille; "I honor you so as a worker that just the knowledge that you were quietly, to the best of your ability, cleaning up your corner of the world, without possibility of corruption or intimidation or exhaustion has sustained me greatly when the men I loved defaulted." This was not to say, she reminded him, that he was not capable, most of the time, of acting like a damn fool. "I did not add wisdom to your virtues. Smart as hell with a machine in your hands—not quite so bright when fingering a heart. Your own, for instance." She ordered him to stop abusing himself physically and emotionally, and recommended that he see a psychiatrist.

> And I think you need a doctor because you're wretchedly unhappy. You're still involved with Kay. I don't mean I think you still love her, not as you did anyway—but you are involved through anger and frustra-

tion and bewilderment. There's a great rage, a fury, in you, and it's making you ill.

Keeler affirmed de Mille's point, even when he misread her. "I thoroughly agree with you that I know very little about hearts, except my own, which often goes into spasm." In the meantime, he had his work. "I expect to be on the rush for some months or even a year or two before my *final* collapse."

During the course of his career hundreds of ordinary citizens wrote to Keeler—"master of the lie detector"—to untangle the knotty heartache of infidelity. A husband in Nutter, West Virginia, wrote to say that he wasn't satisfied with his wife's explanations for various "coincidences"; he wanted to put her on Keeler's machine because he "wanted to know the truth." A farmer's wife in Barton, Vermont, wanted her husband tested, not about his infidelities—she had known about those for years—but about how much money he spent in the local bordello. A young man in Fort Wayne, Indiana, complained that he was in a state of nervous exhaustion: could Keeler confirm his fiancée's repeated assurances that she had never had an illicit sexual affair before the one she was currently having with him?

Double indemnity; the postman always rings twice: everyone knows that fate comes back to bite you. Yet day in and day out, people act as if the script could happen only to someone else, as if they themselves are immune to the operations of destiny. They read the book, they see the movie, and they still don't get it.

The middle-aged editor of the only newspaper in Litchfield, Nebraska (population 400), took out a $500 life insurance policy on his wife, including a double-indemnity clause in case of accidental death, and paid the premium on the spot, in cash, so that the policy took immediate effect. The accident occurred the very next day. She had been at the wheel, he said, as they were returning from a celebration to mark their thirty-fourth wedding anniversary, when the car unexpectedly plunged down a thirty-foot embankment. He had managed to leap clear in the nick of time and crawl back to the road. But his wife, tragically, had died in the crash. How uncanny, his neighbors reported: only a few weeks before, the editor had had a premonition that his wife might die in a car wreck.

Except the car was undamaged. Except the two truckers who pulled over to assist the stalled car were surprised when it suddenly drove into the ditch. Except someone found a heavy iron pipe 150 feet down the culvert. Except the autopsy showed that the cause of death was four violent blows to the back of the head. Except according to rumors in town, the editor had been having an affair with a local woman.

Double indemnity; the postman always rings twice. Everyone knows the story, and nothing can alter the ending. The editor and his wife had been childhood sweethearts; they had a twenty-five-year-old son; they were among the most respected citizens of Litchfield, where the editor had once been mayor. His name was Carl M. Anderson. An Everyman.

But as fate would have it, the assistant attorney general of the county had recently attended one of Keeler's seminars. He knew that modern forensic science could transform an Everyman into a specific person who had done a specific deed at a specific place and time. He offered Anderson a chance to clear his name on the "truth machine," and Anderson agreed. On Monday morning, June 30, 1941, the local sheriff and assistant attorney general flew Anderson from the plains of Nebraska to the vertical heights of Keeler's office on the eighth floor at 134 South La Salle Street. The test began at eleven o'clock in the morning, ten days after the Keelers' divorce became final.

Keeler hooked Anderson up to the machine. He asked if Anderson had killed his wife. He asked if Anderson had killed her with a stone, with a stick, with a fist, with a shoe, with an iron pipe. He asked each question ten times. He went on like this for four hours. And every time Keeler mentioned the iron pipe, the nation's newspapers would report, the "delicate needles of the detector, zigzagging on a graph, wavered violently."

In a break between runs, Keeler informed Anderson that while the questioning had been under way, he had sent the iron pipe out for chemical analysis. The analysis had just come in, and it indicated the presence of human blood, he said. At this, Anderson asked that the questioning cease. "You have all the information you need," he said. "I'll not give you any more."

It was mid-afternoon, and Keeler agreed to take a late lunch break. Carl Anderson—a man of fifty-six, "fairly chunky," nearly six feet tall, weighing 200 pounds—was escorted by the sheriff down to a café in the Loop for a bite to eat. Then they rode the elevator back to the examination room and waited in Keeler's office lobby.

Afterward, everyone agreed that Anderson was a phlegmatic, unemotional man. It was a hot summer day, and he said as much as he flipped through the pages of *Life* magazine. He said he was going to get some air. He stood up and stepped out into the corridor. The sheriff followed close enough to hear him say, "This is just as good a time as any . . . ," before he dived headfirst through the open courtyard window and landed on a roof four floors below.

At the inquest, everyone denied telling Anderson he had failed the test: the sheriff, the assistant attorney general, Keeler's assistant. Keeler himself was not present at the inquest. As his assistant explained, "Mr. Keeler has been suffering from a heart attack, he has a bad heart, and it is possible he has been suffering from that today." The jury returned a verdict of suicide by reason of temporary insanity. Keeler stayed at home—presumably liquored up—for several days.

John Larson (1892–1965), the nation's first cop with a Ph.D. in science, assembled the first working "lie detector" under the auspices of August Vollmer, chief of police of Berkeley, California. This photograph, dated April 24, 1921, shows Larson two days after his first case, the College Hall theft, transformed his life.

2

Leonarde Keeler (1903–1949) was named after Leonardo da Vinci, but preferred to be known as "Nard." He was a Berkeley high-school student and amateur magician when he became entranced by Larson's machine. His patented "Keeler Polygraph" made him the personification of American lie detection.

3

August Vollmer (1876–1955), chief of the Berkeley police department from 1906 to 1932, was the celebrated father of American professional policing. Vollmer saw in Larson and Keeler's lie detector a means to replace brutal "third degree" interrogations with more scientific and lawful techniques.

4

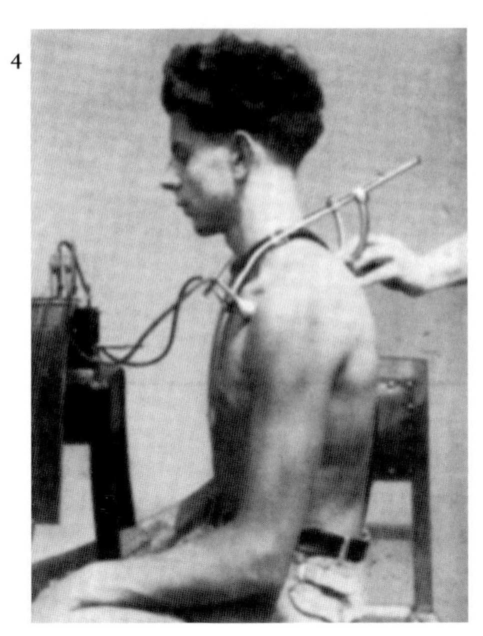

In the second half of the nineteenth century, European scientists like Étienne-Jules Marey recorded the bodily responses of subjects in order to track hidden mental processes, such as stress and fear.

5

At the turn of the twentieth century, Hugo Münsterberg (1863–1916) established a psychology laboratory at Harvard where students transformed their inner lives into the public phenomena we call science. With his techniques, Münsterberg (seated at the head of the table) claimed he could discern when a subject was suppressing the truth.

6

By tracking his fellow students' blood pressure during role-playing games, William Moulton Marston (1893–1947), one of Münsterberg's disciples, claimed he could sort truth-tellers from liars nearly 100 percent of the time. Marston earned a law degree from Harvard and a Ph.D. in psychology.

7

In 1922 a trial judge refused to let Marston (seated at right) testify that his test had exonerated James Alphonso Frye (center), who had retracted a confession of murder. Following this precedent, lie detector evidence has been largely excluded from criminal trials to the present day, although its use has been allowed in criminal investigations and many other settings.

8

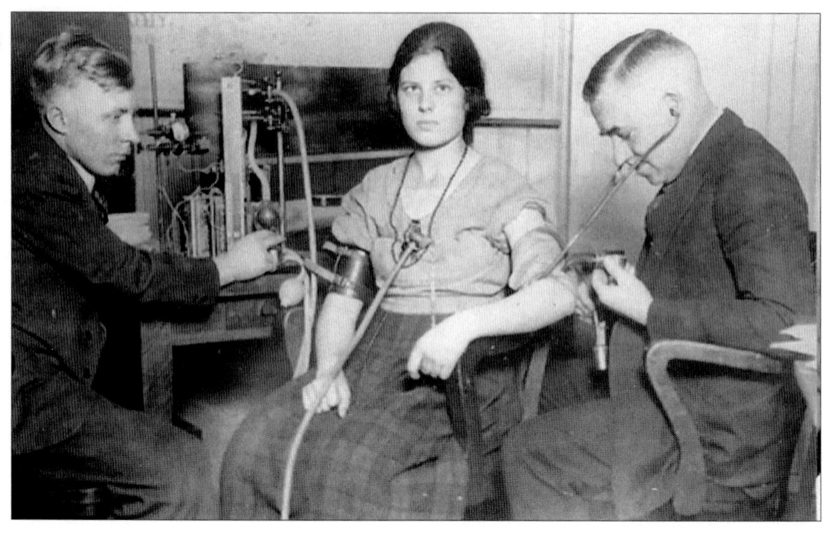

On hearing of Marston's early experiments, Larson rigged a device to continuously record the blood pressure and breathing depth of subjects while they answered "Yes" or "No" to alternating relevant and irrelevant questions. In 1921, Larson found the ideal experimental setup to prove his technique: a theft at an all-woman's college residence, the College Hall case. Here Larson (left) and Vollmer test a Berkeley undergraduate.

9

INVENTOR OF LIE DETECTOR TRAPS BRIDE

Crime Testing Device Utilized by Cupid to Make Expert Prisoner for Life

Dr. John Augustus Larsen, Berkeley criminologist, has lately emerged upon the stage of fame as the inventor of the sphyg—sphygomanom—call it the "lie detector."

Everybody has heard of the "lie detector." Put it on a criminal's arm and ask him a rude question and if he lies the little wings of the machine will flop up and down.

Well, tonight Dr. John Augustus Larsen will take unto himself a bride, Miss Margaret Taylor. She is a University of California co-ed, charming and pretty, as co-eds are wont to be.

Now there may seem, offhand, to be little connection between the two above sets of facts—Dr. Larsen's "lie detector" and Dr. Larsen's triumph in the field of romance.

HERE ARE PLAIN FACTS.

The plain fact is that the Berkeley criminologist won his bride through stumbling on a discovery that bids fair to take the crime out of romance, as it has already taken much of the romance out of crime.

A few months ago the Berkeley women's dormitory, College Hall, was disturbed by an epidemic of petty thievery. Dr. Larsen, being the only "Ph. D." in the Berkeley Police Department, was summoned

Caught in Own Trap

Above is Miss Margaret Taylor, who is to become the bride of Dr. John A. Larsen, inventor of the "lie detector." He placed the instrument on the girl's wrist and the love god manacled him for life.

Larson used the crime victim, Berkeley freshman Margaret Taylor, as his "control" in the College Hall case. The case appeared solved when one of her dorm sisters reacted furiously to being questioned on the machine and subsequently confessed. Over the next few months, Larson invited Miss Taylor back to interrogate her about her feelings for him—in the interests of objective science, of course. A year later, on August 9, 1922, their nuptials were front-page news in the *San Francisco Examiner.*

# NEW CRIME LAID TO HIGHTOWER

5 CENTS — THE CALL — FINAL HOME EDITION

# SCIENCE INDICATES HIGHTOWER'S GUILT

## PSYCHOLOGICAL TEST IN JAIL AT MIDNIGHT BARES HIDDEN MIND

Science penetrated the inscrutable face of William A. Hightower today, revealed that beneath an unruffled exterior is a seething torrent of heart throbbing emotions, and that these emotions indicate strongly that he was the murderer of Father Patrick Heslin of Colma.

During the small hours of the morning, while a dozen or more newspaper reporters clamored for admittance, a scientific "soul test" was administered to Hightower in the San Mateo County jail.

This was the most remarkable bit of psychological research undertaken on this coast, in which an apparatus known as a sphygmomanometer was relied upon to determine the truth or falsity of the statements of a man charged with murder, a man whose life is at stake.

The apparatus, which records respiration and blood pressure, was called on to perform a new miracle of chemistry—the chemistry of psychology—in which a soul was laid bare and subjected to the critical inspection of seekers after truth.

And before the test was concluded this apparatus was to tear the mask of self-control from the unfathomable face of William Hightower and show him to the investigators the man he is—not the cool, calculating, dauntless man he has been regarded, but a man torn with human passions, human fears, human frailties—cringing inwardly while wearing a cloak of iron.

It was the first time such methods had been used in a celebrated case in the history of criminology. The test was arranged by The Call and was carried out under orders of Police

## HIGHTOWER'S HEART AND SOUL UNDER CROSS EXAMINATION

## CONFESSION HINT GIVEN AGAIN BY PRISONER

## PROOF CONCLUSIVE, VOLLMER'S VERDICT

## HIGH PRAISE GIVEN TO TEST BY SWART

This *San Francisco Call and Post* exclusive of August 17, 1921 conferred national fame on the "lie detector," as the press now christened the device. Soon after the device spoke from the front page of the newspaper, a jury convicted Hightower of murdering a local priest.

11

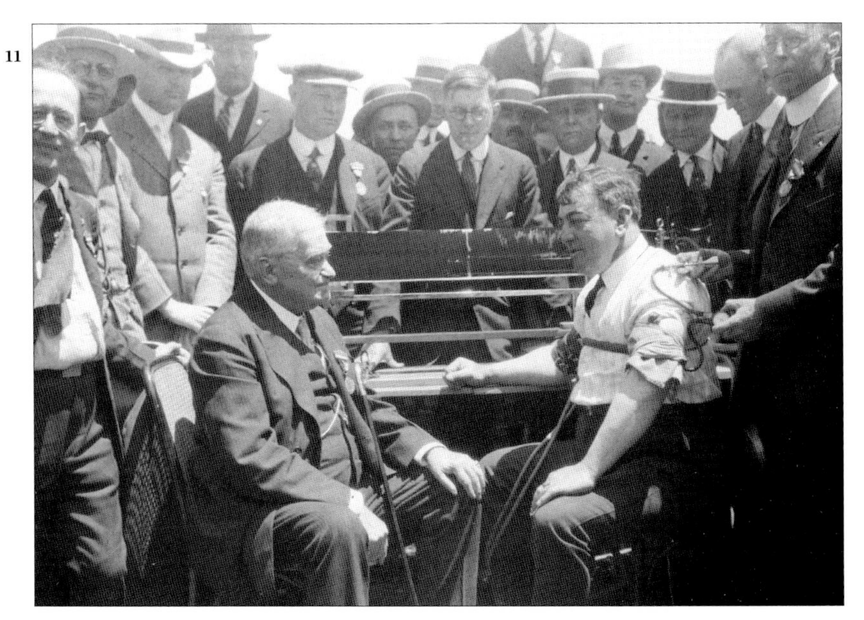

At the 1922 meeting of the International Association of Chiefs of Police—convened in San Francisco by association president August Vollmer (standing, far right)—Larson (standing behind the machine, in the center) watched as the city's police chief, Daniel O'Brien (seated, right), flummoxed the machine by telling "whoppers" to interrogating detective William Pinkerton (seated, left).

12

Charles Keeler of Berkeley—allegorical poet, founder of Cosmic Religion, and father of Leonarde—is pictured here in his "simple home," the natural habitat of the California bohemian bourgeoisie.

13

As an undergrad at Stanford in the 1920s, Leonarde Keeler pursued a patent on a more portable and reliable lie detector. He also refined interrogation techniques such as the card trick, designed to convince subjects he could identify their chosen playing card—though he marked the deck to make sure. Keeler first met fellow psychology student Katherine Applegate (pictured here a few years later) when she beat the card trick by cheating right back. They married in 1930.

14

By 1930, Larson, Keeler, and Vollmer were all in Chicago, America's capital of crime. Larson (left) arrived in 1923 to study criminal psychopathology at the Institute for Juvenile Research. Keeler (middle) followed in 1929 and was soon hired by Northwestern University's Scientific Crime Detection Laboratory, the nation's first forensics lab. And Vollmer (right) came in 1930 to teach police science at the University of Chicago. The woman standing over Vollmer is possibly his wife, Pat.

15

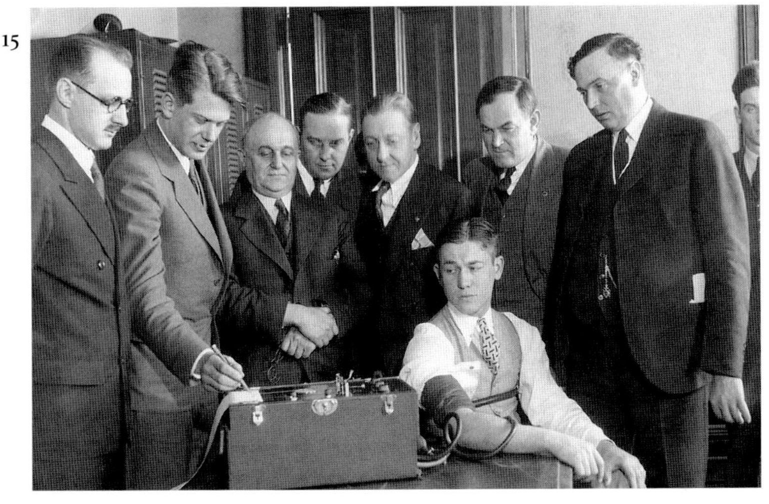

Keeler (second from left) tried his machine in 1931 on Virgil Kirkland, a football star from Gary, Indiana accused of raping and murdering his "sweetheart." Pictured, from left to right: Keeler's colleague C. W. Muehlberger; Keeler; trial judge Grant Crumpacker; defense attorney John Crumpacker (the judge's son!); defense consultant Dr. Orlando Scott (soon to become Keeler's nemesis); and two attorneys. Larson was infuriated by Keeler's brazen effort to introduce the lie detector at Kirkland's trial.

*DICK TRACY—The Lie Detector's Busy Day*

16

Chester Gould, creator of "Dick Tracy," studied with Keeler at the Northwestern crime lab and equipped his square-jawed hero with many of the forensic techniques he learned there, including lie detection.

*DICK TRACY—Starting the New Year*

17

After dispatching Larceny Lu in the final strip of 1932, Tracy used the device to extract the true feelings of his long-suffering girlfriend, Tess Trueheart, in the first daily of 1933—an echo of Larson's courtship of his wife. Popular accounts often emphasized how the machine helped frank and objective men see through the deceptions of collusive women.

18

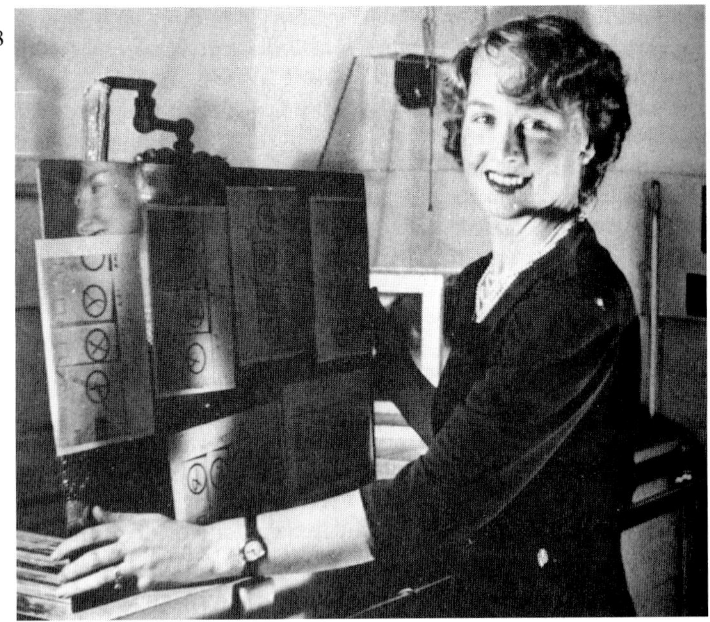

Throughout the 1930s, Kay Keeler battled Chicago's foundational crime: ballot fraud. After the Northwestern crime lab was sold to the Chicago police department in 1937, she started an all-female detective agency. Within a few years, she had divorced Keeler and remarried; he was devastated by her departure.

19

Kay (center) served in the Women's Auxiliary Service Pilots (WASPs) during World War II. She died in a crash in 1944 while flying solo cross-country to halt the disbanding of the WASPs.

20

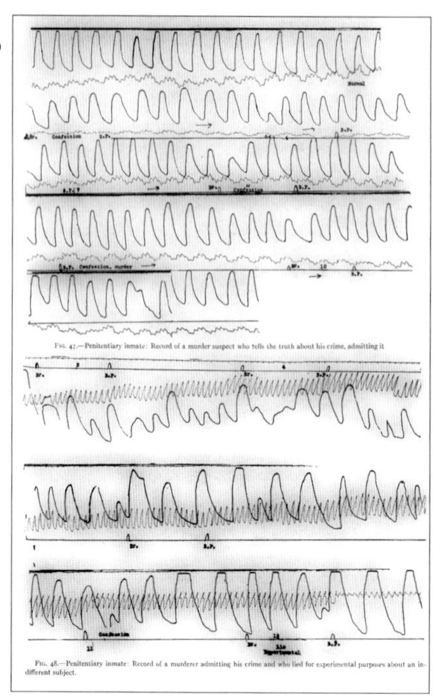


This polygraph chart from Larson's *Lying and Its Detection* of 1932 shows the response of Richard Loeb, Nathan Leopold's co-conspirator in the thrill-murder that shocked the nation. Larson befriended Loeb as part of his investigations into criminal psychopathology. After Larson was fired in the wake of Loeb's prisonhouse murder, he devoted the rest of his career to the treatment of the criminally insane and to the campaign against his own creation, the lie detector.

21

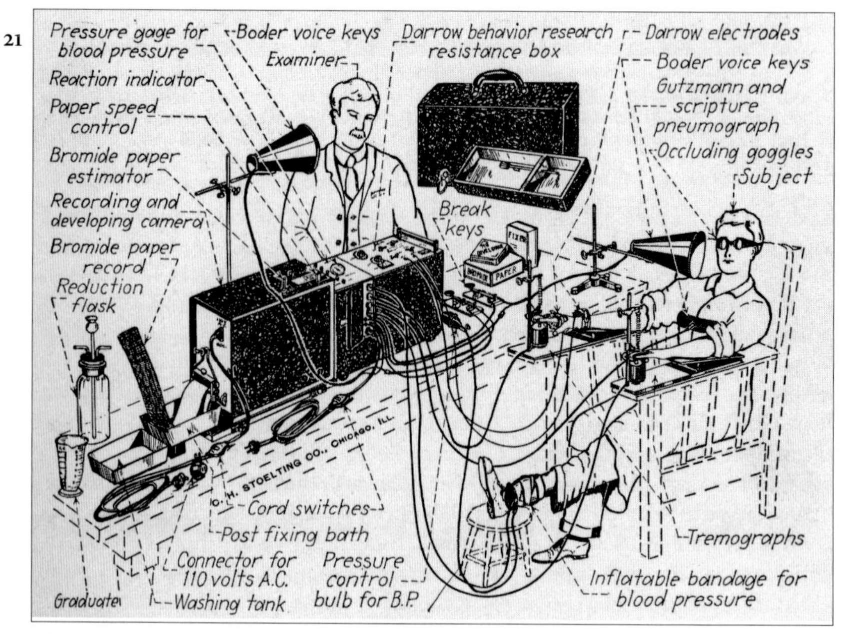


Chester Darrow, Larson's colleague at the Institute for Juvenile Research, developed this Photopolygraph, one of many midcentury devices to study human emotions through their bodily expression.

22

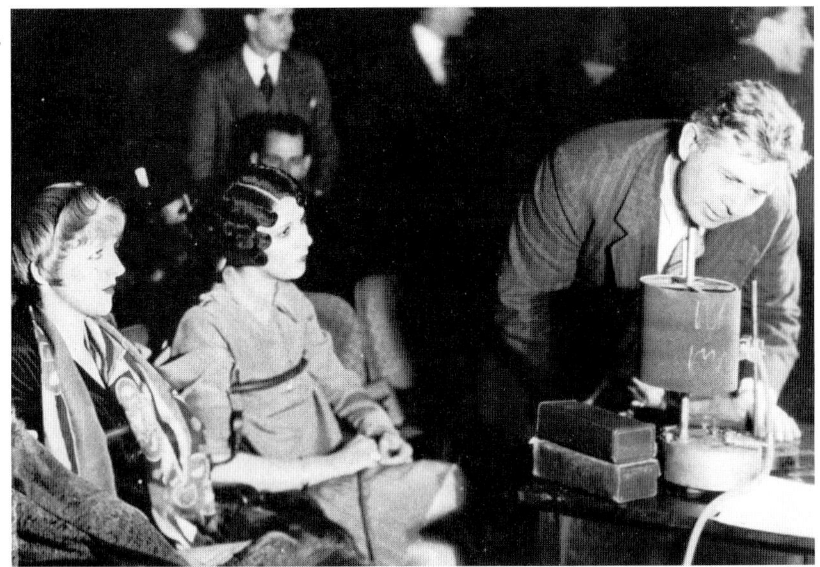

With his showman's flair, William Marston (right) made a comeback in the late 1920s, inviting reporters to the Embassy Theater in midtown Manhattan to watch him use his lie detector to prove that blonds were more emotionally reactive than brunettes. His evidence? The response of starlets to Greta Garbo and John Gilbert making love in *Flesh and the Devil.*

23

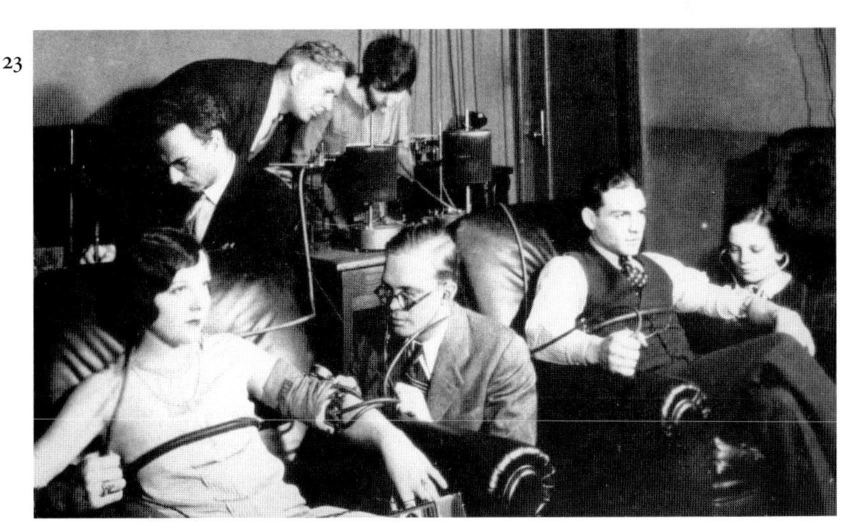

On this basis, Marston was hired by Universal Studios to prescreen movies for emotional content, making him a pioneer in audience testing and studio self-censorship. He can be seen in the back, hunched over the machine, alongside his longtime mistress, Olive Byrne, who lived with Marston and his wife and was the mother of two of his four children.

24

Marston (left) lasted only a year in Hollywood. Behind his mockery of sexual conventions was his own all-encompassing theory of human emotion. Based on experiments at movie screenings and sorority hazing rituals, it posited that dominance and submission were the primordial poles of human feelings, and that both men and women enjoyed submitting to women who were themselves submissive to love.

25

Back in New York, Marston used his device to conduct market research—and vouch for products, as in this 1937 ad. John Larson was appalled by Marston's attempts to make him complicit in these claims.

26

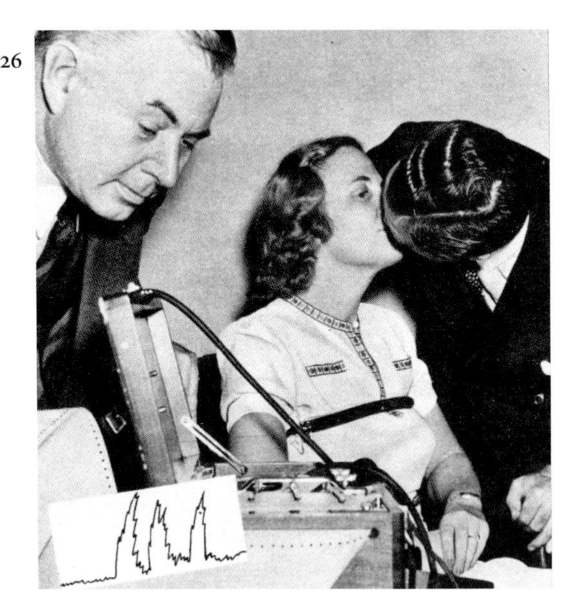

In the 1930s, Marston used his lie detector to dramatize the damage done by the lies men and women told themselves, urging Americans to free themselves from "twists, repression and emotional conflicts." In this *Look* photo spread of 1938, Marston assures a young bride that her marriage can be saved, even though a stranger's kiss is more thrilling than her husband's.

27

Marston's greatest success came in 1940 when he created Wonder Woman, the embodiment of all the psychological principles he had glimpsed in the science of lie detection. In her first solo issue of 1942, Wonder Woman, disguised as Diane Prince, watches as her boyfriend Steve fails to extract the truth from the Nazi Baroness. Wonder Woman later converts the Baroness to the American cause by teaching her to submit to love.

28

In the 1940s, Leonarde Keeler, famous as the inventor of "The Magic Lie Detector," became a founding member of the Court of Last Resort and touted the device as an instrument of exculpation. In 1947 he played himself in a Hollywood movie, *Call Northside 777,* based on the exoneration of Joe Majczek, falsely imprisoned for killing a cop.

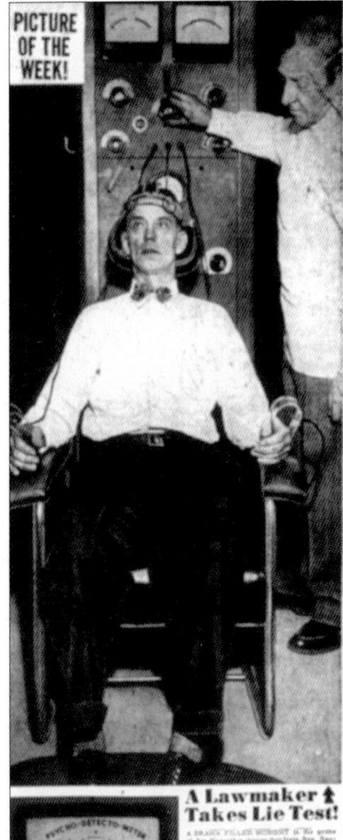

29

In the 1930s and '40s many rivals went into the lie detector business using their idiosyncratic methods. Keeler's nemesis, Orlando Scott, claimed to track "brain waves," but actually measured galvanic skin response (i.e., sweatiness). Yet his terrifying instrument proved effective at extracting confessions. Here he is in 1948 testing an Illinois state representative accused of extorting money from Joe Majczek.

30

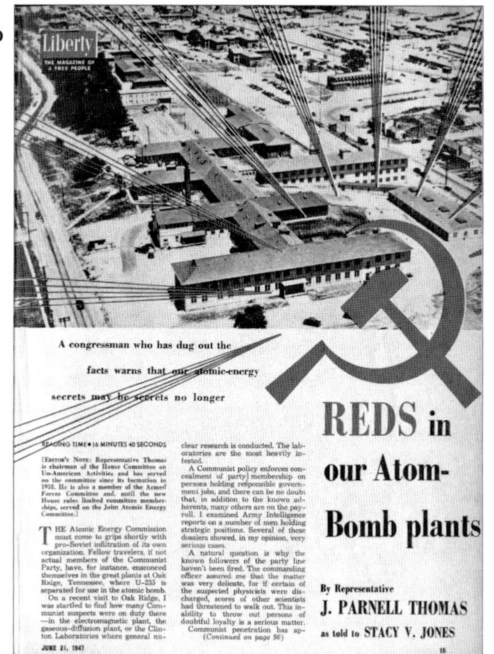

Liberty

THE MAGAZINE OF A FREE PEOPLE

A congressman who has dug out the facts warns that our atomic-energy secrets may be secrets no longer

# REDS in our Atom-Bomb plants

By Representative J. PARNELL THOMAS as told to STACY V. JONES

READING TIME • 16 MINUTES 40 SECONDS

[EDITOR'S NOTE: Representative Thomas is chairman of the House Committee on Un-American Activities and has served on the committee since its formation in 1938. He is also a member of the Armed Forces Committee and, until the new House rules limited committee memberships, served on the Joint Atomic Energy Committee.]

THE Atomic Energy Commission must come to grips shortly with pro-Soviet infiltration of its own organization. Fellow travelers, if not actual members of the Communist Party, have, for instance, ensconced themselves in the great plants at Oak Ridge, Tennessee, where U-235 is separated for use in the atomic bomb.

On a recent visit to Oak Ridge, I was startled to find how many Communist suspects were on duty there—in the electromagnetic plant, the gaseous-diffusion plant, or the Clinton Laboratories where general nuclear research is conducted. The laboratories are the most heavily infested.

A Communist policy enforces concealment of party membership on persons holding responsible government jobs, and there can be no doubt that, in addition to the known adherents, many others are on the payroll. I examined Army Intelligence reports on a number of men holding strategic positions. Several of these dossiers showed, in my opinion, very serious cases.

A natural question is why the known followers of the party line haven't been fired. The commanding officer assured me that the matter was very delicate, for if certain of the suspected physicists were discharged, scores of other scientists had threatened to walk out. This inability to throw out persons of doubtful loyalty is a serious matter.

Communist penetration has ap-

(Continued on page 90)

JUNE 21, 1947 15

In 1946, in response to accusations that Soviet spies had infiltrated America's nuclear facilities, Keeler and his "box" were brought down to Oak Ridge, Tennessee to safeguard America's most potent weapon. He found no spies, but plenty of employees who had fibbed on their applications or believed in federally subsidized housing. In the process, Keeler transformed his device into a psychological deterrent and a gauge of political loyalty.

31

Between 1947 and 1953, some 18,000 residents of Oak Ridge were tested numerous times. Terminated as scientifically suspect, polygraph screening was reinstated at the nation's nuclear weapons labs in 1999 in the wake of the (erroneous) accusations against Wen Ho Lee. These tests continued despite a 2003 report by the National Academy of Sciences suggesting that this sort of security screening is ineffectual, harmful to morale, and a distraction from real security measures. In 2006, the program was being scaled back.

32

This Herblock cartoon from the McCarthyite era (1954) expresses the argument of Edward Shils in *The Torment of Secrecy:* that the era's populist zeal to root out secret conspiracies had overwhelmed the cognitive privacy that is the basis of political conscience. Among other notorious applications, the lie detector was used in this period to "out" homosexuals at the State Department.

33

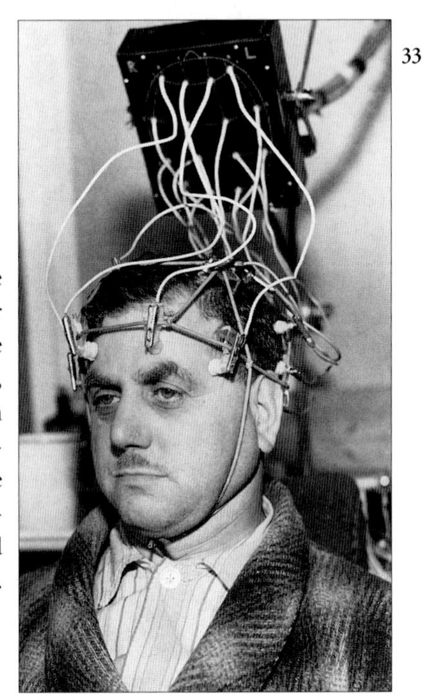

In recent decades—and especially since 2001—deception-testers have directed their attention away from the heart toward the fount of duplicity: the brain. Since the 1940s, some testers have concentrated on brain waves as measured by encephalography (pictured). More recently, neuroscientists have turned to fMRI scans. In fact, these techniques share many of the presumptions—and fallacies—of old-fashioned polygraphy.

PART 3

# *Truth, Justice, and the American Lie Detector*

---

**POPE** (exhausted): *It is clearly understood: he is not to be tortured.* (Pause.) *At the very most, he may be shown the instruments.*

**INQUISITOR:** *That will be adequate. Your Holiness, Mr. Galilei understands machinery.*

—BERTOLT BRECHT, *LIFE OF GALILEO*, 1940

CHAPTER 14

# A Lie Detector of Curves and Muscle

STEVE: *Well perhaps you're telling the truth—will you take the lie detector test?*

BARONESS VON GUNTHER: *No—No! You cannot make me—your American law forbids!*

STEVE [kneeling, straps the monitor to the blond Nazi's leg, while Wonder Woman, in the guise of Diane Prince, takes notes]: *American law does not protect enemy spies in war time—you'll take this test and like it!*

—WILLIAM MOULTON MARSTON, *WONDER WOMAN*, 1942

BORN IN BERKELEY, TRAINED IN CHICAGO, THE LIE DETECTOR was primed for its national debut. Tested on women, blacks, and criminals; deployed on cops and clerks, the machine was ready to take on grander roles. To a nation entering its mid-century struggle against totalitarianism, the lie detector promised to redeem the innocent, scarify the guilty, and ensure political loyalty. It earned a walk-on part in nearly every major scandal of the day: a magnifying mirror held up to America's secret fears and desires. It did not presume to take the leading role, but like Rosencrantz and Guildenstern took its cue from the great players in the drama, an instrument of their larger purposes. Yet for just that reason its performance was all the more revealing.

William Moulton Marston was back for his second act in America. In his book of 1938, *The Lie Detector Test,* Marston asked readers to imagine an America in which everyone sat regularly for the test. He was already optimistic about the instrument's value in taming the criminal mind (thanks, he said, to the work of John Larson) and its value in fostering honest business dealings (thanks, he said, to the work of Leonarde Keeler). Now Marston said he wanted an America in which "truth telling would be as important as dressing decently and washing your face and hands before going to a party or applying for a job." It was a banal enough ambition: honesty as good manners, instead of the contrary. And Marston had a plan for reaching this utopia, as well as a champion to lead America there.

Since his rebuke in the Frye case, Marston had bounced among academic jobs, while parlaying his study of the lie detector into a grand unified theory of the emotions. First in *The Emotions of Normal People* (1928), then in *Integrative Psychology* (1931), cowritten with his wife, he proposed that dominance and submission were the primary human drives, and that women were—or soon would be—the dominant sex. He considered this an optimistic theory of human development. In his scheme, the lie detector became a tool to bring about this necessary adjustment, by fine-tuning the emotional content of movies, by aiding the marketing of consumer products, and by teaching men and women what they really wanted. Marston did not shy away from announcing his conclusion: most people secretly longed to submit to a superior power. His goal was to help them submit to a benign one. For nearly twenty years, in one medium after another, he preached this message, until he finally succeeded in reaching millions.

In 1941, Marston offered America an alluring new champion, a cartoon character he named Wonder Woman. She embodied the benign principles Marston had first glimpsed in the science of lie detection. And where the masculine apparatus of her lie detector proved inadequate, she used her female power to compel love in order to defeat—and even reform—her Nazi foes, such as the Baroness Paula von Gunther, who demanded slavish sexual obedience. She fought for honesty, decency, and human liberation, with female liberation as a necessary prerequisite. She was the ideal mate that Dr. Frankenstein had not dared to create for his monster, a lie detector of curves and muscle and cheerful wit, a female superhero to whom America might safely submit because she herself had submitted to love.

Marston's theory of human emotion identified two primordial feelings: pleasantness, defined as "the free and unimpeded discharge of motor impulses"; and unpleasantness, defined as "conflicts" in neuromotor consciousness. On this basis, Marston reasoned, an external stimulus could produce one of four emotional responses. When the stimulus was stronger than the motor consciousness to which it was antagonistic, the self became "compliant" to avoid unpleasantness. When it was stronger but allied, the self was "submissive." When the stimulus was weaker, but allied, the self practiced "inducement." When it was weak and antagonistic, the self asserted its "dominance." A mixture of these primary emotions, in greater or less degree, in passive or active forms, produced those familiar emotions that poets and others called fear, rage, jealousy, and love. With this theory Marston reinterpreted his lie detector experiments as an attempt to extract compliance from the subject, a contest to establish dominance, rather than a test for fear of being discovered in a lie.

The impetus for Marston's theory had been his discovery that male and female subjects responded quite differently when he wielded the lie detector as opposed to when his young wife wielded it, especially when the topic was sex. Its confirmation came in a set of experiments he conducted in the mid-1920s at Tufts University. In the aftermath of Larson's tests, Marston, with the aid of an undergraduate assistant named Olive Byrne, investigated hazing rituals at the sororities of its women's college. The rules of the "Baby Party" ritual were silly but strict. For one week at the end of the school year first-year sorority sisters were forced to dress up in baby clothes and yield to the sophomore women, who blindfolded their charges, bound them by the wrists, and interrogated them about their misdeeds while threatening to beat them with long sticks if they dared to rebel or escape. Thanks to interviews conducted using lie detection techniques, Marston and Miss Byrne learned that this ritual was exceedingly pleasurable for both the dominant young women and the submissives, especially when their authority was challenged. "From these studies of girls' reactions," they reported, "it seemed evident that the strongest and most pleasant captivation emotion was experienced during a struggle with girls who were trying to escape from their captivity." Yet Marston and Byrne also noticed that many sophomores expressed discomfort whenever the first-year students showed signs of fear or broke down in tears. At such moments, rather than use force, they derived the greatest pleasure from coaxing the younger women to continue. The more docile the final submission, they noted, the greater the "captivation emotion." As for the younger women, they experienced the greatest pleasure in submitting to those who acted selflessly. "A freshman girl reported herself as feeling pleasant passion emotion when compelled to kiss the feet of a girl whom she liked. But this same girl took a large amount of extra punishment rather than shine the shoes of an older girl whom she thought selfish."

By contrast, hazing rituals in fraternities produced very few cases of true "submission passion." There, attempts at resistance just led to more simpleminded punishment. However, Marston saw evidence suggesting that some of the young men would have been only too glad to submit lovingly to young women of sufficient emotional strength. Women, suggested Marston, were superior to men. Men had a stronger sexual appetite and a will to dominate, while women preferred to cultivate the love response. But this seeming submission would eventually enable women to take command of the species. When he polled college students 244 out of 248 men asserted that they would prefer to be an unhappy master, while 162 out of 260 women said they would prefer to be a happy slave. Yet people desired happiness. In surrender lay control; in submission (with proper inducement), dominance.

Olive Byrne clearly had captivation charms of her own. She soon moved in with the Marston household, becoming Marston's assistant and joining him and his wife in a ménage à trois. It was by all accounts a harmonious arrangement. Visitors described a merry home, with Marston the genial leader, six feet tall and bulky, a former high school tackle. "He had a family relationship with a lot of women, yet it was male-dominated," one friend recalled. With his wife, Elizabeth, Marston had two children, Pete and Olive Ann. With Olive Byrne, he had two sons, Byrne and Donn, who were legally adopted by the Marstons. While Elizabeth worked as a lawyer to keep the family in funds, Olive stayed home to mind the children and assist Marston with his experiments. Their sons recall the arrangement as reassuring. An evening's entertainment might include hooking up a guest to the lie detector and plying him with innocuous questions before popping a doozy.

As this arrangement was no more compatible with academic mores than his theories were, Marston looked to the wider world for validation. He wrote up practical suggestions akin to those of contemporary sexologists such as Havelock Ellis and his mistress Margaret Sanger. (Sanger was Olive Byrne's aunt, and Olive was always grateful to the pioneering feminist for teaching her about the world beyond her sheltered Catholic upbringing.) Marston's tips ranged from lovemaking to peacemaking. Genital mechanisms are the teachers of love. Inducement should proceed submission if it is to lead to love. Women should be on top. Men should learn to wait. Nearly half of all female-female love relations were accompanied by "bodily love stimulation," which was perfectly healthy. A normal adult should lead his or her lover to a state of devoted submission. Hate (or racism or "war fever") among individuals or peoples was caused by a failed effort to induce submission and a consequent urge for dominance, so the route to peace was to find terms to which each party could lovingly submit. Marston's solution? It was time for women to take a dominant role in the political and social life of the nation. In 100 years, he predicted, "the country will see the beginning of a sort of Amazonian matriarchy." In 500, there would be "a definite sex battle for supremacy." And in 1,000, "women would rule over the country, politically and economically." This last prediction had just the right frisson to earn Marston hundreds of newspaper notices.

Marston had a knack for blurring the boundary between high and low culture, elite and popular science, provocation and uplift. Part of his charm was some uncertainty as to whether he was kidding. It took a showman to extend the use of the machine to domains where credibility was itself the principal commodity.

In 1928, while he was a part-time lecturer at Columbia University, Marston transformed the Embassy Theater in midtown Manhattan into a scientific laboratory and invited the press to ogle as he cross-checked the emotional temperaments of blonds, brunettes, and redheads. "The air was thick with euphemism," noted the reporter for the *New York Times,* as

Marston strapped down three Broadway starlets, one of each tincture, and monitored their blood pressure while they watched John Gilbert and Greta Garbo make love in *Flesh and the Devil.* His conclusion was that "[b]londes prefer gentlemen, men prefer brunettes, and red-heads prefer a fight." He even maintained his scientific sangfroid when a reporter had the gall to ask whether the blond came from a bottle.

On the basis of these tests, Marston got his big break. Carl Laemmle, Sr., founder of Universal Pictures, hired him "to apply psychology to all departments of the motion picture concern." Marston was to use his lie detector to vet the studio's screenplays for emotional content. "Motion pictures are emotion pictures," he liked to say. His job had as much to do with keeping criticism at bay as attracting ticket buyers. Under pressure from moralists, the Hollywood studios had just subscribed to a voluntary production code to censor their own films for violence and licentiousness. The industry had also begun to sponsor psychological and physiological research to repudiate the claim that movies had a pernicious effect on viewers, including on the dreams of children. Marston's work fit neatly into this agenda.

Marston boasted that his tests would craft movies to touch the "better" emotions of the audience. "No other organization in the world, not even the church, is so powerfully equipped to serve the public psychologically as is the motion picture industry," he wrote. This power had to be tamed and refined for an age of mass audiences. No longer would Marston apply his lie detector to individuals one by one. Through the movies, he proclaimed, "I shall experiment with millions of people instead of hundreds."

In a sense, this alliance of lie detector and moving picture was incestuous: the two devices were siblings separated at birth. In the second half of the nineteenth century, Étienne-Jules Marey, the French physiologist who assembled the first polygraph to record the invisible interior actions of the body with several physiological instruments—the grandfather of Larson's and Keeler's machines—had also sought to record the fleeting movements of the body with sequential photographic images of trotting horses and human calisthenics, which he then re-created with his stop-action chronophotograph, the grandfather of modern cinematography. Both types of instruments enabled scientists to capture bodily actions below the threshold of conscious observation and probe elusive mental states: the polygraph to measure the bodily correlates of fear and anxiety; the cinematic apparatus to capture the fleeting expression of feelings in the human face. Nor was it long before practitioners reversed the influence so that both devices were being used to induce mental states—and consequently bodily states. Keeler used the lie detector to intensify his subjects' fear of being caught, and Larson used it to relieve guilt. As for the motion picture projector, it had become the greatest emotion-generating device in human history. Reunited, the polygraph and the film promised to bring scientific control to the production of human feelings.

Marston's mentor, Hugo Münsterberg, had been the first to articulate how this might be done. In his influential booklet of 1916, *The Photoplay,* he explained how cinematic techniques, like the close-up and flashback, made movies seem an inner process in which the picture appeared "shaped by the demands of our soul." When properly performed, the actors' bodily expressions were then reproduced in the audience as if the audience members were emotional automatons, mimicking those of the performers. "The horror which we see makes us really shrink, the happiness which we witness makes us relax, the pain which we observe brings contractions in our muscles; and all the resulting sensation from muscles, joints, tendons, from skin and viscera, from blood circulation and breathing, give the color of living experience to the emotional reflection in our mind."

In his own how-to book for screenwriters—*The Art of Sound Pictures* of 1930—Marston showed how to reunite this long-separated pair of emotion engines: the dos and don'ts of a gripping story, the physiological gestures that brought the passions to life, plus a thirty-page state-by-state guide to "what censors do to your story" as regards violence, vulgarity, sexually suggestive acts, and criminality.

Meanwhile, Marston was arranging screenings where he tracked the responses of squirming test audiences so that the studio might fine-tune the appeal of the movie, forestall public controversy, and reduce expensive post-release editing. He took credit for the final look of *Show Boat* (1929) and *All Quiet on the Western Front* (1930), helping craft the violent scenes so that they got past the censors. He boasted that he was Hollywood's first on-lot censor, "a big money saver." Marston had been hired not so much to make movies emotionally true as to make them emotionally safe. Still, he did not last out the year at Universal; production executives clearly resented a Ph.D. vetoing their projects. After a brief stint at Equitable Pictures, he returned to New York, as "the best known psychologist in America," but without a job.

Not that Universal had given up on the idea of vetting its pictures for their emotional impact. In *Frankenstein* the studio knew it had a movie which the public (and censors) would find disturbing and which, not coincidentally, would make its fortune. At a preview in Santa Barbara in November 1931, several indignant reviewers walked out visibly shaken and demanding changes.

With Marston no longer on staff, Universal turned instead to Leonarde Keeler. In short order, he had strapped two undergraduates from DePaul University into his lie detector to watch Boris Karloff play the Monster. While the film was projected as flickering images on the screen, Keeler monitored the viewers' physiological reactions on the graph paper unfurling from his box. As the man-made Monster on the screen lumbered into life, pleaded for sympathy, and then turned to rape and murder, the apparatus inside Keeler's box mimicked the viewers' inner palpitations, like a living, breathing machine. And as the resurrected Monster and the young bodies seated in the darkness trembled together, Keeler recorded the connection of sympathy and horror that was the physiological index of the film's truth, its capacity to suspend disbelief and bring the creature to life.

Keeler's report for Universal has not survived, though we know that his instrument was used to adjust the film's shock, as well as to eliminate "emotional . . . dead spots." The studio trimmed a scene in which the Monster drowns a little girl; eliminated three close-ups of Fritz torturing the Monster; and deleted Dr. Frankenstein's blasphemous cry, as the Monster rose from the dead: "Now I know what it feels like to BE God!"

James Whale, the director, considered this creation scene the key to the movie's success. To make it credible, he drenched the lab in electrical razzle-dazzle that awed the characters as much as the audience. As Whale put it, "Frankenstein merely has to believe what he sees, which is all we ask the audience to do." Mae Clarke, who played Frankenstein's young fiancée, agreed. "[W]e actors experienced exactly what future audiences would feel as the film rolled on the screens." In other words, the sets were fake, the images were staged, and the dialogue was censored, but to the audience it all felt true.

On December 4, 1931, *Frankenstein* opened in Times Square to become the country's biggest hit, giving new meaning to the cliché "spine-chilling." When state censorship boards demanded changes, they only fed the publicity. Even the application of Keeler's lie detector to vet the film was used to publicize it.

Marston, back in New York, was already running the same operation for advertisers. His ad agency lasted only a year, but throughout the 1930s he consulted for Madison Avenue. He applied his lie detector to compare the responses of smokers to four brands of cigarettes, and to gauge the satisfaction of 150 drivers who filled their tanks with Texaco Sky Chief instead of a rival fuel. Each stunt won its little notice, which was the point. For his biggest contract, Marston strapped men into "the same scientific instrument used by G-men" while each shaved one side of his face with a Gillette razor and the other with a rival blade. Full-page ads in *Time, Life,* and the *Saturday Evening Post* confirmed that the rival blade made men "grouchy and irritable," while nine of out ten expressed an honest preference for Gillette.

Marston's campaign capped a twenty-year effort by psychologists—again initiated by Hugo Münsterberg—to convince advertisers that psychological expertise could help shape and direct consumers' desires. Politicians might argue that citizens were sovereign; economists might assert that consumers were rational; political theorists might insist that human beings were born free; but psychologists knew Americans to be creatures who responded to conditioned stimuli. All one needed to know was what the lie detector indicated that citizen-consumers believed to be true.

Unfortunately, before expanding their campaign to radio, the folks at Gillette asked John Larson to replicate Marston's findings. Marston had been cultivating Larson for several years, offering to help him write articles, land a job in New York, and wrangle an entry in *Who's Who.* In return, Larson had agreed to write the introduction to Marston's book—even though it was full of the hokum Larson despised—so long as Marston added some oblique slurs against Keeler. But Larson would not let personal favors undermine his scientific integrity. For a $750 fee and a promise to keep his name out of any ads, he used his lie detector to test the emotional response of 100 shavers to seven brands of blades, finding no preference for Gillette. Behind the scenes, according to Larson, Marston begged him to modify his report and even offered him an "inducement" to do so: a cut of the $30,000 he hoped to make from the deal. Larson indignantly quashed the campaign and even ratted out Marston to the FBI.

Undeterred, Marston extended his populist reach in the late 1930s, soft-selling his theories in "pop psych" articles for women's magazines and becoming a scientific pastor of uplift and proper adjustment. "Try Living" was his catchphrase. His most enduring creation was the in-magazine personality test. His particular version let readers assess their urges for dominance and submission, so that they might better adapt themselves at home and office. Today known as the DISC test (for dominance, influencing, steadiness, and conscientiousness), its merchandisers still tout it as the "original, oldest, most validated, reliable, personal assessment, used by over 50 million others to improve lives, relationships, work productivity, teamwork, and communication!"

Through all this, in his private practice, Marston put his lie detector to work uncovering the lies men and women told themselves, thereby freeing them from "twists, repression and emotional conflicts." His gadget, he said, could gauge whether a couple had married for money, the "higher emotions," or "just plain sex." Marston helped one woman whose college roommate had seduced her fiancé; when the seductress was put on the lie detector, she admitted that she didn't want the man and released him to her roommate. One married woman who suspected her husband of having an affair discovered that it was her jealousy that was driving him away. For the readers of *Look* magazine, Marston compared the physiological reactions of a neglected young wife: first when she was kissed by her wayward husband, then when she was kissed by an attractive young stranger; apparently she preferred the stranger, but Marston still had hopes of saving her marriage. "Healthy love adjustment," explained Marston, "requires, first, the cutting away of disguises, the elimination of false expressions of the true emotions underneath."

The culminating act of Marston's career began almost by accident in 1939, when he went on record in *Your Life* and *Family Circle* as a judicious critic of the comics. "Superman" had just taken America by storm, and the moralizers' indictment of the comics ran long: "mayhem, murder, torture, abduction, superman heroics, voluptuous females, blazing machine guns and hooded justice." As a professional psychologist, Marston was awed by their emotional efficacy: 30 million children read 1.5 billion strips a day. Marston's children read them; he himself read them; he even liked them—most of them, anyway. There was nothing wrong with a little wish fulfillment, so long as it wasn't perverted. "Dick Tracy," for instance, struck Dr. Marston as violent and sadistic, but he did not agree with the would-be censors who fretted about female décolletage, while overlooking the comics' portrayals of women as "jealous, mercenary, and moronic." Parents who wanted to bring out the better side of comics should organize a Cleaner Comics League.

This oblique threat caught the attention of M. C. Gaines, a former high school principal turned publisher of All-American comics and discoverer of "Superman." Gaines offered Marston a spot on his advisory board of child advocates. Marston was glad to be co-opted; it was his lifelong strategy. He was soon praising Superman as breathlessly as did the awestruck Jimmy Olsen. It was Gaines's genius, Marston announced, to have seen Superman's "fundamental emotional appeal," his "Homeric inheritance." Marston defended "Superman" from those who complained that the comic starred a vigilante hero who taught the moral superiority of force: in short, a Nazi. On the contrary, Marston argued, the fantasy of using superhuman strength to right wrongs could only be healthy for young Americans at a time when the United States needed to cultivate its national strength to protect weak and innocent peoples abroad.

Once on the advisory board, however, Marston denounced other comics for their "blood-curdling masculinity." But when he proposed a female superhero, the editors snickered; every previous attempt had flopped. This, answered Marston, was because the previous heroines had been weak. "Give them an alluring woman stronger than themselves to submit to," he later proclaimed, "and they will be *proud* to become her willing slaves!" Intrigued, Gaines gave Marston six months to disprove the doubters. In February 1941, under the pseudonym Charles Moulton—a combination of his publisher's first name and his own middle name—Marston delivered the first script for "Suprema, the Wonder Woman."

Wonder Woman became the champion for Marston's worldview. Many elements of her Amazonian mythology are drawn from his life and works. Her magic lasso, which compels obedience, is the ultimate lie detector. Marston called it a symbol of "woman's love charm and allure with which she compels men and women to do her bidding." The Amazonian bracelets that symbolize her submission to the goddess Aphrodite are the source of her strength as well as her vulnerability; she can use them to deflect bullets, but she loses her strength whenever a man attaches chains to them. Apparently Marston's mistress, Olive Byrne, favored large bracelets of this sort. But the details mattered less than the "universal theme" he believed he had tapped: "the growth in the power of women." He warned the editors that although he was willing to consider changes to names, costumes, and story, "I want you to leave that theme alone—or drop the project."

"Frankly," Marston said, "Wonder Woman is psychological propaganda for the new type of woman who should, I believe, rule the world." She certainly took America by storm. One year after her debut in *Sensation Comics* in January 1942, she had her own comic book with a circulation of 500,000. She was quickly inducted into the Justice League of America—as its secretary, sad to say—and by 1944 she had a daily syndicated strip as well. By the end of the war, her circulation topped 2.5 million. Marston wrote all the scripts until his death seven years later, with Harry G. Peter doing the bulk of the illustration.

For dramatic and visual punch, Marston relied heavily on imagery of enslavement and emancipation, with (scantily clad) women led in chains, hypnotized into submission, and otherwise disciplined—only to overcome their bondage, thanks to Wonder Woman and her sidekicks, the Holliday Girls, modeled on the sorority sisters of the Baby Party. Many of Wonder Woman's adversaries were women likewise endowed with sexual and physical powers, except that they sought unbounded domination. By contrast, Wonder Woman reformed her opponents on Paradise Island, where they were put in shackles until they had learned to follow her own "loving submission to authority."

The complaints began immediately. Josette Frank of the Child Study Association of America conveyed her dismay about Wonder Woman's costume, "or lack of it," as well as the "sadistic bits showing women chained, tortured, etc." Outsiders attacked "Wonder Woman" for preaching a "cult of force, spiked by means of her preteniously [*sic*] scanty 'working' attire." Frederic Wertham, an émigré German psychiatrist who would become the nemesis of the comics, perceived a lesbian subtext in "Wonder Woman" and castigated Marston for using psychological science to create a monstrous perversion that could never have arisen "normally" in popular culture. And anxious editors feared the critics were on to something when they received a fan letter from a sergeant in Fort Leonard Wood, Missouri, who relished precisely this aspect of Marston's creation: "I am one of those odd, perhaps unfortunate men who derive an extreme erotic pleasure from the mere thought of a beautiful girl, chained or bound or masked, or wearing extreme high-heels."

To mollify these critics, the staff at All-American recommended making Wonder Woman less "sexy" and more, well, all-American. They also commissioned a female tennis champion to write a monthly piece on real-life "Wonder Women of History."

Marston defended his heroine. Wonder Woman without erotic appeal, he declared, would be like "Superman without muscle." But he denied that she was sexual. None of his "jury" of local kids would stand for anything "mushy." He pointed out that even his own mother, director of the New England Woman's Club, found the costume unobjectionable. This was an outright fib: only a few months earlier Marston's mother had asked her son if the artist might not lengthen Wonder Woman's pants "even a wee bit. . . . And how about an embroidered scarf of red, white, and blue? It might save her from an attack of pneumonia."

Above all, Marston denied the charge of sadism, which he defined as taking enjoyment in the suffering of others. Bondage was not just a symbol of the peril his heroine overcame: it was also (properly understood) the basis for civilization.

> The only hope for peace is to teach people who are full of pep and unbound force to enjoy being bound, enjoy submission to kind authority, wise authority, not merely tolerate such submission. Wars will only cease when humans *enjoy* being bound (by loving superiors of course). Women are exciting for this one reason—it is the secret of women's allure—women *enjoy* submission, being bound. . . . And because it is a universal truth, a fundamental, subconscious feeling of normal humans, the children love it.

Throughout those early episodes Marston contrasted Wonder Woman's cheery sexiness with the dark and sexualized power wielded by fascists, such as her nemesis, the Baroness. He also highlighted America's relatively positive attitudes toward women, without neglecting the continuing struggle for equality at home.

One of Wonder Woman's early foes was Doctor Psycho, a brilliant, stunted psychologist, who had turned misogynist after being spurned by his fiancée, Marva. In the episode entitled "The Battle for Woman-Kind" (June–July 1943), Doctor Psycho sets out to reverse the freedoms won by women, aided by the mythical Duke of Deception, emissary of Mars, the god of war, who is infuriated that 8 million American women are contributing to the American war effort. Using his hypnotic powers Psycho forces Marva to marry him and become his psychic medium; bound and enslaved, she enables him to transform himself into any bodily form he desires. Initially Psycho assumes the form of Mussolini, but he soon hits on a more dastardly plan, better suited to his American audience. Before a large crowd, including the skeptical Wonder Woman, he conjures up President George Washington himself, and in the voice of the great truth-teller warns America, "Women will betray the country through their weakness if not treachery!"

No sooner is this prophesied than a munitions plant staffed by women is sabotaged, and Psycho tricks the female office workers into hiding secret papers in their undergarments so that they must be strip-searched and led away in chains. Even Wonder Woman's square-jawed boyfriend Steve is half persuaded of their perfidy, until Psycho captures him and impersonates his voice and body to lure Wonder Woman into a trap. Thanks, however, to the interventions of the Holliday Girls she breaks free, releases her boyfriend, and liberates Psycho's shackled wife, whom she later interrogates on both a lie detector and her magic lasso:

> **MARVA:** "Submitting to a cruel husband's domination has ruined my life! But what can a weak girl do?"
>
> **WONDER WOMAN:** "Get strong! Earn your own living, join the WAACS or WAVES and fight for your country! Remember the better you can fight the less you have to!"

The episode ends with the Holiday Girls chasing Psycho with a giant paddle. "Catch him kids, give him the Lambda Beta treatment!"

Marston understood that as a piece of physiological apparatus the lie detector was nothing. Its power—like that of movies, advertisements, and fiction of all sorts—derived from its capacity to amplify people's fears and hopes. Marston openly embraced the instrument's ability to compel belief. He removed the lie detector from the realm of science and repositioned it in the more popular (and remunerative) reaches of science fiction. Wonder Woman was Marston's female lie detector: to submit to her authority was to be free. She was the good monster Dr. Frankenstein had not dared create: a creature Marston was proud of. Perhaps this was why he escaped the blowback that afflicted his coinventors. It also helped that he retained his copyright on "Wonder Woman." After he died of skin cancer in 1947, his two widows supervised the cartoon scripts until new editors in the 1950s domesticated Wonder Woman and took her in romantic directions they did not condone. Wonder Woman may have routed the foreign fascists, but she couldn't defeat the hard-hearted men of the 1950s who trumpeted their right to protect America and its women. Not until the 1970s was "Wonder Woman" rediscovered as a feminist icon, albeit without Marston's subversive acknowledgment of the human need for submission. Elizabeth Marston and Olive Byrne continued to live together until their deaths in the 1980s.

CHAPTER 15

# Atomic Lies

**GUILDENSTERN:** *The scientific approach to the examination of phenomena is a defence against the pure emotion of fear.*

—TOM STOPPARD, *ROSENCRANTZ AND GUILDENSTERN ARE DEAD*, 1967

IN HEAVEN, AS IN EDEN, WE WILL PRESUMABLY GO UNCLOTHED. Or perhaps, Saint Augustine speculated, we will be transparent, for in the angelic company, we may be as crystalline as the celestial spheres. Human opacity is a consequence of the Fall. Having eaten the fruit of the tree of knowledge of good and evil, the story goes, the knowledge we acquire is of a mortal kind, and we see the world as through a veil, darkly. Our souls are concealed from one another, and all too often from ourselves, "wrapt up here in the dark covering of uncrystallized flesh and blood." So Adam plucks a fig leaf to cover his nakedness and like a toddler thinks that because he dare not examine his own conscience it is likewise hidden from God's sight. "Adam, Adam, where are you?" begins the world's first inquisition, according to one of Spain's most notorious Inquisitors. But isn't human opacity, fount of mutual mistrust and discord, also the basis of human freedom?

It was Francis Bacon, godfather of experimental science, who updated the medieval millenarians when he boldly promised a return to Eden, not by way of innocence, but by way of knowledge. Though Adam and Eve had filched their capacity for knowledge in violation of God's command—and for this sin been condemned to lives of toil and pain and mutual deception—God had also, in His beneficence, seen to it that this same capacity might, if properly applied, enable their children to reverse the curse. Cultivate that form of knowledge called science, and human beings might ease their common toil, alleviate their daily pain, and perhaps even abate their mutual duplicity. Science offered a superior way to organize human lives, a form of collective knowledge that thrived on reciprocal honesty and shared information. It would even mitigate the deceptions to which thoughts and senses were necessarily prey. But there was always a dark undercurrent to Bacon's thought. What if nature's investigators themselves became corrupt or put their powers to destructive ends? Bacon devoted the final chamber of his imaginary House of Solomon to instruments of deception, though he quickly added: "But we abominate Imposture and Falsehood; insomuch, that all our fellows are strictly forbid, under pain of Ignominy and Fines, to shew any natural Work, or Thing . . . otherwise than pure and simple as it is in itself." Against the dangers of deception, Bacon could only return full circle and urge strict sanctions on the circulation of new knowledge. His ideal scientific society is a secretive outfit unwilling to reveal all it knew.

Since Bacon's day we have learned much about how to manipulate the world to our advantage and our misery. There have even been times when Eden seemed just around the corner: at the dawn of the atomic age, for instance, when a source of unlimited power seemed to promise an end to human toil, even as it threatened to end human existence altogether. Monitoring the balance between peril and hope, some Americans—mostly scientists—urged a universal sharing of nuclear know-how as the surest route to mutual trust and peace. Others, more suspicious, sought to lock down the secrets of the atom. At that crucial juncture, the lie detector was called in to mediate at the boundary between open knowledge and secrecy. For seven years, from 1946 to 1953, the atomic city of Oak Ridge, Tennessee, aspired to be an honest town, a state-of-the-art utopia of crystalline transparency in which there was no need for democracy because there was complete disclosure and public trust. And it was all designed to protect a secret.

The nuclear age began with a desperate plea. In his famous letter of 1939 to President Roosevelt, Albert Einstein begged Roosevelt to take charge of nuclear weaponry. The discoveries in fission trumpeted monthly from Paris, Chicago, New York, and Cambridge had already led German scientists to alert the Nazi government and prompted the Soviet Union to begin atomic research. With western scientists unable to keep nuclear secrets, Einstein and his colleagues wanted the president to clamp down. The challenge for the Manhattan Project was to nurture the open exchange of ideas thought necessary to build the bomb, while building protective walls around ideas and techniques lest others learn how to replicate their feat. Robert Oppenheimer convinced the army that because the bomb's designers were isolated on a high desert mesa in New Mexico, these scientists might communicate amongst themselves without interference.

Secrecy proliferates much as knowledge does. It is, after all, just another form of knowledge, from the stonecutting secrets of the Masonic guilds to the design of atomic weaponry. The latter form of secrecy—classified knowledge—began to breed prolifically during World War II. One historian of science has estimated that in recent decades the United States has been adding five times as many pages each year to the universe of classified knowledge as are added to the Library of Congress.

The boundary between open and secret knowledge is notoriously hard to police. This is not because "information wants to be free," as partisans of the Internet would have it. Nor is it because some knowledge is intrinsically "born secret," as U.S. official classifiers assert of the designs for nuclear weapons. Rather, it is because knowledge is produced in particular settings whose rules are themselves subject to conflicting pressures. Classified knowledge and open science grew up symbiotically. Indeed, many of the features of "open science" that we take for granted—publicly funded scientific research at elite universities, double-blind peer review—emerged in response to the postwar regime of secrecy. When Congress funded the National Science Foundation in 1950, it did so because leading scientists wanted to carve out a domain of autonomous action within the new secrecy regime and because they convinced those who held the purse strings that only a system of open science could train the cadres of scientists who would work on secret research, as well as replenish the stock of public knowledge on which military and corporate researchers relied. The transfer of people and ideas between the open and closed worlds was always part of the plan.

As a nation whose principal military advantage was a weapon assembled from secret knowledge, America at mid-century needed a new kind of barrier against its foes. As an open society whose oceanic defenses offered less protection than ever, it needed a different kind of defensive border.

This would have to be an interior border, a mental border. For these reasons, the U.S. government became the world's largest user of the lie detector. Keeler showed them how.

No sooner had America gone to war than some observers suggested that all citizens employed in sensitive government and industrial posts be polygraphed on being hired and regularly thereafter. Commander McDonald advised his pal Keeler to market his ability to root out "subversives" in industrial plants. To set an example, he hired Keeler to vet all his employees at Zenith for "first, their loyalty to the Government, and second, their loyalty to the company." Within months of the attack on Pearl Harbor, Keeler was offering the army his services for testing personnel.

Originally, he hoped to serve in naval intelligence. But not even the Commander's pull could disguise Keeler's dicky heart problem. When his friends arranged a captaincy in a military crime lab for him in 1942, Keeler scheduled his medical exam for Wright Field, Ohio, where no one knew his history. He "was getting along beautifully" until the physician checked his blood pressure. "It was a terrible bump." The reading was 180 over 100, too high for clearance. Keeler returned on three subsequent days hoping that it would subside. The next day the systolic pressure hit 180 again, then 190, and finally 210. As Keeler knew better than anyone else, the stress of the test itself explained these numbers. "Of course, the 210 reading was obtained when I knew my physical examination was turning out poorly." For once, the lie detector had worked as advertised. Keeler's body had betrayed him.

The rejection for service and Keeler's 4-F status were particularly galling because his former wife had joined the Women's Air Force Service Pilots, one of the few chosen from among 30,000 applicants and by far the oldest. Though the age limit for recruits was thirty-five and she was thirty-seven, Katherine fibbed her way through. Her dashing new husband, meanwhile, had parachuted into occupied Norway to help sabotage its supply of heavy water. He would later drop into Normandy two weeks before D-day. Known as "Commandant Bazooka," he was one of the most decorated soldiers in World War II, honored for single-handedly liberating the city of Thiers.

Against such heroics, Keeler could only ply his ambiguous craft. Ironically, the new governor of Illinois made him chair of the medical advisory board, where he used his lie detector to expose malingerers seeking to evade army service.

But though he never made it into uniform, Keeler sold the army on his technique. His first assignment, identifying the killers of a German POW, led directly to something much bigger: the first mass use of the lie detector to gauge political loyalty. In transforming the lie detector from a discloser of deeds to a disgorger of thoughts, Keeler extended the deterrence logic he had perfected for business to the new world of ideological conflict. The lie detector stepped eagerly into this new role, for which it had long prepared, but which only the U.S. Army could have sponsored, and only wartime could have legitimated.

Keeler's work for the U.S. government began with a criminal case, so it was familiar work despite the new setting. On March 13, 1944, a German POW named Werner Drechsler was found hanged in the bathhouse of the Papago Park internment camp in Phoenix, Arizona. The camp was one of dozens across America holding in all some 372,000 German prisoners of war. Fellow prisoners who had served in U-boats had apparently recognized Drechsler as a former shipmate rumored to have spied for the Americans. The Office of the Provost Marshal General came under intense pressure to solve the murder. J. Edgar Hoover was warning Americans of the "potential menace" posed by the Nazis in their midst. Prominent liberals were accusing the government of letting Nazis bully their fellow prisoners. And nativists were complaining that the military coddled prisoners.

Unable to make headway with the silent prisoners, the army's criminal investigation division called Keeler in. Working his way through 125 suspects who supplied only name, rank, and serial number, he finally got one soldier to admit knowing the identities of the killers, though the soldier refused to divulge their names. At this point, Keeler used his peak-of-tension test and interrogated any prisoner whose name produced a physiological reaction. He identified twenty suspects for transfer to a secret camp in Stockton, California, where they were subjected to methods akin to torture: sleep deprivation, intense heat, strip-searches, and physical abuse. Some were made to wear a gas mask containing an onion. Seven POWs confessed to the murder, claiming to have done what any loyal German soldier would do. Two weeks after the war they were hanged in the last mass execution in U.S. history.

By then the killing of Drechsler had led to a rethinking of the POW problem. Liberal journalists cited the killing when they brought the problem of Nazi rule in the camps to the attention of Eleanor Roosevelt, and thence to President Roosevelt. For the past year the U.S. government had been pondering a POW reeducation program, primarily because the Soviet Union had organized its German prisoners into a Free Germany movement. In anticipation of the coming occupation of Germany, an American program was quickly approved, though it was kept secret to avoid the appearance of contravening the Geneva Conventions, which forbid indoctrination of prisoners of war.

But how were U.S. officers to distinguish "white" Germans amenable to reeducation from hard-core "black" Nazis? One German officer didn't think it would be easy. "The Americans want to know today exactly what we think. But we do not permit him this favor." Also, many Americans doubted that Germans could live by democratic rules. In his book *Is Germany Incurable?* the psychiatrist Richard Brickner argued that Nazism was a symptom of the "paranoid emotional core of German culture." To undo this psychopathology, the reeducation program sought to instill democratic values through open discussion. "Democracy," suggested one organizer, "was a way of life, not just a form of government."

Among the most pressing needs of the new Germany were policemen who respected their fellow citizens. After a decade of Gestapo terror, Germany needed a dose of Vollmer's police professionalism. Shortly after the war in Europe ended, a police school was authorized for Fort Wetherill, Rhode Island. The initial plan was to train 2,000 to 5,000 POWs in two-month courses. The brass wanted to weed out "militarists" (Nazis) and those "with definite communistic backgrounds or tendencies." Preliminary screening yielded 250 candidates for final vetting by Keeler and his team of six assistants.

The school's instructors objected that screening by the lie detector would damage the recruits' faith in the practice of democracy—especially if the examiners weren't fluent in German language or history. Major Maxwell McKnight, the program coordinator, thought Keeler had "sold someone a bill of goods." McKnight was a lawyer who had worked for the

U.S. attorney general and noted that the FBI had considered the lie detector unsuitable. He feared that its use would create a "cross-current of suspicion" among recruits.

But the army's security officers actually wanted to create an atmosphere of suspicion, and not just among the prisoners. Keeler's main sponsor—Colonel Ralph Pierce—saw the tests as a check on the American instructors. One security officer had denounced the camp commander, a former physicist, as a crypto-socialist. Another considered Captain McKnight's sympathies demonstrably "pro-communist in scope."

Keeler arrived in Rhode Island the day before Hiroshima. The three-day tests began on August 15, the day Japan surrendered. Each examiner used a translator to ask about membership in the Nazi Party and belief in Nazi principles; about communism and religious freedom; and about mental disorders, masturbation, criminal acts, atrocities, homosexuality, the elimination of the Jews, and whether Hitler was a great man. These were matters sure to confound men who had lived through a dozen years of Nazi rule. *Was* Hitler a great man? But the ambiguity of such generic questions was intentional; they functioned as what would soon be known as control questions. They highlighted, if only by comparison, a guilty person's response to a more specific question about a crime. In this instance, however, the crime was itself so ambiguous that Keeler's reports slid all too easily from political loyalty to psychological state to physical health: that is, from Nazism to anxiety about Nazism to cardiac disease.

Keeler found that 36 percent of the subjects harbored Nazi sympathies or, more to the point, considered it stressful to be questioned on this topic. Among the 126 recommended for the program, nearly 40 percent denied any affiliation with a political party and exhibited no reaction to questions about current loyalty; about the same percentage admitted membership in the Nazi Party but seemed to have no current loyalty to it; two admitted having been in the SS; and one reacted so as to indicate sympathy for communistic ideas. Of the 96 candidates rejected, half were admitted Nazis with signs of ongoing sympathy; a quarter denied any Nazi affiliation but reacted strongly about current sympathy; 15 percent reacted as loyal to communistic ideas; 10 percent were rejected as mentally unstable; and a few had bad hearts.

Those conducting the test judged it a success, not just for flushing out potential traitors, but for improving morale by enabling "those with a sincere desire to cooperate . . . to weed out the Nazis among their group." In other words, the lie detector fostered mutual trust by encouraging future cops to rat on their comrades. That this procedure did not entirely accord with the program's ethos was noted by some rejected applicants. "We often were told to feel free and we were taught upon the rights of a free man too. I think therefore, that the school should give all students the opportunity of a defense if anybody states something against another student, instead of creating an atmosphere of mistrust." The school's commander dismissed this particular prisoner as suffering from kleptomania, "a habit not desirable in a policeman."

Soon after the testing was complete, the army publicly acknowledged the reeducation program, and Keeler extolled his role to the press. But its achievements proved ephemeral. While only 6 percent of departing policemen still thought of Germans as the master race, 57 percent still blamed the war on the Jews. In the end, the police graduates were given no special role in Germany, in part because the occupation authority forbade any association among former veterans. On the other hand, this testing inspired some columnists to call for lie tests for all returning German prisoners.

For a self-styled pragmatic nation, America took an intensely ideological approach to sorting Germans. The Russians sorted their prisoners not by politics but by rank; they then dosed the soldiers with socialistic propaganda and "turned" the officers by playing to their opportunism. The British separated "whites" from "blacks" on the basis of conformism. Both the Russians and the British focused on group dynamics, whereas the Americans focused on individuals so as to identify and then transform their ideological allegiance. In a sense the lie detector had always been a loyalty monitor at one remove, probing the subjects' conscience for their own belief that they had betrayed the law, their employer, or a spouse. The political struggles of the twentieth century simply reconnected the lie detector to its core competence as a touchstone, not of the truth, but of belief.

Keeler's work screening German POWs landed him the biggest job of his career: a contract to test the workers in America's premier nuclear city. On the surface, this may seem an unlikely alliance: Keeler's humble placebo-in-a-box protecting history's most destructive device. But the lie detector and the bomb actually performed analogous tasks. After all, the bomb's potency derived as much from its symbolic force as its destructive capability. Deterrence, the name for this new doctrine, was the very stuff of which the lie detector was made.

Safeguarding the bomb, after all, meant securing not just its physical matériel, but the idea too. To protect the fissile fuel, its defenders erected guard posts and radiation monitors. To protect the nuclear know-how, they deployed mental guard posts and psychical monitors. For the latter, they brought in Leonarde Keeler, ostensibly to prevent traitors from taking fissile uranium out, but actually to remind those with nuclear know-how that the knowledge they possessed was not theirs to give away.

Whereas Los Alamos was dedicated to the assembly of heterogeneous knowledge, Oak Ridge was a city of purification. Its gigantic industrial plants distilled the tiny percentage of fissile uranium 235 out of the many tons of inert uranium 238. To carry out this task required a heterogeneous array of actors. The federal government set aside thousands of acres of rural Tennessee; major chemical corporations invested vast sums; thousands of scientists, engineers, and technical workers were recruited from far-flung urban areas; and tens of thousands of semiskilled laborers were lured from the rural South. This agglomeration was strictly managed. The races were segregated; management and labor were sharply divided; unnecessary shoptalk was prohibited. During the war, the town was denied political representation, a free press, and labor unions. When the physicist Richard Feynman visited Oak Ridge from Los Alamos, he violated security protocol when he told production engineers what they were doing so that they could do it right.

The bombing of Japan and the formal disclosure of the existence of Oak Ridge intensified this obsession with atomic secrecy in some quarters, although it convinced others that nuclear know-how ought to be made transparent. The U.S. Army wanted to control atomic know-how, and its congressional allies demanded that the government protect America's nuclear secret at all costs. The scientific elite and their allies in the Truman administration wanted to assert civilian control over atomic power; they warned that there were no atomic "secrets" as such: in five years, or fifteen at most, the Russians would master the atom. Given this inevitability, the United States should share its nuclear knowledge through the United

Nations and control the proliferation of weaponry. Nativist Congressmen denounced these internationalists as naive and disloyal. Then rumors of leaks began. In January 1946 the press learned that the Russians had acquired nuclear information from Canadian spies. The House Un-American Activities Committee hinted that a comparable spy ring was operating at Oak Ridge.

To assuage these fears, the Army Corps of Engineers brought Keeler and his team to Oak Ridge in February 1946 to determine "insofar as possible, the loyalty, integrity, reliability, mental stability, and suitability" of its nuclear workers. Keeler unearthed no spy ring, although of the 690 individuals he tested—at $13 apiece—nine admitted having "stolen product material," and seven knew someone who had done so. In each case, however, the infraction proved minor: a tiny chip of uranium removed as a souvenir or a practical joke. More common (3 percent) were unreported radioactive spills. Far more common were the shenanigans that always surround job sites: 10 percent had lied on their job application, 12 percent had stolen tools, 3 percent had used an alias at some point in their life, and so on.

Soon after, the army signed a contract with Russell Chatham, one of Keeler's team, to test workers at uranium separation plants run by the Carbide and Carbon Chemicals Corporation. Keeler did not bid on the contract; he would not have wanted to give up his high-flying image and move to Tennessee. In 1947, when the civilian Atomic Energy Commission (AEC) wrested control over nuclear power from the military, it expanded on Chatham's testing program so as to assure the nation of its diligence. The chairman of the House Un-American Activities Committee had again generated huge headlines by announcing that Oak Ridge was "heavily infested" with communists, that nuclear secrets had been stolen, and that scientists had allied themselves with the workers' union to protect traitors and keep the military out. Fighting public leaks of secret information with leaked announcements about security arrangements has since become standard procedure in American security politics, leaks being the pressure valves in the continual conflict between the public's right to know and the demands of the security regime.

By the early 1950s, Chatham and his team of polygraphers were periodically testing 5,000 scientists, engineers, mid-level managers, and laborers at Oak Ridge—including one another. Employees were tested at hiring, at regular intervals thereafter, and on departure. Employees who objected were sent to a superintendent who threatened to revoke their security clearance, making the tests voluntary in name only. Only two employees dared refuse, and they were transferred to other jobs. Because of the sheer volume of tests—1,700 in June 1951—most exams lasted twenty minutes, with two sets of questions for each subject separated by an interval during which subjects could explain ambiguities and make confessions. This mass screening was designed not so much to uncover deceit as to enforce a new form of loyalty.

> Q: Have you belonged to any organization whose purpose is not for the government of the U.S.?
>
> Q: Do you have any acquaintance or relatives which you know of who are connected in any way with any organization which is un-American?

Only a small number of subjects disclosed "derogatory" information: 2 percent in 1952, of which one-third were tagged as having "friends or relatives associated with organizations considered un-American." Yet on closer examination, those who "sympathized with the Communist movement" were citizens who supported federally subsidized housing and the Tennessee Valley Authority. It was such "liberal thoughts" as these that caused one subject to wonder if "deep down he could be sympathetic toward Communism." In his case the polygraph operator indicated that the subject had probably confused his "isms." The majority of those affiliated with un-American organizations were patriotic members of the Ku Klux Klan. The managers of Carbide and Carbon Chemicals Corporation were far more pleased by the reduction in stolen tools and other items, especially Kleenex, that were in short supply during the postwar period. More generally, the tests helped keep the labor force quiescent during the CIO's union drives, and damped down protests against hazardous releases of radiation and toxic mercury.

That the test uncovered so few "traitors" suggests that Chatham had set a relaxed or "friendly" threshold of suspicion. Polygraph operators have many ways of adjusting the trade-offs between false positives (wrongly accusing truth-tellers of lying) and false negatives (wrongly exonerating liars). So long as the percentage of spies was assumed to be small, setting a severe threshold for false negatives would hardly increase the odds of catching one, and would badly damage morale. This is an intrinsic feature of any probabilistic search for needles in a haystack. Suppose there are ten spies among 10,000 workers (surely a sufficiently paranoid guess). And suppose the polygraph has an accuracy of 90 percent (higher than any field study has found). Even if operators set out to trap half of the ten spies as deceivers, the test would net 345 "guilty" individuals, among whom the five spies would still have to be fingered, leaving 340 innocent workers wrongly accused and the five other spies entirely undisturbed. If interrogators decided to get even tougher and trap eight of the ten spies, they would need to accuse 1,606 individuals of lying, among whom interrogators would still have to pinpoint the eight true spies, while still letting two spies go scot-free.

An opinion poll conducted for Chatham confirms that the tests for the AEC aimed more at deterrence. To be sure, the fifty-nine responses are hardly to be taken at face value (the subjects' names were listed). But they do suggest what security-minded employees thought security officers wanted to hear. Several subjects praised the lie detector as "a major deterrent." They said the test kept employees "on the ball," and made a person "think twice" before doing or saying anything to jeopardize security. One officer said that the tests cut "loose talk" by "70 percent," a number Chatham blithely quoted as a demonstrable fact. The test also gave employees confidence in one another. Said one engineer: "Passing the test makes me feel better personally and feel that the organization must be an honorable one." Several wanted the test to be required of all Oak Ridge employees and to be given at other nuclear sites as well. Then the same individuals praised the machine for clearing them of guilt in their own minds too. They thanked the test for giving them "the confidence that they have not done wrong." One physicist phrased his satisfaction in charmingly physicist-like terms: "I have a personal satisfaction in passing each test, to have a recalibration so to speak." In a world permeated by suspicion, with shifting presumptions about guilt, the lie detector provided "mental relief from worry" about whether one could count oneself among the honest.

By 1952 some 50,000 tests had been administered to more than 18,000 individuals, making Oak Ridge a sort of Tennessee Eden, purifying the world's most potent product, and staffed by America's most self-assured, trusting, honest technicians.

---

Why not extend this solace to a nation anxious about nuclear annihilation? The first Americans to whom this assurance was offered were those expected to be first to experience this destruction personally: American soldiers. Because General Douglas MacArthur was eager to use tactical nuclear weapons on the Korean peninsula, the army needed to convince Americans (and America's foes) that U.S. troops had the stomach for nuclear war. At the tests code-named Desert Rock in the fall of 1951 American troops were put through a set of showy pseudo-secret maneuvers at a nuclear test site in Yucca Flats, Nevada. Their goal, according to army psychologists, was to ease the irrational fear of radiation that is "almost universal among the uninitiated." To assess (and ease) this fear, the paratroopers who carried out this first advance into a zone of nuclear devastation were subjected to before-and-after polygraph exams by Keeler's associate Paul Trovillo. Trovillo accompanied the men through their education sessions on radiation and as they marched to within 500 yards of ground zero, where they found charred earth, blasted equipment, half-blind sheep, and not much else. His results compared the soldiers' psychophysiological reactions to this experience (their "private" views) with their verbal reactions (their "public" views) and their observed behavior as recorded on film.

The findings heartened those who believed American soldiers could share the battlefield with nuclear weapons. True, the polygraph revealed that a majority had been fearful before the maneuver, though only half verbally acknowledged their fears (in language that shocked veterans of World War II). On the other hand, the troops exhibited less fear than a comparable group of paratroopers preparing for a drop, and thirty-eight out of forty-five said they'd be grateful for "atomic support" in any future combat. The tests suggested that the troops' concerns could be assuaged, at least in this stage-managed "attack."

This same therapy was then extended to the nation. The most cheery of all postwar atomic propaganda was undoubtedly a CBS radio documentary, "The Sunny Side of the Atom." Over uplifting music, the narrator (Agnes Moorehead) described a nuclear tomorrow, with "men standing straight and tall and confident, facing, without fear, their future, urged forward by a new hope, by the infinite wonder and possibility of a new life." To make sure the message got across, CBS monitored a select group of listeners on polygraphs while they pressed red or green buttons to signal their conscious approval or disapproval. This sort of audience testing, pioneered by Marston, would soon become routine practice in mass media. And it confirmed the program's success: apparently, as regards nuclear weapons, the broadcast assuaged the fears of 46 percent of the group, with only 3 percent becoming more fearful.

But every Eden has snakes who whisper of freedom. In 1953—just as Senator Joe McCarthy was threatening to use the lie detector whenever he encountered an obstacle to his search for the truth—the AEC pulled the plug on the polygraph. Though few employees at Oak Ridge had publicly vented their displeasure with the polygraph, several surveyed by Chatham had obliquely noted that "others" were hostile. One engineer admitted that more than half of the technical people felt very strongly that it was "lousy" to use the test to prosecute petty security violations, like leaving classified papers on a desk in a secure office. Some denied that the machine could read human intentions accurately; others thought that it violated the constitutional protection against coerced self-incrimination.

These scientists, moreover, were organized, and their skills gave them leverage. Shortly after Hiroshima, atomic scientists in Tennessee had formed the Association of Oak Ridge Scientists and Engineers, which joined the Federation of American Scientists to press for a more open discussion of science. While the association was grateful to President Truman for his assurance that atomic scientists would not be subject to political intimidation, it noted that several scientists had lost their security clearance for frivolous reasons, such as a chemist punished for "raising a fuss . . . over the Negro question." Partly as a result of hassles over security, the association said, 40 percent of senior physicists and chemists had quit Oak Ridge in 1947–1948. As for the lie detector, no sooner had Keeler begun his investigation than the association's national publication noted that the instrument was scientifically discredited, had been rejected by the U.S. courts, and was chiefly used as "an instrument of third-degree intimidation."

Tellingly, after years of praising the technique, Carbide and Carbon Chemicals Corporation suddenly turned against the lie detector in the early 1950s, when the company assumed responsibility for recruiting scientists for the research lab at Oak Ridge. The lie test put them in "an unfavorable and sometimes embarrassing position with respect to the other contractors," all of whom were universities.

Privately, many administrators at the AEC were also worried. Many were New Deal liberals, concerned that "violent opposition" among scientists would cause the best scientists to quit government work. America's strength lay in the open exchange of ideas. As one administrator put it, the AEC did not want to "get stuck in a Maginot line" with regard to nuclear knowledge. He feared the lie detector scared off the sort of "minds which will keep pushing back the frontiers of knowledge." One administrator described the agency's aim as "assur[ing] that in this essentially secret and totalitarian work—and I use this word advisedly—[you] . . . should bring to bear as much of the traditional democratic process as you possibly can and still retain final authority where it belongs, in the agency itself."

In December 1951 the *New York Times* published the first report that the instrument was regularly used by the CIA, intermittently by the defense and state departments, and on a quasi-voluntary basis at Oak Ridge, and cited both its potential to reduce the backlog of security clearances and the repugnance it aroused in some circles. Within a month Senator Wayne Morse of Oregon, a political maverick and a former professor of constitutional law, took the Senate floor to denounce the machine as "un-American . . . , repugnant, foreign, and outrageous," citing a prominent civilian who had been denied a post at the Department of Defense on the basis of his supposed reaction to an innuendo-laced inquiry about his sexual preference.

Within weeks the AEC had secretly convened a panel of experienced polygraph operators to evaluate the program. Though all of them praised Chatham's skill, they disparaged his mode of mass screening. Chatham, they complained, was using the machine mechanically! Fred Inbau, Keeler's close colleague, would have none of this: "the examiner draws upon the sum total of his experience. . . . Someone cannot sit down and measure [lying]." Moreover, gauging "dispositions" and intentions was particularly difficult, especially among subjects who were themselves experts. "When you get an expert," remarked another panelist, "he can never say yes or no, really, to anything, which is one of the measures of an expert." Early in 1953—just as the lie detector became Washington's symbol of the new machinery of denunciation—the AEC terminated the "experiment" at Oak Ridge.

In his brilliant tract of 1956, *The Torment of Secrecy*, the social scientist Edward Shils offered a searing portrait of McCarthyism as a crisis in the American equilibrium between public scrutiny and private autonomy. Public scrutiny kept the governing elite honest, while privacy protected the autonomous conscience at the foundation of democratic decision making. Shils saw the scientific community as a model for this sort of polity; a community dedicated to finding consensus, yet committed to autonomy of thought. America's democratic balance had been upended by McCarthyism, a movement which had redirected populist mistrust of political and financial conspirators against (among others) those atom-age longhairs who hoarded the secret knowledge that simultaneously underpinned the nation's safety. These experts also threatened the authority of the old-style governing elite. That was why McCarthy's obsession with secrecy fed on a surfeit of publicity: congressional exposés, impatience with the Fifth Amendment, and loyalty oaths.

A loyalty oath does not guarantee loyalty any more than a lie detector guarantees truth. What traitor would hesitate to swear an oath? Rather, an oath is both speech and act, a public avowal and an invitation to legal jeopardy—and for that reason, it has been shunned as a threat to the autonomy of conscience since the Elizabethan age. The oath publicly affirms society's mastery over private conscience, and binds the oath-taker to the body politic, as the pledge would have it: "[O]ne nation indivisible" (with the phrase "under God" added in 1954 at the height of anticommunist sentiment). To oblige a person to take a lie detector test served much the same purpose: affirming the subordination of the inward self to the public's demand for conformity. "In many respects," one security officer noted, "a sworn statement has the same psychological effect as a lie detector test." Like an oath, the lie detector functioned as an intensifier: heightening the subject's self-consciousness in hopes of prompting disclosure, putting the legal person at risk of further prosecution if he or she misspoke, and placing the souls of believers in mortal danger. Just as an oath disrobed the believer before God—while the secular authorities watched—so the lie detector, for those who believed in science, disrobed the subject before science, while the operator watched. In that sense, the Bible and the lie detector serve the same role in the court: as placebos, which work to the extent the subject believes they work; that is to say, as psychological "intensifiers," which raise the stakes of each utterance, the one by reference to the all-seeing eye of God, the other to the all-seeing eye of Science.

CHAPTER 16

# Pinkos

*Such a hypothetical Mind Resonating Organ, by adjusting itself to the Fields emitted by other minds, could perform what is popularly known as "reading emotions," or even "reading minds," which is actually something even more subtle. It is but an easy step from that to imagining a similar organ which could actually force an adjustment on another mind.*

—ISAAC ASIMOV, *SECOND FOUNDATION*, 1953

THE WATCHWORD OF THE EARLY COLD WAR WAS VIGILANCE. The great fear in those grim years, among both conservative nativists and liberal internationalists, was that America would be lulled into complacency. The cold war was a battle of wills. This was a manly form of combat. The will to wait—containment. The will to take a stand—dominos. The will to absorb a blow and strike back—deterrence. And above all, the will to keep secrets while uprooting conspiracies—J. Edgar Hoover. It was an age anxious about the discrepancy between surface and depth and the demands of self-mastery. There were so many ways the nation's self-control might be sapped, not just by covert enemies but also by unacknowledged personal and sexual weaknesses. No wonder America fretted about its subconscious. This was a golden age of pop psychology. The social critic Dwight McDonald dubbed it the "lie-detector era."

Nearly 400 years had passed since Queen Elizabeth I had declared that her majesty did not "make windows in men's souls," even though the sectarian conflicts of the Reformation and Counter-Reformation were pushing the sovereign powers of Europe to assert their mastery over inward belief—

extending the definition of sin from deed to desire, and redesigning confession to elicit a new sort of conscience. So too, under the strain of their twentieth-century conflict with Nazism and communism, Americans began to monitor their fellow citizens for signs of secret deviance. In such a battle, how to distinguish friend from foe? The enemy might come disguised as a foreign ally, as a golden exemplar of American success, or even—who knew?—as one's own unacknowledged self. Here the lie detector came into its own: a psychological strip search for America. At stake were such mundane matters as domestic political advantage and control over the direction of U.S. foreign policy. One unanticipated consequence was to transform personal identity—including sexual identity—into a new kind of political claim.

Running parallel to the atomic arms race of the cold war was a psychiatric "minds race." If the lie detector was a mirror in which America faced its own fears, then when cold warriors looked in the mirror, they saw Soviet mind control. At the CIA, where each employee was required to take a polygraph test, the agency suspected the enemy of developing techniques of scientific interrogation that bore an uncanny resemblance to those America was developing: the lie detector, sodium amytal, LSD, hypnosis, electroshock treatments, lobotomies.

The CIA's covert Bluebird and Artichoke programs of the early 1950s used intensive polygraph exams as a "cover" to initiate more coercive interrogation techniques with suspected enemy agents. Initially justified as a way to help CIA operatives defend themselves, these programs quickly expanded to include offensive methods. Polygraphers were to double as hypnotists and monitor the effects of LSD and other drugs on the truthfulness of subjects. Under the guise of these programs the CIA reached beyond the lie detector to create a highly classified arsenal of new but related techniques to bring acute psychological pressure on subjects, including such methods as sensory deprivation, self-inflicted pain, and sleep deprivation. Designed to break the personality of the subject, rendering him dependent and compliant, these forms of psychological torture constituted a crucial component of the CIA's interrogation tool kit during the cold war, and have been quickly adapted to the current "war on terror."

At the same time, Americans were given chilling accounts of how advances in psychological science let communist regimes extract dramatic public confessions from their prisoners and former opponents. Popular books like *Brainwashing*—written by the journalist and covert CIA operator who invented the term—described how communist psychologists had advanced from studies of Pavlov's salivating dogs to new techniques of hypnotic mind control. In reality, neither the Soviet Union nor Nazi Germany before it saw any need for the lie detector—as the CIA secretly acknowledged. Totalitarian governments brook no impediment to their control over the bodies of their subjects, not even the minimal restraint that justice appear fair and machinelike. Thus it was not until the U.S. occupation that the German military tentatively adopted the lie detector. Similarly, the Russians never used the lie detector until the Soviet Union fell. The only place the lie detector turned up in Soviet Russia was in spy movies when steely KGB agents outwitted the American device.

Ironically, in both the Soviet Union and the United States during the 1950s, psychologists favored behaviorist explanations of human action that gave little or no scope to an autonomous will. As one British author pointed out, the American lie detector elicited confessions much as the Soviet Union converted people to its cause, or as religions did: by commanding loyalty through control of emotional responses. Of course, a common psychological theory need not produce equivalent outcomes. Psychiatric hospitals in the Soviet Union were horror houses where dissidents were regularly detained and tortured until they confessed. Nothing quite like that existed in the United States, though the CIA would soon be covertly exporting its methods of psychological torture to Vietnam and Latin American regimes. The parallel was there. Civil libertarians like Stewart and Joseph Alsop warned that the lie detector test would erode psychological privacy to the point where "we begin palely to imitate the system we fear."

In that sense, the history of the lie detector during the cold war shows how closely its fate was bound to a conflict within American society over the balance of civil liberties and national security. The lie detector is a kind of hypocritical homage that democracy pays to justice: a public facade of machinelike fairness, which partially constrains the unequal treatment within. This, as George Orwell noted, is the minimal constraint that public ideals impose on justice in liberal democracies, as compared with totalitarian regimes: "An illusion can become a half-truth, a mask can alter the expression of a face. . . . In England such concepts as justice, liberty and objective truth are still believed in. They may be illusions, but they are very powerful illusions. The belief in them influences conduct, national life is different because of them."

Americans were so uncertain about their ability to distinguish foreign friend from foe that they even began to distrust their own interpreters. In 1950, in the midst of the Korean War, the army decided to bring in Keeler's former partner, George Haney, to test its translators for communist sympathies. The presumption here, as with Keeler's German POWs, was that the lie detector, because it read underlying physiological reactions which transcended the subtleties of language, would enable Americans to assess otherwise indecipherable foreigners. Not that anyone had tested the ability of a white American man to pull this off. After evaluating the honesty of twenty-five translators—using, of course, a Korean translator to pose questions and interpret answers—Haney recommended that several be discharged. He also recommended that a team of fifty polygraph experts "sort" through the massive influx of enemy prisoners, following the model of Keeler's operation with German POWs. (The polygraph would be also used periodically in Vietnam.)

By 1953 the army was polygraphing its own members on a regular basis—something it had done only in criminal cases during World War II. It began with personnel at nuclear facilities, then expanded to those in intelligence work and certain administrative jobs. After Senator Morse's outcry, the Department of Defense agreed to ban the device, only to backtrack and reinstate it. Soon the department was buying the Stoelting Company's Deceptographs in bulk. This expansion meant big business. Every employee at the army post exchange in Washington was tested, in order to stop pilfering. Every sailor on the *U.S.S. Pritchett* was tested, to learn who had damaged onboard machinery. By the late 1950s the army was using the lie detector in the way Keeler had pioneered at Chicago's banks, to ensure honesty and loyalty in a hierarchical organization.

The paranoia of this era took in more than foreign friends and one's own comrades; Americans even began to doubt whether they could trust themselves. America's anxieties about its own wobbly will were intensified by the brainwashing scandal of the Korean War—the inspiration for *The Manchurian Candidate.* After the cease-fire of 1953 many Americans were shocked to learn that one-third of the 4,000 American POWs in communist hands had apparently collaborated with the enemy. Conservatives claimed that this capitulation was unprecedented, and they saw it as evidence that tolerant democratic brains lacked the mental toughness to resist totalitarian mind control. In fact, follow-up studies found that the collaboration rate had been much exaggerated. Many GIs had just been "playing it cool," feigning compliance to outwit their captors. At the time, however, Americans were given reasons to doubt that they could trust their own minds.

By chance, the American prisoner who became most celebrated for resisting indoctrination was Major General William F. Dean, the highest-ranking officer captured in Korea, a winner of the Medal of Honor, and a former college cop in Berkeley who had been the best man at John Larson's wedding. Dean had been the first fellow officer Larson had trained in the art of lie detection, yet even he had signed an ambiguous statement asking his own side to confine its attacks to military targets. On his return from Korea, he publicly supported Larson's efforts to end abuse of the technique. By then, however, politicians had discovered the instrument's value.

Richard Nixon deserves credit for first discovering the potential of the lie detector as a prop of political theater. The instrument made its debut—with Keeler waiting in the wings—during the high drama of the cold war known as the Alger Hiss affair. The controversy began in 1948, when Whittaker Chambers, an unsavory editor at *Time* magazine who was a repentant former communist, accused Hiss, a former golden boy of the State Department, of being a member of the Communist Party. Hiss angrily denied the charge, and President Truman bluntly dismissed it as a "red herring."

Ordinarily, Chambers—a rumpled, overweight ex-communist (and rumored homosexual)—could not have matched the credibility of Hiss, a dapper diplomat who had been Roosevelt's aide at Yalta and currently served as president of the Carnegie Endowment for International Peace. Yet after listening to both men testify, one member of the House Un-American Activities Committee concluded that "whichever one of you is lying is the greatest actor that America has ever produced." This was to underestimate a first-term congressman from California. Richard Nixon had found a tool to unmask those traitors who seemed to have reputation and demeanor on their side. To probe beneath the surface of establishment respectability took courage—and a flair for populist politics. This was a job for the great equalizer: the lie detector.

Nixon set his trap carefully. In later years he would claim that he had been on the verge of dropping the investigation when the journalist Bert Andrews privately suggested challenging both witnesses to a lie-detector test. On August 7, as Chambers's closed-door testimony was ending, Nixon dramatically asked him if he would be willing to repeat his accusations on a polygraph.

> CHAMBERS: Yes; if necessary.
> NIXON: You have that much confidence?
> CHAMBERS: I am telling the truth.

Only then did Nixon place a call to Leonarde Keeler in Chicago. He queried him about the polygraph and Keeler agreed to "fly out and assist the committee" if that became necessary.

A week later, when the committee brought in Hiss, Nixon asked him the same question. Hiss temporized, lawyerlike, demanding to know in advance who the examiner would be. This may have been good lawyering, but it was bad political theater. Nixon responded that Leonarde Keeler, "the outstanding man in the country," had agreed to fly to Washington, D.C., to conduct the tests. Hiss agreed to get back to the committee with a reply.

The next morning Nixon's challenge, which had been leaked, was reported on the front pages. The day after that, Hiss affirmed his refusal to take the test, explaining that whatever else the machine was, it was "*not*, however, a 'lie detector'"; that it had not been scientifically validated; that no federal court had accepted it; and that the FBI doubted its validity.

In the next weeks, elite liberal opinion rallied behind Hiss and waxed indignant against trial by polygraph. An editorial in the *New York Times* dismissed Keeler's polygraph as little more than a gauge of human emotions and cited his nemesis: "In the opinion of Dr. John A. Larson, the lie detector is not to be trusted." The *Washington Post* likened it to a medieval trial by ordeal, noting that even Keeler considered his own machine only "82 percent accurate."

Feelings in the conservative camp ran just as high. In later years Nixon recalled that Hiss's evasiveness had swung him against the accused. Certainly this evasiveness helped dramatize the conflict in populist terms, pitting unrehearsed frankness against elitist shilly-shallying. Decades later Nixon could be heard on the White House tapes urging his subordinates to use the lie detector to flush out leakers: "I don't know anything about polygraphs, and I don't know how accurate they are, but I know they'll scare the hell out of people." As for Chambers, he was so eager to be vindicated that he arranged to be interviewed on television while hooked up to a polygraph. Only with difficulty did the FBI and his lawyers dissuade him.

Nixon's strategy launched the political career of the lie detector. Nixon had discovered that no one actually had to take the test for it to generate a headline. The box worked even when no one was hooked up to it. By the time Hiss was convicted of perjury in January 1950, Nixon's innovation had been adopted by Senator Joseph McCarthy.

Soon after he delivered his notorious speech of February 1950 in West Virginia, alleging that 205 communists worked for the State Department, McCarthy began to issue challenges indiscriminately, urging use of the lie detector. Witnesses who presented conflicting testimony to Congress should take a lie-detector test, as should anyone who sought a "sensitive" job in the federal government; and so on, until his final showdown with the army, when he took the theater of ordeal to its logical conclusion. For years critics had taunted McCarthy by saying that he himself ought to submit to the lie detector. Now he offered to do just that. He had complete confidence in the test, he said, having used it to resolve cases since his days as a county judge in Wisconsin in the 1930s (a plausible claim, given Keeler's successes in the Appleton area at the time). As usual, though, McCarthy was bluffing; he never took the test. But while he bluffed his way to ignominy, McCarthyism took root in America.

Arousing fears of secretive internal enemies has long been a potent political weapon. But denouncing communist infiltrators had one major drawback; notwithstanding certain causes célèbres, honest-to-goodness American communists were actually quite scarce in the government. Even expanding the witch hunt to include "communist sympathizers" did not promise to snare many more. Far more promising was the campaign against sexual deviants. The so-called "lavender scare" of the late 1940s and early 1950s intensified when such allies of McCarthy as senators Kenneth Wherry and Styles Bridges sought to bolster his campaign by alleging that the U.S. government was riddled with homosexuals. This apparent non sequitur had one huge advantage: there actually were homosexuals working for the U.S. government. Furthermore, this was a witch hunt the federal bureaucracy was willing to pursue. Thus, even though Deputy Undersecretary John Peurifoy indignantly repudiated McCarthy's charge that the State Department harbored 205 card-carrying communists or had shielded Hiss, he proudly admitted that it had recently purged ninety-one homosexuals as "security risks."

Denunciations of homosexuals played on many of the same fears as denunciations of communists. Homosexuals were said to pass unnoticed, collude by secret signs, conspire to recruit others to their cause, and sap the moral strength of the nation. Some even suggested that homosexuals, like communists, possessed unusual mental abilities that they had turned to evil purposes. In suggesting that homosexuals had infiltrated the federal bureaucracy, the nativists were attacking the civil service as effete, ineffectual, and decadent. This was grist for down-home voters concerned about the rising power of mommy-knows-best government. It also allowed them to impugn the manliness of the patrician "striped-pants" elite of Truman's foreign policy establishment, led by Dean Acheson. If the cold war could be won only by force of will, then America was as vulnerable to the depredations of limp-wristed government as it was to deliberate treachery. Many high-profile accusations of political disloyalty also involved thinly veiled accusations of homosexuality.

The national chairman of the Republican Party, writing in support of Senator McCarthy, denounced "the sexual perverts who have infiltrated our government" as "perhaps as dangerous as the actual Communists." To be sure, a homosexual was not a traitor so much as a "security risk," a hazy category which nominally embraced anyone who might betray the nation's secrets because of some moral defect: blabbermouths, alcoholics (because they were blabbermouths), and those susceptible to blackmail because they engaged in illicit sex. Ostensibly, the last group included heterosexual adulterers. In practice, though, only homosexuals were purged, despite the fact that no one could name a homosexual who had ever been blackmailed into betraying a state secret. As some commentators noted, if homosexuals were vulnerable to blackmail, it was because of the purges.

Lending urgency to these accusations was the discovery that America was overrun by homosexuals. The Kinsey Report of 1948 had presented some jaw-dropping numbers: while "only" 4 percent of American men were exclusively homosexual, 10 percent had been homosexual for at least three years, and 37 percent of all males had experienced at least one homosexual act since puberty. It was enough to make the novelist John Cheever wonder, "Is he? Was he? Did they? Am I? Could I?"

In fact, the expansion of the civil service since the 1930s had attracted many homosexuals to Washington. Lurid exposés like *Washington Confidential* (1951) warned that in addition to the "90 twisted twerps in trousers . . . swished out of the State Department" there were 6,000 homosexuals living covertly on the government payroll. When the son of the Democratic senator Lester Hunt was arrested in Lafayette Park on a morals charge, the Republican senator Styles Bridges threatened to publicize the case if Hunt did not halt his campaign for reelection; Hunt vacillated before committing suicide in his Senate office.

But purging the U.S. government of these deviants would be no simple matter. Even the authors of *Washington Confidential* admitted that it would not be possible to "stop at every desk and count people who appear queer." How would perverts be identified? What were the telltale signs? Vigilant citizens had their own methods. Blanche Blevins, an unmarried fifty-five-year-old secretary in the State Department, suspected her own boss of lesbianism because the boss's best friend had "peculiar lips, not large but odd shaped." Miss Blevins also tipped off security officers to men at the State Department with a feminine complexion or a girlish walk. But were these reliable signs? Indeed, what made a person a homosexual, anyway? Was homosexuality a matter of deed or desire? And if deed, how often, and at what age?

To answer these pressing questions, the U.S. Senate, under pressure from Republicans, authorized an investigating committee in 1950. Behind closed doors Senator Margaret Chase Smith asked psychological experts if there was a "quick test like an X-ray that discloses these things?" She was discomforted to learn that experts did not consider homosexuals deficient in any visible way. Indeed, the committee report would conclude, "Most authorities believe that sex deviation results from psychological rather than physical causes, and in many cases there are no outward characteristics or physical traits that are positive as identifying marks of sex perversion." Nonetheless Senator Karl Mundt seized on a navy psychiatrist's admission that the lie detector might offer a way to peer beneath the surface of ordinary behavior to catch the essential difference (not that the psychiatrist advocated such tests). In fact, the test had already been used at the State Department to "out" recalcitrant gays. One applicant in 1947 who aroused suspicion because of his "mannerisms and appearance (use of perfume etc.)" initially denied that he was a homosexual; but when he was put on the polygraph, he confessed to homosexual activity and was promptly denied the post.

During the Truman administration, however, the professionals in the State Department cautioned against such methods. "Unless scientific tests can be devised and adopted (this is not a recommendation), it should be expected that the problem [of homosexuality] will continue to be with us." But after Eisenhower took office in 1953, R. W. Scott McLeod, an ex-FBI man with close ties to senators McCarthy and Bridges, was detailed to the State Department to intensify the purges. He put two of his security officers to work full-time on the problem. These investigators were not to halt their inquiries just because a staff member was married or professed "ardent Catholicism." Homosexuals were known to adopt conventional social and sexual guises. Instead, investigators assembled a full dossier, interviewing all male applicants, inquiring after "hobbies, associates, means of diversion, places of amusement." They were to take note of any unusual traits of speech, appearance, or personality, including a "jelly hand shake" or "feminine complexion." The officers then combed through the applicants' school records, police records, and lists of acquaintances. "Look into his period of studying costume design," noted the file of one junior diplomat stationed in Vienna. On the assumption that homosexuals recognized one another, admitted homosexuals were pressured into outing others. Once a damning dossier had been assembled, the individual was confronted with the charges, whereupon, according to the snoops, 80 percent confessed. Those who continued to deny their perversion were invited to "clear their name" with a lie detector test. Thanks to these methods, McLeod's team soon boasted that they were firing one pervert a day.

The lie detector had been used in sex cases from the beginning. In Berkeley in the 1920s, Larson had used it to extract confessions from homosexuals, though these were men accused of specific criminal acts. In Chicago in the 1930s, Keeler had deployed it to help a local sociologist chart the homosexual world, though less to intimidate subjects than to map sexual mores and press for legal reform. In forcing closeted homosexuals to declare themselves, security officers took advantage of the instrument's ability to blur the distinction between deed and desire, and the question of whether once was enough. This left the definition of homosexuality to be filled in by the subject—a matter of self-proclaimed identity, albeit one shaped by the subject's fear of what he believed the examiner believed to be the case.

Keeler was among those who adapted his technique to this new quarry. There were two ways to trap homosexuals. One was for the interrogator to discuss homosexuality calmly with the subject in advance of the test, explain that the question was merely for "security" reasons, then invite him or her to volunteer any information up front. From there, it was relatively easy to lure individuals into confessing complicity by degrees. If the subject denied being a homosexual, the operator could point to a bump in the chart and ask for a clarification. The subject would often then recall an experience he "just happened to remember," which occurred when he was, say, seven. The operator would then readminister the test, and show the subject where the telltale bump had reappeared. Suddenly the subject recalled that "darn if it didn't happen when he was seven and a half." This pattern was simply repeated until "you work him on up until last week and there you are."

Equally insidious techniques could be used when the interrogator wanted to snoop for homosexuals "without mentioning sexual perversion." Keeler trained his students to use his "peak of tension" method to expose homosexuals by slipping among their questions a word "loaded" for gay men in a way it wasn't for "normal people." For instance, if the subject was read the series "buy . . . , sell . . . , trade . . . , donate . . . , borrow" and reacted to "trade," he was probably gay because "trade" was a code for casual homosexual pickups. Of course, he might also trade stamps; or work at the Chicago Board of Trade, where he feared being accused of insider trading;

or simply be familiar with homosexual slang and worried that this line of questioning meant he was suspected of homosexuality.

Other scientific probes sought to expose sexual dissimulation in an analogous manner. The Royal Canadian Mounted Police experimented with a device that monitored the diameter of the subject's pupil while he was shown pictures of naked men. Unfortunately, simply shuffling the pictures dilated the pupil more than the most alluring image. So in the 1960s, the Mounties skipped straight to the source, fastening a fluid-filled tube around the subject's penis to register tumescence while he viewed lurid images of men, women, and children. To this day, the penile plethysmograph is used throughout North America to assess sex offenders, and U.S. courts have allowed prisons to make release or parole conditional on passing such tests. This polygraph of arousal, for all its crudity, simply inverts the assumptions and fallacies of the standard lie detector, so that bodily reaction equals desire, and that desire equals criminal propensity. This sort of reverse inference is a species of what Alfred Whitehead identified as the "fallacy of misplaced concreteness," as if a specific response to a particular stimulus can be read as implying a specific cause or deed, when multiple factors could be responsible—for instance, the experience of having something strapped around one's penis.

Yet this scientific aura gave a pretense of respectability to purges that were intended to shape American foreign policy. When Eisenhower appointed the Sovietologist Charles Bohlen, a man of ambiguous sexuality, as ambassador to the Soviet Union, the McCarthyite "primitives" in Congress counterattacked by accusing both Bohlen and his brother-in-law, Charles Thayer, of homosexuality. Bohlen refused McCarthy's challenge to take the test, and the administration backed his decision, not wanting to risk failure. But Thayer, a patrician foreign service officer, had already taken an exam, during the course of which he had made one of those partial admissions that the lie detector so readily induced: an Afghan boy "might" have once performed a sex act on him in the 1930s. Knowing that this information was in J. Edgar Hoover's file put Thayer in a state of "anguish, revulsion, and emotion." He resigned, knowing that if he persisted, McCarthy would use the report to "make mince meat" out of him.

The State Department insisted publicly that all tests were voluntary. But internal memoranda explained how easy it was to corner personnel into taking them. Security officers in the Miscellaneous Morals unit boasted that no subject had "beat" the machine, and that seventy-four out of seventy-six tests had investigated "morals" charges (i.e., homosexuality). Of the hundreds of men and women forced out of the State Department between 1945 and 1956, only one refused to resign after confessing to homosexuality, and he was forcibly terminated. In all, some 1,000 accused homosexuals lost their jobs in the State Department, and perhaps as many as 5,000 in the federal government as a whole.

Surprisingly, the most potent objection to the lie tests came from the FBI. J. Edgar Hoover had a long-standing antipathy to the lie detector. This was not a matter of hostility to scientific police work in general. On the contrary, Hoover had followed Vollmer's lead, brandishing ultramodern forensic science to underscore the incorruptibility and efficiency of his G-men. And certainly Hoover considered the ongoing struggle against political subversion to be a battle against lies. "More than anything else Communists fear truth—for deceit is their strategy, and lies and chicanery are their tactics." Moreover, he worked strategically behind the scenes to expose homosexuals. Yet Hoover dismissed the lie detector as little more than a "psychological aid" in public hearings before Congress, in private audiences with U.S. presidents, in handwritten memorandums to his staff, and in correspondence with Joe McCarthy. Why?

Since the mid-1930s Hoover had kept close tabs on the device he always called "experimental." In 1935, he ordered the special agent E. P. Coffey to spend a week in Keeler's lab training on the Keeler Polygraph and observing bank cases. (Hoover had always been envious of the Northwestern University crime lab for getting the jump on his own.) Initially skeptical, Coffey had returned to Washington persuaded that the instrument produced results, albeit only in the hands of a skilled operator such as Keeler. This conclusion confirmed Hoover's suspicion—corroborated by his correspondence with John Larson—that the lie detector gave the operator wide discretion over the findings.

Still, the FBI did purchase a Keeler Polygraph and Coffey conducted experimental tests on his fellow agents, with mixed results. During the 1940s the FBI even secretly began to use the test in selected cases of embezzlement and sabotage—though it deliberately concealed this fact from both the public and other government agencies, lest they come to expect an exam in all cases, or worse, demand to know if an exam had exonerated a person the bureau wished to charge. Then, in the early 1950s, the bureau abandoned the machine altogether. Hoover thought its results particularly misleading in screening for political loyalty, a situation where many emotions came into play. After all, to the extent the machine was believed, it offered an alternative to Hoover's own machinery of accusation, which he preferred to operate outside public scrutiny. Tellingly, he always insisted that the machine was ineffectual against sexual deviates. "I personally would not want to accept solely the evidence of what the operator of a lie detector says the lie detector shows in proving that a man was or was not a sex deviate." And no wonder, given the rumors, whether true or false, of his own proclivities.

In the end, the "lavender scare" ruined more lives than the "red scare," and proved an equally potent weapon in the battle to shift America's domestic politics and foreign policy toward the machismo right. The historian Robert D. Dean has plausibly suggested that the foreign policy establishment was so eager to display its manliness that it marched the nation into Vietnam. And the historian David K. Johnson has suggested that one unintended consequence of the lavender scare was to push the American homosexual community to organize itself politically for the first time. It is no accident, he notes, that Frank Kameny, founder of the nation's first publicly active gay group, was both a scientist and a civil servant who lost his job in the purge. The lie detector was, of course, no more than a tool in these larger struggles, and an ambiguous one at that. The instrument bred suspicions as fast as it assuaged them, creating suspect foreigners, suspect friends, suspect selves. In the process, the lie detector blurred the differences between formal crimes and personal sins, political views and sexual identity. In the end, everyone was a potential subject of suspicion: after all, we're all pink inside.

CHAPTER 17

# Deus Ex Machina

*He was of no real importance, of course. Just a human being with blood and a brain and emotions.*

—RAYMOND CHANDLER, *THE LONG GOODBYE*, 1954

THROUGHOUT HIS CAREER, LEONARDE KEELER HAD SERVED authority. Though he never wore a uniform or drew pay from the government, he had consulted for police and prosecutors, seconded corporate managers, and subcontracted himself out to the U.S. Army. Not an organization man himself, he had labored at the behest of organization men, with this one proviso: he was free to exonerate as well as condemn. Now, in the postwar period, in business for himself, Keeler began to offer a new service: absolution. From across America, the letters poured in.

Some of the writers were troubled by a nameless guilt. "No one has ever accused me of doing anything wrong," confided one seventy-four-year-old woman, "but for a long time I have felt that my movements were watched with suspicion." She dated her troubles to the day, thirty years earlier, when she had "picked up an empty glass case by mistake." Now she felt under arrest in her own room. Might the lie detector absolve her? An elderly man wanted to take a lie detector test to prove to his psychiatrist that he had never worked for the Secret Service.

For decades petitioners had sought Keeler's exoneration for crimes, real and imaginary. Now their fellow citizens longed to hear tales of vindication. To assuage that longing Keeler joined in 1949 with two new pals, Erle Stanley Gardner and Raymond Schindler—America's top-selling novelist and its most famous detective—to found the "Court of Last Resort," an exculpation service for a Kafkaesque age. When police and prosecutors presumed your guilt, when law courts favored the rich and powerful, when politicians ignored your plight, you could still appeal to the Court of Last Resort and its famous lie detector.

Leonarde Keeler had always operated the lie detector like a deus ex machina, an oracle whose judgments descended on the scene of human entanglement like a mechanical god descending from the rafters: condemning the guilty, absolving the innocent, and making retrospective sense of the drama. The lie detector resolved ambiguity by shoving it into a box. This form of justice has never satisfied the critics. Aristotle hated the device of the deus ex machina. He held that the resolution of a drama had to emerge from within the logic and conflicts of the plot itself. The courtroom trial is one such drama, in which the audience—the jury—sorts through the crosscutting motives and actions to render final judgment. But courtroom justice is expensive, delayed, and untidy. On behalf of millions of Americans disappointed by legalistic wrangling, the Court of Last Resort summoned two gods of our own day—science and publicity—to render final judgment. In this capacity Keeler won his greatest public fame, without himself receiving absolution.

Keeler's first step toward the summit of celebrity began with the case of Harry Oakes. In the midst of World War II, the press barons nevertheless found room on their front pages for the scandals of the rich, famous, and aristocratic. In 1943, hundreds of correspondents descended on the Bahamas to cover the trial of Count Alfred de Marigny, accused of bashing in the skull of his multimillionaire father-in-law, Sir Harry Oakes. Oakes had been the owner of the western hemisphere's wealthiest gold mine, and an intimate friend of the duke of Windsor, formerly King Edward VIII, who had abdicated the British throne and was now royal governor of the Bahamas. Erle Stanley Gardner, creator of Perry Mason, sent by the Hearst papers to get the "behind the scenes stuff that other writers can't get," called it the "greatest murder mystery of all time" and "the most intense drama of the century."

On the morning of July 8, 1943, Oakes's friend and real estate agent Harold Christie had discovered Oakes in bed, his head crushed and his body ritualistically burned. The duke of Windsor delayed the public announcement long enough to assign the case to a pair of detectives from Miami—his sometime bodyguards—who pinned the blame on Marigny, a twice-divorced French roué who had married Oakes's svelte, redheaded daughter Nancy two days after her eighteenth birthday. Father and son-in-law were known to detest each another.

Yet Nancy stood by her husband. She hired America's most famous detective, Raymond Schindler, who in turn hired Keeler. The sleuths lodged with Nancy's friend from boarding school, the baroness Marie af Trolle, where they entertained reporters. Nancy wanted the detectives not just to hunt for scientific clues but to "turn around the bad publicity . . . and see that the truth is told."

Gardner's job was to guide millions of readers through the legal stratagems, "just as Perry Mason would do if he were summing up the case." The challenge for the defense, as Gardner saw it, was to explain Marigny's fingerprint in the Oakes mansion, which Marigny claimed not to have visited in two years. In his first dispatch, Gardner promised an interview with Professor Leonarde Keeler, "THE MAN WHO READS MINDS WITH A BOX." Keeler, announced Gardner, had come to test Marigny on his lie detector—in open court! Of course, as the British courts did not accept the lie detector as evidence, Keeler would simply offer to test Marigny in open court, a challenge that Marigny would surely accept.

Keeler never issued his challenge. John Larson, hearing of this plan for "grandstanding," spent several hours on the phone, briefing the prosecutors on the instrument's flaws. But Keeler's other forensic skills did help win the day for the defense. He proved that the copy of Marigny's fingerprint taken from the crime scene could not have been lifted from a Chinese screen, as the American detectives claimed, since the image lacked the telltale background design.

The lie detector had been denied its day in court, but the court of public opinion would not be disappointed. Like a gun waved conspicuously in the first act, the device had to go off before the final curtain. Two days after Marigny's acquittal, at a victory celebration at the home of the baroness af Trolle—by now passionately in love with Keeler—the machine was brought out for its delayed performance. Keeler was playing the card-guessing game with the ladies, the newspapers were informed, when

Marigny, seized with the desire to clear his name, overruled his lawyer's scruples and volunteered to be put on the machine. Wonder of wonders, reported the newsreels, he was cleared by the box.

The case lifted Keeler to a new level of American stardom. It was the apotheosis of the lie detector, the moment when it rose above its quasi-scientific origins and ascended into the realm of pure publicity known as celebrity. Back on the mainland, Keeler received the full three-part treatment in the *Saturday Evening Post.* "The Magic Lie Detector"—by the first writer to win a Pulitzer Prize for science journalism—was an ode to the miraculous potency of the polygraph, a triumph for science as superstition. Then Keeler reached the apogee of fame: he was asked to play himself in a Hollywood movie.

In 1948, Twentieth Century Fox released *Call Northside 777,* starring James Stewart. It was based on the real-life vindication of Joe Majczek, a Chicagoan who had spent eleven years in Joliet for the murder of a police officer during a holdup at a delicatessen. Vera Walush, the proprietress and sole eyewitness, had identified Majczek and his partner Theodore Marcinkiewicz as the killers; the jury disbelieved their alibis; and in 1933 the judge sentenced both men to ninety-nine years in prison. Eleven years later, the reporter James McGuire of the *Chicago Times* learned that Majczek's mother was scrubbing floors to assemble a $5000 reward for evidence to prove her son's innocence. What began as a human interest story became a newspaper crusade to exonerate Majczek.

McGuire learned that the trial judge (since deceased) had had doubts about the case; that no other witness had identified the two men; that the "delicatessen" was a speakeasy; and that Walush, who had been hiding in a closet during the robbery, had twice failed to pick out the two accused men in a police lineup before changing her story the next day, apparently under pressure from the police. Indeed, the police had altered the date on Majczek's arrest sheet to cover up this switch. Furthermore, the killing had occurred before the Chicago world's fair, so the cops themselves were under intense pressure to solve the case quickly. Still, there was no proof that Majczek was innocent until McGuire arranged with Keeler for Majczek to take a polygraph test in November 1944. The morning headline announced the machine's judgment: "Lie Detector Clears Joe."

It was the 1930s all over again—the lie detector versus police corruption—but this time with the lie detector playing defense. On August 15, 1945, Majczek was pardoned by the governor and was given $24,000 in compensation, plus $1,000 from Fox for the rights to his story. The movie took the form of a film noir documentary, advertised with the line "Every word is true." Shot on location, in rancid city neighborhoods, it touted the superiority of the independent investigator over the sclerotic formality of the law, and impartial scientific evidence over self-interested testimony. The movie gazed into the shadowy mirror of noir paranoia: the story of an innocent man who is incarcerated—and finally exonerated, though no thanks to the official guardians of the law.

Keeler's lie detector makes its entrance at the fulcrum of the plot. The scene opens with a close-up of a bulky apparatus of knobs and dials—"a formidable looking-machine," says the screenplay—while a husky voice explains that "[t]he only thing the machine is for is to record the emotional reactions of an individual." The camera then rises toward the large blond man in the wide lapels who is instructing the reporter while he lovingly adjusts the machine with supple fingers. Keeler is not movie-star handsome, but real-life handsome, if a tad haggard, with a broad body, a reassuring Nordic face, and hair flipped neatly to one side. His explanation done, the reporter leaves and the prisoner is brought in. He is a small man, visibly nervous and obviously "ethnic," but not shifty-eyed: a decent man who has gotten a bad rap. With clinical courtesy, Keeler invites the ethnic man to take a seat and relax. Then he hooks him up to the machine and asks him to pick a card, any card.

As the interrogation begins, we watch the prisoner's chest rise and fall, his eyes dart left and right, his nostrils flare, his fingers twitch, and the machine's needles dance across the page. It is an intimate moment, and the camera delicately averts its gaze. By the time it returns to the interrogation room, the session is done. Keeler lights a post-interrogation cigarette and passes it to Majczek before releasing him from the chair and dismissing him. The two reporters rush in.

FIRST REPORTER: What's the verdict?

KEELER: Well, there's the record.

FIRST REPORTER: What's that jump there?

KEELER: Well, he reacted in all three curves. Very specifically. He lied to that question.

SECOND REPORTER: Is that where you asked him if he killed [the cop]?

KEELER: No. "Are you married?"

FIRST REPORTER: Well, but he didn't lie. He isn't married. He's divorced.

KEELER: Yes, but he's a Catholic and he still thinks he's married and he feels within himself that he's married. So he reacted with deception.

It was as if the machine had read Majczek's soul. The next scene has honest Jimmy Stewart typing the only possible conclusion: "PASSES LIE TEST."

But in real life, Majczek was back in trouble, and once again calling on the services of the lie detector. This is the problem with summoning a deus ex machina: let it descend from the rafters but once, and there will be no end of human mysteries it will be asked to resolve. According to Majczek, the state legislator who had sponsored his restitution had demanded a $5,000 cut for his pains. "[A] lawyer would ask for a bigger fee," he'd allegedly said. This the legislator denied, and demanded that Keeler give him a lie detector test to clear his name. When Keeler refused, the legislator turned to Keeler's nemesis, Orlando Scott, who posed the legislator in his "brain-wave detector" for an exclusive in the *Chicago Herald American*—and found him truthful too. This prompted Majczek to turn to Keeler for another test, to be published in the *Chicago Times,* to prove he was telling "the entire truth"—whereupon the legislator responded that he "didn't believe in the lie detector" anyway.

Then Theodore Marcinkiewicz, Majczek's partner, still in prison, demanded that Keeler test him too. His test indicated that he was lying, even after he admitted stealing cars, robbing government trucks, casing Walush's speakeasy, and pulling a previous crime with Majczek (who, it now turned out, may have done time for an earlier robbery). As none of these episodes meshed with the reporter's story line, none was reported. Two years later Marcinkiewicz was also freed.

---

The publicity given the Majczek case—and Erle Stanley Gardner's own role in a reprieve from death row—inspired Gardner, Schindler, and Keeler to found the Court of Last Resort, a chance for convicted men and women to appeal to two authorities even more sacrosanct than the law. Ordinary Americans deserved access to the same scientific defense as Marigny. And every American ought to be free to present a case to the true court of last resort, the American people. Working with the publisher of *Argosy,* a men's monthly devoted to detective stories and manly adventures, Gardner explained his court's plan to place scientific expertise before the public, or at least before the next best thing, the American magazine subscriber. He urged subscribers to become "a militant body who are willing to fight for justice." "Remember," he told them, "this is *your* Court. We are merely investigators."

The Court of Last Resort was not bound by the legal maneuverings whereby prosecutors bamboozled jurors, experts faced humiliating cross-examinations, and persnickety judges barred the door to valuable techniques like lie detection. This was the route of the men's magazine to swift and certain justice. Gardner boasted the Court of Last Resort gave voice to those who believed that state power had run amok. It planned its own fleet of mobile forensic labs, one for each state, with teams of crime scene investigators.

First, however, it had to decide which cases to take on. As Gardner put it: "We had to know the men we were talking with were telling the truth." This was a job for the lie detector. "[W]hen Keeler pronounced it as his opinion that a man was guilty, he usually had the definite proof to back up his contention."

Public officials responded warily to the Court of Last Resort. Some prosecutors dismissed it as a publicity stunt. Others cooperated in hopes of assuring the public that no innocent man was incarcerated. Yet when prosecutors reopened cases, they discovered miscarriages of justice.

Though Keeler helped on two early cases, he never made the switch to defense. As he told his collaborators, with every prisoner in Joliet clamoring for a test, officials in Illinois now insisted on preapproving each one. Besides, his own health was erratic. In his place, the Court of Last Resort turned to two of his disciples. Even so, the court was swamped with pleas, and its achievements were mixed. By the mid-1950s its subscriptions were dropping. In 1957–1958 it enjoyed a brief run as a TV drama opposite *Dragnet.* When Gardner finally dissociated himself from the Court of Last Resort, it was handed over to the American Polygraph Association, which rechristened it the "Case Review Committee." By the 1960s, the lie detector had its own television show, in which polygraph tests were run "live" before millions of viewers. This strategy of using scientific investigations to publicly expose wrongful convictions would not be taken up again until the 1990s, when DNA evidence helped reopen old cases to public scrutiny.

In the years after Kay left him, when he wasn't pursuing criminals, exonerating innocents, and drinking at the city's hot spots, Keeler let himself be seduced by a series of beautiful, spirited women—none of whom trusted him. Or at least, all of them tried to improve him, and none succeeded. Through everything, Keeler maintained his outward cool. The gossip columnists figured him for the "still-waters-run-deep" type. Girls were warned to watch their alibis when the "lie detector man" was around. He was still known to use his lie detector machine as an aid in seduction, as he had done since high school, with his pal the Commander tipping him off about "good ones" who expressed interest in a "personal demonstration." (Among them, perhaps, was the Commander's wife; the two men broke off contact after the McDonalds' divorce, and Keeler was later accused of being her lover.)

In the Bahamas Keeler was pursued by his exuberant, glossy hostess, a twenty-one-year-old baroness, originally Marie Goodwill of Montreal, married to a Swedish earl of forty, whom she called "Papa" and with whom she had a one-year-old daughter. By her own admission she was a possessive mistress who feared discovery but defied it. "My darling," she scrawled in one mid-morning missive to Keeler, "I wonder how I can live without someday betraying myself to others in my looks at you, for when, like us, the fruit is forbidden, then must a glance, a touch, tell so very much." Not that she cared whether her husband disapproved or not (apparently he didn't). Nor did she care what the censor thought. "After all, if the Imperial British censor cannot smile to himself and keep a secret, what has become of Imperial British censorship[?]"

Back on the mainland, Keeler took up with Sarah Elizabeth Rodger, a former debutante who had already published a raft of stories for women's magazines, a book of poetry (*If I Cry Release*), and two novels (*Strange Woman* and *Not with My Heart*). When she met Keeler in New York in 1943, she was divorced and had a five-year-old son. In Chicago they cruised on the lake, drank at the Pump Room, and canoodled in his office while his secretary was on break. Rodger was a woman who knew enough not to trust men, yet knew how little ground they had to trust women, who had little reason to trust one another, despite how famously they all got along. She sent a copy of her title poem, "If I Cry Release," to Keeler: "Oh hold me closely, say the lovely lie/ And then have done. The little rapture slips/ To nothing—even underneath your lips."

In her fiction men and women smolder under society's conventions. They blush, flush, and otherwise betray their secret desires. Romance fiction had this in common with the lie detector: both were based on a sentimental appreciation of the human body, the idea that the beating heart and heaving breast spoke a truth deeper than words—though those words were inevitably confessed in the end. Keeler, she recognized, was that rare man, "one you can tell anything to." Yet she also recognized the dangers of such seductive listening. It made him the sort of man a woman could read anything into, a mirror of her feelings, an enchanting lie detector. When his letters to her proved insufficiently heartfelt, she wrote herself love letters on his behalf (with parenthetical commentary) for him to mail back to her.

> I just don't know what girls like to hear (you liar, you knew all about girls while I was still wearing pajamas with the feet in them!) so all I can write is what I am thinking—and most of my thinking—aside from the lab!—is about you. (A whopper, but I would probably eat it up and beg for more.)

When she pressed too hard for marriage, the relationship ended; she married someone else soon after.

Most decisive of all was Franja Hutchins, the eighteen-year-old daughter of Robert Hutchins, the formidable chancellor of the University of Chicago. When Franja was fourteen, her mother, an artist, shocked the faculty by illustrating her annual Christmas card with a drawing of the pubescent Franja in pigtails, stark naked, holding seasonal candles. By the time she was sixteen the society pages were aflutter over this "stunning young woman," so "tall and poised, with dark brown curls that frame her pretty face." By the time she was a freshman at Bennington she was romantically linked to Keeler, then thirty-nine. She reassured him that the age difference meant nothing. A twenty-one-year-old friend of hers had married a man of fifty, "so take heart, young man." One gossip columnist imagined them engaged. Privately, Franja sent Keeler overwrought poems, hinting at secrecy and betrayal, including,"We cannot love/ or trust."

Her ambition for him made her nag. But Keeler never put any more effort into his women than his business. He lived comfortably in his bachelor's apartment in the pseudo-Renaissance villa on Dearborn, two blocks south of Lincoln Park in one of the city's most elegant districts. He had a maid to clean his apartment and cook his meals, including breakfast. What more did he need?

The public Keeler was everything the private Keeler could never be. In public, he was gregarious, charming, a pal to all the world: one of the manly men who got things done. Wherever he went, the women cooed, the alcohol was poured. "My hobby is collecting friends," he once said. As one eulogist would comment: "Many times he neglected his business for his hobby." Privately, he was wounded, wary, mistrustful. He considered himself a failure and bemoaned the narrowness of his achievement. Alone, he hated himself, and so he hated to be alone. Leonarde Keeler was a man whose most intimate relations took place in a soundproof interrogation room where men and women bared their souls—at some risk to his own.

Even the ordained priest taking confessions in the name of an all-knowing power runs an acknowledged risk. Confession manuals since the Middle Ages have guided confessors on how to conduct themselves, both for the sake of the souls of the penitent and for their own. Priests were warned not to inquire too directly into the sex life of married couples lest they develop a prurient interest, or into any matter out of idle curiosity. Inquisitors had to be at least forty years old and of strong character: honest, prudent, virtuous, and versed in theology. Moreover, priests and inquisitors could always turn to one another for absolution. Where could the operator of the lie detector turn?

The operator of the lie detector claimed to draw on the preeminent authorities of the modern age: science and the law. But Keeler knew better. What comfort did science or the law offer him? The lie detector was condemned by respectable science, and only half acknowledged by the law. Where then could he turn for absolution, except perhaps to alcohol, the universal solvent? Even psychiatrists see psychiatrists. What about lie detectors?

The damage done by the mechanical lie detector—like the destruction wrought by the third degree or judicial torture—is not confined to its subjects, or even to the legal system that bears the burden of unreliable confessions and distorted justice. The interrogator also pays the price of perpetual mistrust and human isolation, like Dostoevsky's "Grand Inquisitor." According to the head of the polygraph unit for New York City, all the polygraphers of Keeler's generation were alcoholics.

In her diaries Agnes de Mille left a searing record of one of Keeler's grand meltdowns. She was in Chicago at Christmas 1943 to celebrate the opening of *Oklahoma!*—though a bit blue because her bridegroom had been shipped off to Europe. She wanted to elicit some sympathy from Keeler; but the first time she tried, he had a blond singer on his arm, and the second time he had a redhead. Finally, he invited her to his apartment for what she thought would be a companionable New Year's celebration. When she arrived, a party was in full swing. Keeler's personal physician, himself well liquored up, pulled de Mille aside to warn her not to let Leonarde drink too much. "This is the anniversary of Kay's leaving, and he's never anything short of desperate. Watch him."

She tried. She followed the party to the Double Eagle, where dancing Russian waiters served flaming shish kebabs. But she lost Keeler in the roar that always greeted him. When she next sighted him, he was stationed at the bar beside a pretty blond and sagging at the knees—but not from love or drink. Something was badly wrong. By the time she got there, he was pushing past her and out the door—only to return, moments later, flashing his gold police badge and flanked by two large men.

The coat-check girl gave her the lowdown. Apparently Nard had ordered a boorish lout to stop pestering the blond, and the drunk had hit him "low, very low." But Keeler hadn't struck back. He had walked away, then returned to arrest the lout. Now the drinking could begin in earnest. Keeler was lit up, manic.

By the time she coaxed him back alone to his apartment his charm had turned to melancholy. Half sobbing, he recited his bitter catechism: "How Kay had betrayed him, how he had put a detective on her and found out the extent of her infidelity, how nobody cared for him, how he despised his daily occupations." During an interval, de Mille hid the whiskey. "He was the greatest detective in America and he couldn't find his own bottles of whiskey right under the sink." Keeler slammed his fist into the door, called her a bitch, then crumpled.

She was stroking his damp hair when the doorbell announced another set of revelers, including a glittery brunette who quickly located the booze. De Mille pleaded, "He's been drinking for forty-eight hours non-stop. No food. He's on the point of collapse. I'm going to send for a doctor and first aid. You mustn't do this. It's wicked."

"Who are the hell are you?" sneered the brunette. "His New York lay or something?" Even Nard shoved de Mille aside. "Sorry, honey, but I don't think you know about these things."

Instead of the doctor, it was the doctor's wife who arrived, a small, white-haired woman in her sixties, as blotto as the rest of them. "Didn't expect me, did you?" she giggled, her speech slurred like a child's. She ordered the men to carry Nard into the bedroom, then ordered them out, gave de Mille a wink, and shut the door behind her. Silence.

> After a while I rapped sharply and pushed. Nard lay on the bed giggling. The doctor's wife sat beside him her dress down to her waist. She was snickering and giggling too and her little reduced breasts with their pale pink points looked like promises on some obscene child. Nard was fingering hers in a matter of fact way, more as something expected of him than with any real interest.
>
> "He's lovely," Mrs. Doctor lisped. "I don't understand why Kay left him. She walked out on him. Real mean!"

Nard sat up long enough to convey the doctor's wife's malicious verdict: that de Mille was just acting the war widow; that what she needed was a screw. De Mille slapped him hard. Nard kissed her hand, then toppled over.

When the doctor finally arrived with two medics and a stretcher, Nard fled to de Mille to beg for absolution. He spent the first three weeks of January in the hospital tortured by the "bitter memory of that massacre of December 31, 1943."

Eleven months later, Katherine Dussaq (formerly Kay Keeler) was killed in a nighttime airplane crash outside Patterson Field, Ohio, one of forty women pilots who died in service during World War II. She had been flying around the country trying to reverse the army's plan to disband the WASPs and had run out of fuel in bad weather. The news of her death drove Keeler into a new paroxysm of grief. Her mother wrote to comfort him: "Kay loved to live dangerously and at last chance caught up with her and perhaps she would not have had it otherwise."

In 1948 Keeler spent several months in a sanatorium in California getting treatment for severe hypertension. His blood pressure had hit 248 over 154, and the doctors ordered him to change his ways. A single cigarette would push the pressure up twenty points, and his drinking didn't help. His friends tried to intervene. Vollmer offered him a kind of sinecure as a professor of criminology in Berkeley. But Keeler turned it down. He had spent twenty years building up his business, he told Vollmer, and still hoped to make Chicago the world center of polygraphic studies.

Yet behind the bravado, Keeler, Inc. was struggling. Total income had risen from $17,000 in 1947 to $20,000 in 1948, barely enough to pay salaries. Privately, Keeler admitted that he was a "complete wash-out" as a businessman. He didn't even keep books. There were times when he was so drunk he couldn't conduct business. There were many lost weekends.

Then on Memorial Day weekend in 1949, while visiting a disciple in Michigan, Keeler had another "blowup." He developed double vision and discovered blood in his urine. He spent five weeks in a hospital in Chicago on a low-sodium diet of rice and fruit, followed by two weeks at the Mayo Clinic. The bed rest wore on his nerves. The doctors recommended a vacation, preferably "on a wharf with my big toe dangling in the water as fish bait." He announced that he was granting himself a year's leave from lie detection.

That summer his systolic blood pressure hovered around 190. In September he made one last trip to visit his old friends the Wilsons at their summer cottage in Sturgeon Bay, Wisconsin. On the night of September 7, he suffered a major stroke. For several days he lay unconscious in Door County Memorial Hospital, while friends urged him to put up a good fight. He never got the message. He died on September 20, 1949, at the age of forty-five.

His sisters divided Leonarde's bric-a-brac; sold Keeler, Inc. to an employee for $10,000; and granted his manufacturer sole rights to design and sell the "Keeler Polygraph," subject to a 7 percent royalty fee. The newspapers noted his passing—"INVENTOR OF LIE TEST DIES"—and led off his obituary with his early sorority case. His faithful secretary, Viola Stevens, set up a charity for high blood pressure in his name. And his beloved sister launched a radio series, "The Hidden Truth," which dramatized (and embellished) his greatest cases. Though it was soon surpassed by rival firms, the Keeler Polygraph Institute continued to train operators and conduct personnel work for several decades.

Death is the ultimate deus ex machina. However expected, its arrival always comes as something of a surprise, and is rarely aesthetically pleasing. Death appears to resolve all doubts, but only by shoving the ambiguities—the mind-body problem included—into a box.

The anthropologist Claude Lévi-Strauss tells the story of a skeptical young Kwakiutl Indian from the Pacific Northwest who decided to study with the local shamans to find out whether their cures were "true or only made up." Indeed, young Quesalid soon discovered that, among the other tricks of the trade, the shamans often concealed a tuft of eagle down in their mouth, which they then spat out as a "bloody worm," pronouncing this to be the source of their patient's sickness. Yet Quesalid also found that the ill greatly desired his aid and that the trick with the bloody worm seemed to heal the sick. In time, he even came to disparage other cures as inadequate or fake. In the end, Lévi-Strauss concludes, "Quesalid did not become a great shaman because he cured his patients; he cured his patients because he became a great shaman." Leonard Keeler was such a shaman.

## CHAPTER 18

# Frankenstein Lives!

FRANKENSTEIN: *They? I am the one to find him. . . . I'm the one who's guilty. I created the Monster—I must be the one to destroy him!*

—*FRANKENSTEIN*, UNIVERSAL PICTURES, 1931

CHIEF VOLLMER FELT KEELER'S LOSS ACUTELY. "I REGARDED him with the affection that a father would a son and I am certain that he had a feeling of loyalty to me that is only found in father and son relationships." For more than twenty years the Chief had shared Keeler's joys, as well as the secrets that gnawed at his heart. "I was his leaning post when he felt too weak to stand alone." Yet he had no hesitation in pronouncing his young protégé "one of the finest men that I have ever known."

Imagine, then, the Chief's response when John Larson, after a two-decade silence, took the occasion of Keeler's death to write him a five-page letter filled with innuendo, reproach, and slurs. Over the years, Larson had kept close tabs on Keeler's successes and blowups, noting with special satisfaction the collapse of the Northwestern University crime lab and the refusal of the Chicago police to accept Nard. In private letters, in unpublished manuscripts, in private conversation, and even in print (albeit with more circumspection), he expressed his withering contempt for the man whom he had trained, and who had betrayed their common purpose. It was the one error he acknowledged: how in a moment of weakness—as a personal favor to Vollmer—he had agreed to train a single lay operator, that "high school kid in short pants," Leonarde Keeler. And how he regretted it! In Keeler, he had created a monster: a "salesman," an "exploiter," a showman who had "prostituted" his technique by exacerbating, not alleviating, his subjects' feelings of guilt. Thanks to Keeler's example "each pseudo-expert carries about his 'lie box,'" the result being "a brainwashing of the public by quacks."

By contrast, Larson, said Larson—referring to himself in the third person—had "always refused to commercialize his method, patent apparatus, or train lay operators." Privately, he gave Keeler this much: the young man had been a master showman whose "flamboyant self-confidence undoubtedly struck terror in many of the guilty subjects" and got them to confess.

Keeler's death seemed to offer Larson a chance to kill the monster once and for all. When Keeler's former manufacturer asked to use Larson's name to promote its device, he spurned the offer. He enlisted old colleagues to verify that it was he, not Keeler who had invented the lie detector—not that any one had "invented" anything, he hastened to add. Some, like the old cop in Berkeley, Frank Waterbury, fed him stories about Keeler's misdeeds.

When Larson wrote to Vollmer in 1951, his pent-up resentment outran his prose. He was writing, he informed Vollmer, to set the record straight. The "many years of exploitation by Lee, Keeler and many others" had allowed the "so-called lie detector" to be turned into a "psychological third degree." This had been disastrous from a scientific point of view. "We have found that the error in the interpretation of records may range as high as 40%." By contrast, Larson had spent forty years assembling the records that would tell the true story of the lie detector's birth and its fall from grace: how the instrument that he had developed at Berkeley had been debased by Keeler.

The Chief's reply was courteous but curt.

> Maybe I have been asleep, but I did not know that there was any doubt about who developed the old deception technique in Berkeley. . . . What you started in Berkeley—and which was resisted by many policemen—is now recognized as one of the most valuable investigative tools in possession of the cop. The fact that a dozen or more people have been called "Inventor of the Lie-Detector" cannot rob you of your contribution to police science. *The written record cannot be obliterated.* . . . Greetings to you and your nice wife.

The Chief—recently named "America's greatest cop" by *Collier's*—was suffering from Parkinson's disease and had been diagnosed with throat cancer. His wife had died a few years back. Yet his mind was as lucid as ever. He told his friends that he would never be an invalid. One afternoon he asked his young successor to bring him his service revolver from his old desk. A few weeks later, on November 4, 1955, after helping his housekeeper make the beds, he stepped into the hallway, told her to call the police, and shot himself.

His death was of a piece with his life: unsentimental, but with a human touch. In the previous month, he had systematically purged all embarrassing correspondence from his files. The contents of a policeman's desk are not for public consumption. Since his appointment as Berkeley's chief of police fifty years ago, his program of professionalization had become the lodestar of American law enforcement. In 1960, his disciple O. W. Wilson was appointed police chief of Chicago.

But neither Keeler's death nor Vollmer's brought Larson any nearer to his goal of caging the monster he had unwittingly unleashed. On the contrary, his frustration grew as he realized that the machine had outlived Keeler, outlived Vollmer, would outlive them all. Where there had once been a single Keeler, now there were hundreds, each touting his own device. The machine had begun to breed, proliferate, mutate. As Larson shifted jobs, he obsessively tracked the monster's wreckage, collecting news clippings, scientific reports, and sensational police cases for his final refutation.

The medical troubles that caused Larson to be rejected by the army—bad eyesight, prostate trouble, kidney stones, arthritis, ulcers, and possible growths in his left ankle—never reduced his capacity for work, not even when those cancerous growths cost him his leg in 1947. Two days after his third operation he was back at work, putting in ten- and twelve-hour days, six days a week, without holidays or a day off, as he hop-skipped with a wooden prosthesis around the country, working at a dozen different posts, each cut short by that insidious force known as "politics."

Trading Chicago for Detroit had not meant an escape from corruption: the judges in Detroit were just as beholden, the bureaucrats as craven, the criminals as well-connected. Larson took a one-year position at a psychiatric clinic on Long Island before quarreling with the director. Then he worked seven months in Seattle, at an asylum, only to leave because of "their disorganized methods of thought." Then he worked for seven months at a drying-out clinic in Blythewood, Connecticut, but quit because he refused to toe the Freudian line. Next came New Mexico, where he was chief psychiatrist at the state asylum; the pay was execrable and the patronage politics were fierce. A year later he was medical director of the Arizona State Hospital for the Insane, a post he held for two years, "cleaning up an awful mess," until he was denounced by "political snipers" in the local press and forced out of office. He had dared to suggest that the asylum's most famous prisoner, Winnie Ruth Judd, the alluring trunk-murderess, was the secret paramour of several of Arizona's leading politicians.

He managed to hold on from 1949 to 1957 as medical director of the Indiana state mental facility in Logansport. There he banned the strait-jacket, introduced penicillin for syphilitic patients, and opened a successful chapter of Alcoholics Anonymous. Annual operating costs rose, but Larson's outpatient treatments freed beds and saved the cost of a new facility. He also used a lie detector—which he now called a "Reactograph"—to test lobotomy patients, diagnose mental disease, and continue his efforts to standardize testing for deception. His reforms were recognized in 1951 by the American Psychiatric Association, which granted Logansport second-place honors for most improved institution. A glowing write-up in *Life* magazine was as close as he would come to public fame.

During his time in Logansport, Larson also founded the International Society of Police Psychiatry and Criminology, a gathering of like-minded criminologists from Indiana. Despite his incipient paranoia, Larson always inspired loyalty from those who admired his integrity. Captain Robert Borkenstein, head of the Indiana state police lab, accepted Larson's offer to be a coauthor of his book, and the two published a brisk defense of Larson's team approach to clinical polygraphy and offered to train operators in their methods. They had no takers. Increasingly, Larson supported anti-polygraph laws, sending rambling letters to politicians in states considering a ban on testing.

Yet even at Logansport, Larson's scrupulousness bred resentment. When he uncovered a scheme to line the pockets of one of the governor's chief aides, he was promptly fired. Recommending Larson for another job, one friend explained: "[Larson] is one of the most honest and hard working men I have ever known and he cannot tolerate dishonesty or inefficiency in any form."

Larson resumed his nomadic ways, taking five more jobs in as many years. At the Tennessee Maximum Security State Penitentiary, he fired thirty-five asylum workers and brought in an undercover attorney to investigate horrific conditions: patients remained naked in cells, forgotten except for the guards who sold them alcohol and beat them. Two years later he was on his way again. Besides, Marge, who had hated the cold weather in Chicago, found Tennessee too hot and humid. In 1955 she was diagnosed with a bowel obstruction, and the surgeon removed two feet of intestine. Thereafter she needed constant doses of glucose and suffered occasional convulsions. She died on December 12, 1960, at the age of fifty-seven. Without her, Larson was disoriented.

When one young staff psychologist recognized his boss as *the* Dr. John Augustus Larson of lie-detector fame, he found Larson much diminished: one-legged and nearly deaf, with poor eyesight and faulty comprehension. Though the psychologist estimated Larson to be well into his eighties, he was only sixty-seven. Yet he was still putting in "fifteen-to-twenty hour days," and a pastor praised his compassion for each patient. At the same time, even Borkenstein recognized his colleague's "eccentricities." Borkenstein explained: "Sometimes his idealism and unwillingness to compromise for financial gain has led him to be misunderstood by many people. He is relentless in his defense of his convictions." After so many years of fierce battles against dishonesty Larson seemed to have become somewhat loopy. He had become a kind of Dr. Dippy, the asylum keeper as mad as his patients.

The final years of Larson's career saw him working as psychiatric director at mental institutions in Montana (ten months), Iowa (two years), and South Dakota (one year), before retiring to Nashville in 1963, now seventy, living on Social Security and a $200 annuity.

Rarely had he collected the fees his patients owed him. He was too trusting, his friends told him—too trusting and too paranoid. Yet his chaotic energy showed no sign of tapering off. His mind had always raced ahead of his ability to convey his thoughts, jumping from idea to idea until the initial point was lost. His writing had always been an illegible scrawl, his syntax mangled, his typing erratic. Early in his career, he had dictated letters to his wife, who cleaned up his prose for publication. A succession of secretaries at various asylums had organized his correspondence. Now he had no help. He rented rooms in the Nashville Colony Motel Court, where he devoted one room to his voluminous files.

For forty years, he had carted around his accumulated correspondence, polygraph records, data sheets. The time had come for him to get all this "whipped into shape" for his "magnum opus." In the 1940s his manuscript had been provisionally titled "Unmasking the Lie Detector." By the 1950s, it had acquired a more noble title: "Psychobiology of Detection and Removal of Stress during Interrogation, with Special Reference to Larson's Cardio-Pneumo-Psychogram Test." Pitching the book to the managing editor at the University of Chicago Press, he explained his ambition this way [*sic* throughout]:

> Our text from the book after presenting an objective, realistic account in which I am not defensively or aggressively assaultively is to give a historical description from the scientific literature of the procedure. . . . There is a difference in defensiveness in attacking those who did not use suitable technique in the proper fashion and in crusading for the proper uses of available technique integrating where possible other technique of value with my original but discarding cumbersom, impractical truck leads of Rube Goldberg's equipment.

The book, he promised, would pick up where the first left off, covering the "Alpha and Omega of the Cardio-pneumo-psychogram." Its central feature would be extracts from his personal file of 900 letters—especially the correspondence between himself, Keeler, Vollmer, and Marston. Or, as he put it in his final will and testament [again, *sic* throughout]: "the entire work plusveryhot letters 890 showing Keeler's criminal record; Marstons Million racket case we blew up for the Gillette co. and give it if kept intact." These documents would expose Keeler's duplicity: the way he had stolen the ideas of others; affixed his name to a machine he had done nothing to create; and bungled a whole string of cases, including the Santa Barbara Klan case, the Mayer case in Seattle, and Rappaport's execution. It would be a history based on original sources, not on "cheese-cake type news interviews." He did not mention the fact that it was now 9,000 pages long.

There is no record of any reply from the press.

In the early months of 1963—after several months in Chicago recuperating from an infected prostate; and a short break for a nostalgic visit to Berkeley where he was honored for his innovation—Larson moved back into the Colony Court Motel to resume work. On September 23, 1965, he was sitting on the patio, sorting through his manuscripts with the help of his wife's cousin, when he had a sudden heart attack and died, age seventy-four.

As instructed in Larson's will, Robert Borkenstein went through Larson's papers to prepare the manuscript for publication. He found thousands of pages in no particular order, scramble-typed by Larson, along with thousands of letters, bundled lie detector records, tattoo records, fingerprint records, and modus operandi files. Borkenstein searched valiantly for a "key to this puzzle," but found himself unable to locate a "cohesive vein of material that I could call a manuscript."

Larson had failed to slay the monster he had created, or contain its destructive fury. "Beyond my expectation," he wrote shortly before his death, "thru uncontrollable factors, this scientific investigation became for practical purposes a Frankenstein's monster, which I have spent over 40 years in combatting." There is no simple mechanism that can undo the process of creation, just as there is no route back to the prelapsarian innocence we enjoyed before the Fall. And Larson had no doubt as to who was responsible for his own fall from grace. It was Keeler, he said, who had "fostered a Frankenstein's monster in the form of a psychological third degree." But what of Larson's own responsibility?

Larson always considered his instrument benign because he judged his own motives to be pure: to end police corruption, diagnose the criminal mind—and who knows—perhaps even cure it. But the ambition for knowledge is never wholly pure, nor is its realization perfect, as Dr. Frankenstein learned. From the start Larson been entranced by the power of the machine. Why else had he invited young Margaret Taylor back to his lab, if not in the hope that the machine would make her more likely to comply with his desires? And why else had he paraded his discovery before the noir newspapers of San Francisco, if not to boast of having been the first to open a window on the soul? A creator always has unexamined motives, and perhaps for just that reason, something always goes awry with his creation. Indeed, it is these unacknowledged motives that give the creature its hideous strength, just as it is the creator's disgust with those imperfections that launches the creature on its autonomous journey.

Once out of nature, every human creation takes on a life of its own, animated by the imagination of its audience—or, in certain cases, by its nightmares. The lie detector was an instrument designed to feed on the fears and desires of those hooked up to it, as well as the public who observed its operations from afar. So long as the people fear the lie detector—even if they suspect it is hokum—they will never succeed in killing it. There will always be a sequel. *Frankenstein, Bride of Frankenstein, Son of Frankenstein, Blood of Frankenstein, Revenge of Frankenstein, Young Frankenstein....*

This was the final indignity of Larson's life: that his name, his honest labors, and his scientific aspirations would be forgotten, and only the monster would survive. The lie detector had become the hero of a life story that contained so much more: his fingerprint work, his modus operandi system, his collections of tattoos, his asylum reforms, and the thousands of patients he had comforted. It was the fate of Dr. Frankenstein, so often identified with his monster that readers habitually referred to the monster as Frankenstein. Larson feared that the monster would upstage him. And so it did.

Yet this oblivion is, in its own way, just. After all, if Larson had not invented the lie detector, someone else would have. Indeed, all the men formerly famous for having "invented" the lie detector have been forgotten, except one. Only William Moulton Marston, creator of the lie detector and of Wonder Woman, has endured: happy in his ménage à trois, ebullient in his showman's patter, preaching his gospel of psychological uplift, and dedicated to the coming dominion of women. Marston embraced the lie detector for what it was: a figment of the popular imagination and an emblem of our dangerous suggestibility. He was proud of his creation, and never seemed to suffer for it.

CHAPTER 19

# Box Populi

*"Nobody will ever believe you if you write a thing like that," I told myself. . . . So, for the sake of plausibility, I lied about her!*

—DASHIELL HAMMETT, "LETTER TO THE EDITOR," *BLACK MASK*, 1924

OVER THE COURSE OF THE PAST EIGHTY YEARS, LIE DETECTION has been perhaps the most investigated forensic technique. Hundreds of investigations of the polygraph have been conducted, only to be reviewed in review essays, studied in metastudies, and analyzed by committees of scientific analysts. Their conclusion? A review by the National Research Council in 1941, a meta-analysis by the Congressional Office of Technology Assessment in 1984, and a survey by the National Academy of Science in 2003 each concluded that the techniques of lie detection, as used in investigative work by polygraphers, do not pass scientific muster. Yet lie detection lives on.

The lie detector cannot be killed by science, because it is not born of science. Its habitat is not the laboratory or even the courtroom, but newsprint, film, television, and of course the pulps, comic books, and science fiction. To put it in the more sober language of economics: lie detection is demand-driven. For more than eighty years a polygraph of pulsing tubes and swelling diaphragms has supplied the requisite "science." Now, at long last, the era of the polygraph seems to be drawing to a close, not because the science has changed—that happened decades ago—but because the public has come to believe in a different kind of science. The results have been mixed.

While corporate use of the lie detector has sharply declined, the courts are under pressure to allow polygraph evidence, even as many in the U.S. government are rushing to embrace new lie detector technology.

The one constant is the machinery's role in political theater. For the past several decades nary a public scandal has gone by without its polygraph moment. In his testimony before the Warren Commission, Jack Ruby's first request was to be tested on the lie detector. Watergate generated its own industry of affirmation: Chuck Colson took a polygraph test to prove his innocence, and passed; Jeb Magruder took one to confirm his accusation that John Mitchell had lied, and passed; even F. Lee Bailey took one just for representing someone, and passed. Apparently everybody was on the up-and-up, until Senator Sam Ervin, chair of the congressional hearings, dismissed the technique as "twentieth-century witchcraft."

James Earl Ray took one. And O.J. Simpson went so far as to be hooked up to a polygraph machine, though he later denied in court that he had actually "sat" for the lie detector test or had called it off when it seemed to be going poorly, as was widely reported. Anita Hill took a test to avoid being accused of avoiding one; Clarence Thomas refused the test and disparaged its value. In the great fan dance of Bill Clinton's impeachment, Monica Lewinsky offered to take a lie test if the special prosecutor, Ken Starr, offered her immunity from perjury charges, then changed her mind. As for Paula Jones, whose public accusations got the show rolling, she later let herself be strapped into a machine on the nationally syndicated television show *The Lie Detector,* for the edification of the public.

> Q: Did President Clinton, when he was governor of Arkansas, expose his private parts to you?
> A: Yes.

When the public doubts your sincerity, there's nothing like a private date with a friendly polygraph operator. The parents of the murdered pageant princess JonBenét Ramsey took a test that pronounced them innocent. Jeffrey Skilling of Enron arranged for an exonerating polygraph. Representative Gary Condit, rumored to have knowledge about the death of his intern Chandra Levy, refused to take the FBI's polygraph until a national poll indicated that 83 percent of Americans thought he should take one; whereupon he took one privately and announced the results, as he presumably would not have done had he failed. Whereas a lie detector wielded by a prosecutor feeds on stress and pushes for confession, a lie detector in the friendly hands of a consulting polygrapher provides reassurance and exoneration.

In this context, the lie detector is not a hoax so much as a mirror of make-believe. Since the 1950s America's sci-fi religion, Scientology, has used the E-Meter—a simple galvanometer—to gauge the spiritual health of initiates as they verify their ascent to "clear." In the 1960s Cleve Backster—Keeler's disciple and the founder of the CIA's polygraph program—hooked a lie detector to his household plants and discovered that they reacted dramatically when he threatened to burn their leaves or killed brine shrimp in a nearby tank; this was evidence, he said, of a "primary perception" that linked all living things. In the 1980s and 1990s, the *National Enquirer* relied on lie detector tests to confirm tales of alien abduction.

For the past half century the debate over the polygraph has gradually deviated from the pattern laid out by Keeler and Larson. Despite Larson's prodding, the American Psychiatric Association went no farther in 1944 than "cautioning against advertising" the lie detector, lest the public think a mechanical device could replace the association's brand of expensive excavation: "Whereas there is no conscience robot and no diagnosis robot." In 1958 Pope Pius XII praised modern psychology in general, but warned against methods that entered a person's "interior domain" against that person's will, among them lie detection. Then, during the 1960s, the Warren Supreme Court discovered a right to privacy in the interstices of the Bill of Rights, and some legal scholars suggested that the lie detector, even when "voluntarily" submitted to, violated citizens' psychological integrity. The device, these civil libertarians reasoned, probed not by means of the ordinary senses but like an electronic listening device that pierced the walls of a person's home. Hence, even when the subject agreed to a test, the operator might unexpectedly turn his attention to areas the subject had not agreed to discuss (as Keeler often did). This view, however, never found favor with the courts. Once the subject had given explicit assent to the exam, they ruled that anything the subject said could and would be used against him.

In the political arena, resistance to the lie detector fared somewhat better. In the late 1960s, the union representing government employees, with the help of the American Civil Liberties Union, pressured Congress into ending interrogations into sexual practices and private acts. The Civil Service Commission added restrictions on the instrument's use in preemployment screening of nonsecurity government personnel. The AFL-CIO spoke out against lie detection in the workplace. Abuses of the instrument became a favorite object of congressional inquiry, with politicians damning the machine as an "invasion of the mind." And in the late 1960s, California, Oregon, Alaska, and Massachusetts restricted employers' rights to polygraph workers at will. By the mid-1980s, twenty-two states had done so; all of them had high rates of unionization.

Yet the same period saw a tremendous expansion in the use of the lie detector. In 1959, the National Labor Relations Board ruled that private employers could require being tested on the polygraph as a condition of employment. That same year the FBI trained its first operators, and they were testing nearly eighty cases a month by the end of 1960. When a report by the Pentagon cast doubt on the scientific legitimacy of the polygraph in 1964, the report was immediately classified. In 1965 the American Bar Association unapologetically defended the technique as a "useful tool" of interrogation in cases where consent had been given. By the middle of the 1960s, some 1,000 operators were at work, a tenfold increase since the 1950s; and half of the nation's police departments were making use of the technique, along with thirteen federal agencies, and some 40,000 businesses. Then the use of the machine increased tenfold again, from 200,000 Americans a year in the 1960s to 2 million by the 1980s. By then, one-fourth of all U.S. firms used the technique for preemployment screening or systematic checkups on honesty.

In those same decades the polygraph finally began to make modest inroads outside the United States, particularly among its close allies. Canada, West Germany, Japan, South Africa, and Israel have all adopted the machine for their military police and some criminal investigations. Most foreign courts still officially reject the technique, however, and the United States remains by far its largest per capita user. Despite the growing use of the technique across cultural and linguistic barriers, there has been little research published on its effectiveness in these circumstances. Since Keeler inaugurated his program with German POWs, American polygraphers have interrogated Korean and Vietnamese combatants, Chinese spies, and Arab terrorists. But almost all polygraph operators are still white Christian males, and there is scant evidence that the lie detector can minimize cross-cultural misreadings.

Occasionally, polygraph experts have urged their colleagues to set rigorous protocols for interrogation and establish licensed training schools. In fact, only cursory standards have been adopted, and the reason is simple enough. Keeler's style of polygraphy works best when the examiners are not constrained by norms.

In the past half century, public debate over the government's use of lie detection has generally pitted the executive against the legislative branch. The executive branch (often in Republican hands) has typically favored the instrument as a means of controlling the federal bureaucracy. The legislature has alternated between howling for more lie detection whenever the executive branch failed to protect national security, and howling for curbs when constituents complained.

Ironically, a push by the Reagan administration to expand the use of polygraphs in the 1980s finally led to federal restrictions. The administration wanted to end the disclosure of classified information, nominally in the interest of national security, but mostly to halt embarrassing revelations about the true cost of Reagan's military buildup. The president complained that he was "up to his keister in leaks" and authorized polygraph exams of presumed leakers. But a study by the Congressional Office of Technology Assessment concluded that lie detection was, at best, an uncertain science and unsuited for "dragnet" screening. Then, after his reelection, Reagan ordered random polygraph testing of all government employees with access to classified information, placing nearly 200,000 bureaucrats under the lie-detector regime. Secretary of State George Shultz interpreted this as an attack on the integrity of his department. When his private pleas failed to get the order rescinded, he publicly announced that he would sit for a test at the president's request, but would resign "the minute . . . I am told that I'm not trusted." The president quickly modified the order to cover only those suspected of espionage. This snafu helped jar Congress into passing the Polygraph Protection Act of 1988, which forbade private businesses from requiring their employees to take polygraph exams under most circumstances. Under pressure from civil libertarians on both the left and the right, Reagan signed it. The law did preserve some notable exemptions. The polygraph could still be used in criminal investigations; when employers had grounds for suspecting a crime (as Keeler often pretended); and on federal, state, and local government employees, including police officers. Still, the law of 1988 led to a dramatic drop in use of the polygraph.

Businesses quickly made their peace with the ban. Whatever its former role as a deterrent, the polygraph had ceased to make sense in the new economy of the late twentieth century. Introduced early in the century to vet employees in the hierarchies of managerial capitalism, the polygraph was too crude a tool to enforce loyalty in the horizontal firms of the new service economy, where "knowledge workers" operate almost like subcontractors, with quasi-entrepreneurial roles inside the firm. The sort of loyalty that once came with lifetime employment is rarely expected or delivered.

Ironically, just as corporate America was giving up on the lie detector, the era of the Frye rule was ending, potentially opening criminal courts to the polygraph. In the middle decades of the twentieth century, the judiciary's ban and strictures against police coercion had simply given investigators more reason than ever to deploy the lie detector to extract pretrial confessions. Likewise, the famous *Miranda* ruling of 1966 simply inoculated the police against more intrusive regulation of their activities, much like the printed health warning on packages of cigarettes. All along, of course, polygraphers continued to press for the courts' recognition as qualified experts. In the 1950s, Russell Chatham at Oak Ridge sponsored a survey, which found that twice as many polygraphers (75 percent) as psychologists (36 percent) believed reactions on the instrument were a sign of deception. In the 1980s and 1990s, proponents and opponents of lie detection conducted rival surveys, contending for the right to speak in the name of the "relevant" experts.

Meanwhile, the courts had come under pressure to define more clearly what counted as scientific evidence. In 1977 new federal rules of evidence declared that courts hear relevant scientific evidence whose probative value outweighed its potential to prejudice the outcome. Then in *Daubert v. Merrell* (1993), the U.S. Supreme Court explicitly replaced the Frye rule with a multipronged test. Henceforth, trial judges were to let jurors hear scientific evidence that had been (1) subjected to a program of scientific testing, (2) published in peer-reviewed journals, (3) assigned a known error rate, and (4) found acceptable by the relevant scientific community (the old Frye rule). Hailed at the time as a victory against "junk science," the *Daubert* decision led some polygraph operators to believe that their day in court was coming at last. After all, decades of rejection had spurred polygraph researchers to conduct laboratory and field studies, publish in peer-reviewed journals, and announce their low error rates. Some appellate courts agreed to let criminal courts consider polygraph evidence, which was, after all, already approved for parole cases, civil suits, and national security purposes.

In 1998 this contradictory landscape came to the attention of the U.S. Supreme Court in *Scheffer v. United States,* a case involving a military officer who had been denied the right to introduce polygraph evidence that seemed to exonerate him for failing a drug test. But in the end, *Scheffer* was decided on narrow grounds. Writing for an eight-to-one majority, Justice Clarence Thomas—perhaps recalling the imbroglio with Anita Hill—upheld the prerogative of the appropriate rule-maker (here the president in his capacity as commander in chief) to formally ban the polygraph so long as the rule-maker provided some minimal rationalization. Yet only three justices (forming a conservative minority) agreed with the remainder of Thomas's opinion: that the polygraph threatened to usurp the jury's role in making assessments ("the *jury* is the lie detector," quoted Thomas) or his fear that the machine's "aura of infallibility" would overwhelm their deliberations. The four partially concurring justices and the lone dissenter (forming a liberal majority) were troubled by America's double standard for the polygraph and defended the jury's ability to assess all forms of evidence, including scientific evidence. Indeed, in their eagerness to assure the court that jurors would not be flummoxed by lie detector evidence, advocates of the polygraph had submitted briefs arguing that the results of polygraph tests now had a minimal influence on jurors' judgments. It would be ironic, indeed, if the polygraph were finally admitted into court because most people no longer believed in it.

The muddled ruling left lower courts free to admit the polygraph, as some have cautiously begun to consider doing. In the courts, as elsewhere, the debate over the polygraph turns not so much on its reliability as on other questions. For conservatives, the debate turns on whether the machine is used in conjunction with institutional authority. For liberals, it turns on the right of the individual citizen to mount a full defense, especially when the accused seems likely to lose a swearing contest with the police. As always, with the polygraph what counts is less its reliability than the circumstances that surround it.

To those circumstances we must now add the resurgent national security regime and its "war on terror"—and the domestic posturing these campaigns enable. The CIA has long required its agents to take polygraph tests even as it trained its operatives on how to beat them. From prison, Aldridge Ames, the CIA agent who betrayed his country for twenty years—despite being repeatedly tested on the polygraph—now mocks the machine. "Like most junk science that just won't die (graphology, astrology and homeopathy come to mind), because of the usefulness or profit their practitioners enjoy, the polygraph stays with us." After the FBI agent Robert Hanssen was also discovered to have betrayed his country for twenty years, the Bureau publicly announced that it would polygraph those agents who have access to intelligence information. Thus, to great fanfare, the polygraph lock was ceremonially affixed to the barn door.

In 1999, forty-six years after the AEC shut down the Oak Ridge screening program as unscientific and counterproductive, its successor agency, the Department of Energy (DOE), reinstated the ritual of political purification at the nuclear weapons labs. The impetus was again a politically motivated leak alleging that nuclear secrets had been leaked. That spring, with the Republican Congress excoriating the Clinton administration for supposedly kowtowing to China, the *New York Times* published a story based on anonymous sources suggesting that the Chinese had miniaturized nuclear warheads using a design possibly stolen from America. Behind the scenes, a panicked search for a culprit had already led security officers to fasten their suspicions on Wen Ho Lee, a Taiwanese-born American scientist who had worked at Los Alamos for twenty years and had been suspected of meeting furtively with Chinese scientists. One hitch was that a polygraph operator subcontracted by the DOE had judged Lee "not deceptive," and it was three weeks before the FBI reinterpreted Lee's polygraph record as inconclusive and demanded a redo. This time the FBI's examiner implied that Lee had done poorly, but there was no judgment until the article in the *Times* asserted that the government's chief suspect was an unnamed Chinese-American weapons scientist who had taken a polygraph exam and was "found to be deceptive."

In a follow-up interrogation the next day, the FBI's special agent Carol Covert accused Lee twenty-six times of having failed his polygraph. She also threatened him with the loss of his job, and with the same fate that befell the Rosenbergs. Two days later, the *Times* revealed that the DOE's principal suspect was named Wen Ho Lee and that he had "stonewalled" investigators. When Lee was also alleged to have transferred to an open computer a library of software codes for simulating nuclear tests, he was placed in solitary confinement and shackled. Within a year, however, the government's case against Lee had unraveled, and almost none of it has ever been substantiated. Among many other misrepresentations, the FBI agents had lied when they accused Lee of failing his polygraph exam. On the first exam he had received a high score for honesty. The second had been arranged solely to accuse him of failing. Many Americans were shocked to learn that such ruses were "not uncommon." In the end, the Justice Department had to withdraw the bulk of the case against Lee and release him.

Yet the DOE's solution—its public-relations solution—was to make polygraph tests obligatory for lab personnel. This time around, the weapons scientists noisily expressed their disgust. At public hearings in Livermore, Albuquerque, Los Alamos, and Washington, D.C., they condemned the tests as unfair, demoralizing, lacking empirical validation, and as yet untested by the open debate on which science thrived. For purely political reasons, they said, they were being subjected to tests that "would never pass muster in a high school science fair." In 2003 a survey by the National Academy of Sciences of decades of lie detection studies criticized dragnet screening as useless, warning that it did no good in itself and undermined national security by producing a false sense of security. Later that year the DOE seems to have misread the report to justify an increase in the number of exams and forced out a physician-physicist who protested vehemently. For more than five years, some 20,000 employees have been subject to mandatory exams, a return to the situation of the early cold war—

although in late 2006 DOE seemed to rescind polygraph screening and this flawed approach to national security.

September 11, 2001, which was supposed to change everything, has reprised every bugaboo concerning lie detection. The FBI has been using the lie detector on suspected terrorists with mixed results. When Abdallah Higazy, an Egyptian citizen who had been staying at a hotel across from the World Trade Center on 9/11, returned several months later to collect his possessions, he was held as a material witness. A security guard said he had found a pilot's radio stashed in the safe of Higazy's hotel room. In the course of a four-hour interrogation on the polygraph, the FBI's operator screamed at Higazy, repeatedly told him he had failed the exam, and warned him that Egyptian security forces would investigate his family in Egypt (this being tantamount to threatening the family with torture). Sobbing with fear, Higazy admitted that the radio was his. He was placed in solitary confinement until the radio's owner turned up to retrieve it a few days later. The security guard was later convicted of lying to the FBI, even though he too had previously passed a polygraph exam.

As usual, it is the innocent who are tripped up by polygraphs; the bad guys aren't fazed. Documents seized by U.S. forces in Afghanistan show that jihadists have been trained in countermeasures against the lie detector, which they consider a perversion of the methods used by the medieval Islamic philosopher-physician Avicenna, who monitored the pulse to determine his patients' true feelings. Iraqi jihadists have posted this information on the Web.

In the meantime, American interrogators have hastily reassembled the coercive techniques of the cold war. The road to Guantánamo and Abu Ghraib may have been laid in the month after September 11, 2001, when President George W. Bush issued secret orders exempting detainees in the war on terror from the Geneva Conventions and from the protection of U.S. laws and international treaties against torture. But in charting this course the president was reviving long-honed techniques of psychological torture used by the CIA in Vietnam in the 1960s and its Latin American proxies in the 1970s and 1980s. As the historian Alfred McCoy has shown, nothing better dramatizes that legacy than the iconic photograph of the hooded Iraqi prisoner standing on a box, his arms outstretched, with fake electrical leads trailing from his fingers. Widely read as evidence that the interrogations at Abu Ghraib were the handiwork of a few perverted guards, the image actually documents the return of the CIA's techniques of psychological torture, including sensory deprivation (the hood), self-inflicted pain (the outstretched arms), and the dire threat of physical pain (the fake electric leads).

In important respects, of course, polygraph techniques differ radically from these coercive methods. The polygraph can be used only on subjects who are ostensibly cooperative, and resistant subjects can always disrupt a test if so inclined. The physical discomfort experienced during a polygraph exam is nothing like the pain caused by forcing prisoners to assume stress positions. The anxiety of taking a polygraph exam is nothing like the terror of simulated drowning known as "water-boarding." Although the practice of lie detection may follow the *logic* of torture, it is not torture, not even psychological torture. To define torture "downward" until it encompasses such procedures would diminish our abhorrence for coercive techniques that cause acute physical or mental suffering.

Yet the similarities too are revealing. Although polygraph tests are nominally voluntary, prisoners may well have little choice but to take them. And once strapped down, few will refuse to answer all the polygrapher's questions, even when examiners stray from agreed-upon topics. Even subjects who clam up are still supplying information involuntarily through their bodily responses. Both psychological torture and polygraph interrogation are predicated on total control over the subject's environment. Keeler carefully designed his interrogation room as a kind of theater-for-one; the Darrow Photopolygraph monitored all the subject's sensory inputs. Moreover, both sets of techniques deliberately rule out brute force lest it engender resistance or prompt false confessions. Like polygraphers, interrogators who use coercive techniques are told to remain cool, detached, and scientific. In both cases, the goal is to exert psychological pressure until a confession is forthcoming. It is telling that these new coercive methods were first developed in the 1950s under the cover of polygraph exams.

Finally, even though many polygraph operators and professional military interrogators have publicly denounced America's use of torture, some polygraph subjects conflate their experience with psychological torture.

And no wonder. For instance, the Pentagon has currently deployed large numbers of polygraph operators to Iraq to examine prisoners and informants. Exam conditions there are far from the scientific ideal, sometimes conducted with the aid of balky interpreters or in the bathrooms of bombed-out buildings. As one examiner acknowledged, "Some examinees were familiar with the American 'truth machine' while others were scared to death of having electrical wires attached to them."

Until recently, most Americans, when they gave any thought at all to the distasteful topic of torture, imagined that it involved the infliction of brutal, if metered, violence—as typified by the iron maidens of medieval dungeons. And most Americans—to judge by the popularity of television dramas like *24* and *NYPD Blue*—think even good cops sometimes have to break the rules and hurt bad guys because that is the only way to get the necessary information. Unfortunately, these false impressions have been reinforced by the ambiguous behavior of the U.S. government.

In the first place, according to the UN Convention of 1984, the definition of torture encompasses severe mental, as well as physical, suffering when "intentionally inflicted on a person for such purpose as obtaining from him or a third person information or a confession." Yet the United States, when it ratified the Convention in the 1990s, appended a set of little-noticed reservations that largely preserved the CIA's clandestine psychological techniques. And since the attacks of September 11, 2001, the government has repeatedly denied that it has authorized torture, even while authorizing techniques such as sensory deprivation, stress positions, disorientation, and water-boarding that clearly constitute torture as defined by U.S. laws and treaties.

In the second place, there is reason to doubt that psychological torture generates information any more reliably than physical torture. As recently as September 2006, U.S. officials asserted that extraordinary techniques were the only way CIA interrogators had been able to pry crucial information from captured al-Qaeda figures like senior aide Abu Zubaydah, who confirmed that Khalid Shaikh Mohammed was the mastermind of the September 11 attacks. Yet these official claims were contradicted by reports which suggested that all the reliable information from Abu Zubaydah had been obtained in the initial, conventional interrogation by the FBI and that the crucial intelligence that led to the capture of Khalid Shaikh Mohammed was in fact obtained from a tipster who collected a $25 million reward. Other accounts have suggested that Abu Zubaydah is no longer psychologically stable.

The passage in 2005 of the McCain detainee amendment to a military appropriations bill and the subsequent redrafting of the Army Field Manual in 2006 would seem to prohibit many of the harsh interrogation tactics previously authorized for enemy combatants. Yet the Military Commissions Act, signed by President Bush in October 2006, for the first time opens the door to formally authorize CIA interrogators to use these techniques on suspected terrorists and permits the introduction in special military tribunals of evidence obtained with these methods. It would seem that the U.S. government has decided to actually sanction these techniques both as a matter of law and as a clandestine practice.

Under the circumstances, it is understandable that America has once again turned to science for a solution to the problem of how to coax reliable information from recalcitrant human beings. But a half century of controversy and failure seems to have finally eroded the public's faith in the traditional polygraph. Surveys of jurors in civil cases suggest that the public has become skeptical about lie detectors that consist of pulsing tubes and dancing needles. When science gets old, it's time to get some new science. In recent decades researchers have floated many new schemes for lie detection. Some have listened for the pitch of the human voice under stress; others for heat around the eyes.

At the medical school of the University of California at San Francisco, Paul Ekman has compiled a vast library of facial expressions, sorted by emotional type. The faces of liars, he says, involuntarily telegraph their deceit, sometimes in muscle reactions too quick to be consciously observed—unless you know what to look for or have the right equipment. Ekman's methods build on a series of assumptions about the links between human biology and society: that thanks to evolutionary pressures certain human emotions—such as fear—produce a common and specific reaction in the human body; that their visible signs serve some role in social signaling and hence can be read by others, even if not always consciously; and that, with sufficient attention and training, the fear of being caught in a lie is one of the reactions that can be detected. But it is just as plausible to say that evolutionary pressures have made people good at getting away with lies as it is to say that evolution has made them good at catching falsehoods. And subsequent studies have shown—and Ekman himself admits—that even with training very few people do much better than guesswork, and some do worse!

The latest techniques offer an alluring combination of high-tech equipment and a direct assault on the fount of all duplicity: the brain. In place of the old correlation of the body and its emotions, neuroscientists promise to lead interrogators directly into the factory of our consciousness.

One technique is based on the electroencephalograph (EEG), which measures the electrical activity of the brain. The EEG has the advantage of being relatively easy to record, though it gives an aggregate picture of neuron activity, not readily localized. It was first proposed for lie detection in the 1940s, when it was used to evaluate a variety of mental illnesses; John Larson, for instance, used it to assess lobotomy patients. One recent adaptation looks for a rise in the P300 wave said to occur 300-plus milliseconds after a subject encounters a familiar, but rare, stimulus—such as when someone suddenly calls out a former lover's name amid a stream of babble. Lawrence Farwell, its chief promoter, has patented an algorithm that records this and other signals to produce a "brain fingerprint" that he says can detect guilty knowledge 99.9 percent of the time. In the early 1990s Farwell received $1 million in funding from the CIA, although the agency soon ceased to support his research—as it explained in the months after September 11—because it decided the technique could not readily be adapted to screening purposes and because a full evaluation of the technique's effectiveness was impossible so long as Farwell refused to divulge crucial aspects of his algorithm. At the time Farwell was being named one of the "TIME 100: The Next Wave, the 100 Innovators Who May Be the Picassos or Einsteins of the Twenty-First Century." In fact, Farwell is poised to become the Orlando Scott of our time. And like Scott, he gets results by "art rather than science." His own mentor says the technique suffers from many of the problems of traditional polygraphs, such as assuming that the subject's memory is a storage bin from which exact matches can be drawn, rather than an active and creative faculty. The fear is that examiners may interpret brain activity as a sign that the subject has recalled a rare event which the person has never in fact experienced.

A more recent—but technically challenging—method is based on functional magnetic resonance imaging (fMRI). These are real-time scans that enable neuroscientists to gauge the level of activity in various regions of the brain by tracking the blood (oxygen) flow to each region. The basic assumption is that when subjects tell a lie, they draw on distinct cognitive abilities located in identifiable regions of the brain. Taking the images is laborious and requires full cooperation from the subject, who must lie supine inside a confining tube for nearly an hour. Yet researchers using these techniques have been able to reproduce Keeler's card trick and catch volunteers who have "stolen" small sums of money.

For all their state-of-the-art plausibility, however, these new tests are plagued by many of the same ambiguities as old-style polygraphy. The studies still assume a link—via some unspecified mediation—between deception and bodily reaction, though of course they squeeze the "body" from the viscera to the brain—as if the brain's "feelings" were necessarily more authentic, or less easily faked, than those in the lowly heart or stomach. (Not that these humble organs have been entirely neglected; one new lie-detection technique zeros in on the autonomous response of the stomach muscles.) More specifically, most fMRI researchers, when testing for guilty knowledge, look for extra activity in the prefrontal cortex—thought to be responsible for inhibiting responses—on the assumption that honest recollection is the default mode of human beings and that deceit involves a distinctive pathological behavior to cover it up. But as Marston discovered a century ago in his evaluation of Münsterberg's word-association tests, some people seem to enjoy telling lies and do so without inhibition. Indeed, other fMRI researchers have blundered on what Montaigne bemoaned four centuries ago: that lies may assume a hundred thousand shapes—for a start, those we come up with on the fly and those we stake out in advance—and these varieties seem to draw on different regions of the brain, depending on the mix of memorization and invention. And what about those cases in which subjects feel guilty or defiant or simply refuse to play along with the experimenters' "interesting scenarios" in return for a $20 reward? Besides, isn't the human mind capable of fabricating memories, or feeling guilty for no good reason, or in some cases calmly and pathologically lying? Who knows, might we not even be evolutionarily equipped to lie, the way other animals (unconsciously) deceive predators?

On closer inspection the fMRI technique seems to return us to the era of phrenology. By their own admission many researchers in the field engage in "blobology," in which they reason back from their colorful images to the sorts of cognitive processes that "might" have been involved when the subject performed some task. These sorts of inverse inferences have always plagued lie detector research. The fMRI deception studies conducted thus far have been confined to laboratory tests, amalgamating the responses of half a dozen subjects asked to lie about an incident with little ambiguity and low consequences, lest their emotions swamp the processes involved in cognitive deceit. Even under such artificial conditions, fellow researchers caution, the results barely cross the threshold of significance. And the challenge for lie detection, as always, is to get good results in the messy world of tangled tales and fearful innocents.

Then there are the problems that plague any attempt to detect deception. Drugs, mental discipline, or other countermeasures might foil the new tests. False memories and other tricks of self-deception will always cast doubt on their conclusiveness, as might strong personal commitments. And finally, these new techniques all retain the old software of interrogation: the stimulation question technique, the guilty-knowledge test, etc. Thus they seem poised to renew the classical "misdirect" of the magician's art, focusing the subject's attention on the newfangled and intimidating equipment, when it is actually the examination ritual that has produced the results.

But a drummed-up need is the mother of marketing, and in the years since 9/11 many millions of dollars and considerable brainpower have been devoted to this field. In 2006 two commercial firms began to offer fMRI deception testing based on patented algorithms. These firms are offering their services to the Department of Homeland Security and the director of National Intelligence, as well as to civil courts, employers, advertisers, and movie studios interested in gauging viewers' emotional responses to "media information." Leonarde Keeler would be proud, John Larson horrified, and William Marston amused.

Of course one can never rule out the possibility that researchers will someday find relatively reliable methods of distinguishing genuine belief from deliberate deceit. Nothing in this book suggests that such an outcome is impossible. To be sure, one may doubt that it would be welcome; ethicists are already on the case, worrying that the new brain techniques will actually succeed in invading the final realm of human privacy. By that time, of course, as the judge in the *Frye* case said, we may all be dead. In the meanwhile, the more worrisome and likely outcome is that new techniques, no more reliable than Larson and Keeler's old-fashioned methods, will slip neatly into the role the polygraph once played. That is how badly we believe in science.

It would be comforting to think that science could sanitize the messy business of extracting reliable information from recalcitrant human beings. How much easier it would be if we could just ship off the problem of human deception to a remote and spotless laboratory, and get the report back overnight. The temptation is not new and it is perfectly understandable. But we cannot expect science to bear burdens that we ourselves won't shoulder. Science does not reside outside the entanglements of human history, and it is when we delude ourselves into thinking that it does—by, say, treating people as bodily specimens when their lives and liberty are at stake—that we create monsters like Frankenstein's.

# Epilogue

*If the fixture of Momus's glass in the human breast . . . , had taken place,— . . . nothing more would have been wanting, in order to have taken a man's character, but to have taken a chair and gone softly. . . , and looked in,—viewed the soul stark naked;—observed all her motions,—her machinations;— . . . then taken your pen and ink and set down nothing but what you have seen, and could have sworn to:—But this is an advantage not be had by the biographer in this planet;— . . . our minds shine not through the body, but are wrapt up here in the dark covering of uncrystallized flesh and blood; so that, if we would come to the specific characters of them, we must go some other way to work. . . . I will draw my Uncle Toby's character from his* HOBBY-HORSE.

—LAURENCE STERNE, *THE LIFE AND OPINIONS OF TRISTRAM SHANDY*, 1759

OUR OBSESSIONS DEFINE US. THEY TUG AT THE RHYTHMIC trace of our life like the secret pull of our heart and the stifled hitch of our breath. They are the actions—both commonplace and idiosyncratic, willed and determined—that we cannot but choose to repeat. Our obsessions make us who we are. At this historical remove, we cannot of course interrogate men like Keeler, Larson, and Marston or make them answer to their own device. There are no polygraphs for the dead. Instead, their portraits have been assembled according to the patient methods of the old-fashioned private eye: by collecting and corroborating the scattered traces their obsession has left us—their public pronouncements, personal correspondence, private diaries, and secondhand slanders—and reading each piece of evidence for its own emotional slant. And in the gap between their public acts and private selves we catch a glimpse of that equivocal form of self-consciousness that used to be called the soul.

To deceive is human. There has never been, nor ever will be, an honest society. And so long as we lack the means to quantify lies or weigh hypocrisies, we have no basis for supposing any society more dishonest than any other. Rather, what distinguishes a culture is how it copes with deceit: the sorts of lies it denounces, the sorts of institutions it fashions to expose them. Only in America has the campaign to expose lies taken a techno-scientific turn. The polygraph is a banal assemblage of medical technologies, a concatenation of physiological instruments available throughout the developed world for more than a century. Yet only in America has it been repurposed for interrogation.

The lie detector has thrived in America because the instrument played into one of the great projects of the twentieth century: the effort to transform the central moral question of our collective life—how to fashion a just society—into a legal problem. To do so, it drew its legitimization from two noble half-truths about our political life: that democracy depends on transparency in public life, and that justice depends on equal treatment for all. As a nation founded on an explicit political contract rather than a common history or shared kinship, Americans have aspired to resolve social conflicts with explicit public rules—regardless of any chicanery taking place behind the scenes. And lest anyone protest that these rules themselves are rigged, we have often tried to justify them in the name of science, itself considered the least arbitrary and most transparent form of rule making. Hence, perhaps, our propensity to treat deceit, the original sin of the social contract, with a redoubled dose of science.

These noble half-truths about democracy have dovetailed with two popular half-truths about science: that it offers an unerring method for piercing the mysteries of nature, and that it does so by eliminating the personal predilections of the investigator. In the case of the lie detector, these half-truths have been coupled to a novel twentieth-century assumption: that as creatures of nature, human beings express their thoughts and feelings in bodily terms. On this basis, the proponents of lie detection have packaged their technique as a mechanical oracle that can read the body's hidden signs for evidence of deceit—while they sidestep the skeptical interpretive labor that scientists ordinarily demand of such claims. The lie detector and its progeny have been repeatedly denounced by respectable science—but since when has that stopped millions of Americans from believing in something, especially when the public media breathlessly extol its successes? To a nation eager for justice that is swift and sure, it hardly matters that the lie detector succeeds by pretense. The lie detector "works" and that is enough. It resolves cases, extracts confessions, assures fidelity, and underwrites credibility. In short, it provides answers—and nothing feels better. The lie detector is less a "technology of truth" than a "technology of truthiness."

Americans, it would seem, still concur with Saint Augustine and the authors of romance fiction in believing that the truth resides in the heart and that deceit—like self-consciousness—is produced in the gap between our inner feelings and our calculated speech. Even the advocates of the new neuroimaging techniques still try to detect deception by prying apart the liar's divided self, although instead of pitting the autonomic heart against the willful brain, they set the various regions of the brain against one another. The implication is that an honest person does not just match word to deed, but to sentiment as well. From the era of the Puritans to the Age of Aquarius, Americans have not been satisfied with proving our worth with worldly deeds. We have demanded authenticity too: often from our neighbors, always from our leaders, and sometimes from ourselves. In sum, we are a nation of sentimental materialists. When Father Brown, G. K. Chesterton's ordained detective, first encountered the American lie detector, he nearly laughed out loud: "Who but a Yankee would think of proving anything from heart-throbs? Why, they must be as sentimental as a man who thinks a woman is in love with him if she blushes."

We believe in the lie detector for all sorts of elevated reasons: because we long for a form of justice that is swift, certain, and noncoercive; because we can't imagine how the soul could not be manifest, somehow, in the body; because we expect that science can and will pierce the veil of earthly appearances. But perhaps, in the end, these are all just excuses. We believe in the lie detector because it promises us a chance to peek through Momus's window. What will we see when we look within? We believe in the lie detector because—no matter what respectable science says—we are tempted.

# Note on Sources

This book is based on primary sources, most of them unedited manuscripts held in public archives and private collections. Only this type of source enables us to peer behind the scientific facade of lie detection and discern the motives and mores of the men who made the machine their hobby horse. As J. Edgar Hoover said, it is the polygraph operator who is the true lie detector. Of course, much like a polygraph chart, these archival records provide a kind of inky evidence whose trace must be read for its emotional slant and corroborated by other evidence. In this historians are like detectives—as well as voyeurs of a sort.

The family papers of Leonarde, Charles, and Eloise Keeler are in the Bancroft Library at the University of California-Berkeley, although portions of Leonarde Keeler's papers are also at the Department of Defense Polygraph Institute and the Northwestern University Archives. I thank the archivist at Northwestern, Patrick Quinn, for his help. The letters of Katherine Applegate are held by her family, and I thank Joyce Schwartz, Thomas "App" Applegate, and Penelope Eckert for their generosity in sharing them. As none of these repositories include Keeler's criminal cases or the correspondence between Leonarde Keeler and Katherine Applegate, we may assume that these papers were destroyed.

The bulk of John Larson's papers are also in the Bancroft Library; I am most grateful to the archivist, David Kessler, for helping me access these materials. Important additional papers were shared with me through the generosity of Larson's son, Bill Larson; and his former assistant, the psychiatric nurse Beulah Allen Graham. I am extremely grateful for their generosity. Another set of Larson's papers are in the Johns Hopkins University Archives.

The papers of August Vollmer and the administrative files of the Berkeley Police Department are likewise in the Bancroft Library. The case records of the Berkeley Police Department are held by the department, and I am deeply indebted to Sergeant Michael Holland for help in locating them. Some of William Moulton Marston's papers are in the Smithsonian Institution, but the bulk are still in the possession of his family; I thank his granddaughter, Margaret Lampe, who allowed me to consult her portion. I am very grateful to Penelope Eckert, Fred Inbau, Len Harrelson, and Bill Larson for granting me interviews.

In this book, I have drawn on the work of many scholars, some of whose works are cited in the Notes and Selected Bibliography. Although this is the first history of the lie detector, two related works guided my thinking. Tal Golan's fine book of 2004—*The Laws of Men and the Laws of Nature*—covers some of the same ground as my Chapter 4. I also admire the excellent dissertation by Geoffrey C. Bunn, "The Hazards of the Will to Truth: A History of the Lie Detector" (1997), though my approach differs from his. I first laid out my own thinking on the lie detector in two academic articles: "To Tell the Truth: The Polygraph Exam and the Marketing of American Expertise," *Historical Reflections* 24 (1998): 487–525; and "A Social History of Untruth: Lie Detection and Trust in Twentieth-Century America," *Representations* 80 (2002): 1–33.

The biography of Leonarde Keeler by his sister Eloise Keeler, *The Lie Detector Man: The Career and Cases of Leonarde Keeler* (N.p.: Telshare, 1984), is based on documents and recollection, although silent about Keeler's shortcomings. A candid look at Keeler's personal life can be found in the unpublished biographical portrait by Agnes de Mille; I thank her son Jonathan Prude for letting me quote from it. There are, of course, many articles and books about the polygraph for both scientific and lay audiences. These vary widely in their purpose and quality. The most cogent summary of the technique's merits (or lack thereof) can be found in a report of the U.S. National Academy of Sciences, *The Polygraph and Lie Detection* (2003).

Throughout the text, I have silently corrected misspellings and grammatical errors in original quotations, except to convey the idiosyncrasies of John Larson. To refer to the generic subject of a lie detector test, I have (roughly) alternated between the pronouns "he" and "she," while reserving "he" for the operator, who was almost always a man. I have provided pseudonyms for polygraph subjects whose personal information I gathered solely from archival sources; by contrast, I have supplied medical, personal, or psychiatric information about the leading interrogators when I acquired that information through material which they or their heirs made publicly available. Scholars seeking access to documents on the use of the polygraph at the FBI and Oak Ridge obtained through the Freedom of Information Act should contact the author.

The archives and libraries I consulted are listed below by abbreviation. In the notes, I provide citations for direct quotations, quantitative claims, and some factual matters. The numbers that follow the archival collection refer to the box number. All dates are in the month/day/year (or month/year) format, with all years from the 1900s unless otherwise indicated.

## Abbreviations of Archives and Libraries

ADMP Agnes de Mille Papers, New York Public Library of the Performing Arts, New York

AVP August Vollmer Papers, Bancroft Library, University of California at Berkeley, BANC/MSS/C–B 403

BPHS Berkeley (California) Police records, Berkeley Police Historical Preservation Society

BPP Berkeley Police Department Papers, Bancroft Library, University of California at Berkeley, BANC/MSS/72/227c

CCCA Chicago Crime Commission Archives, Chicago, Illinois

CCCP Cook County Court Papers, Chicago, Illinois

CKP Charles Keeler Papers, Bancroft Library, University of California at Berkeley, BANC/MSS/C–H 105

DDRS Declassified Document Reference System, Thomson-Gale Databases

EFMP E. F. McDonald Papers, Zenith Corporation, Lincolnshire, Illinois

EKP Eloise Keeler Papers, Bancroft Library, University of California at Berkeley, BANC/MSS/93/121c

FBI FBI polygraph materials (responsive to FOIA request, documents in possession of author)

FGP Fred Green Papers, Northwestern University Archives, Evanston, Illinois

| | |
|---|---|
| FIP | Fred Inbau Papers, Northwestern University Archives, Evanston, Illinois |
| HMP | Hugo Münsterberg Papers, Boston (Massachusetts) Public Library |
| ISA | Illinois State Archives, Springfield |
| JLP | John Augustus Larson Papers, Bancroft Library, University of California at Berkeley |
| JLP-BG | John Augustus Larson Papers, personal collection of Beulah Allen Graham |
| JLP-JH | John Augustus Larson Papers, Chesney Medical Collection, Johns Hopkins University, Baltimore, Maryland |
| JWP | John Wigmore Papers, Northwestern University Archives, Evanston, Illinois |
| KAP | Katherine Applegate [Keeler, Dussaq] Papers, personal collection of Penelope Eckert |
| LKP | Leonarde Keeler Papers, Bancroft Library, University of California at Berkeley, BANC/MSS/70/40c |
| LKP-DoD | Keeler Papers, Department of Defense Polygraph Institute, Fort Jackson, South Carolina |
| NARA | National Archives and Records Administration, Washington, D.C. |
| NRCP | National Research Council Papers, National Academy of Science, Washington, D.C. |
| ORP | Oak Ridge Papers, Atomic Energy Commission, Oak Ridge, Tennessee (responsive to FOIA request, documents in possession of author) |
| RNP | Richard Nixon Papers, Richard Nixon Library, Yorba Linda, California |
| TWU | Katherine Applegate [Keeler, Dussaq] Papers, Texas Women's University, Denton |
| UCA | University of Chicago Archives, Chicago, Illinois |
| UICA | University of Illinois at Chicago Archives |
| WJP | William James Papers, Houghton Library, Harvard University, Cambridge, Massachusetts |

WMP William Moulton Marston Papers, personal collection of Margaret Lampe

WMP-SI William Moulton Marston Papers, Dibner Collection, Smithsonian Institution, Washington, D.C.

## Abbreviations of Newspapers and Journals

*AJPS* *American Journal of Police Science*

*BG* *Berkeley [Daily] Gazette*

*CDN* *Chicago Daily News*

*CT* *Chicago [Daily] Tribune*

*JCLC* *Journal of Criminal Law and Criminology* (changes title several times)

*LAT* *Los Angeles Times*

*NYT* *New York Times*

*SFC* *San Francisco Chronicle*

*SFCP* *San Francisco Call and Post*

*SFE* *San Francisco Examiner*

*WP* *Washington Post*

## Abbreviations of Proper Names in the Notes

A.V. August Vollmer

C.K. Charles Keeler

E.K. Eloise Keeler

J.L. John [Augustus] Larson

K.A. Katherine Applegate (first marriage to Leonarde Keeler, second to René Dussaq)

L.K. Leonarde Keeler

W.M. William [Moulton] Marston

# Notes

### PREFACE

xii *"all the man's thoughts":* Lucian, "Hermotimus; or, The Sects," in *Translations from Lucian,* ed. Augusta M. Campbell Davidson (London: Longmans, Green, 1902), 72.

xii *"Who's Lying to You":* R. Don Steele, *Body Language Secrets: A Guide During Courtship and Dating* (Whittier, California: Steel Balls Press, 1996), back cover.

xiii *In 2006, one review:* Bond and DePaulo, "Accuracy of Deception Judgments." DePaulo et al., "Accuracy-Confidence Correlation." The lie-detecting ability of the best professionals—federal officers—topped 70 percent, but that was for liars already screened for having exhibited behavioral clues; see Paul Ekman, Maureen O'Sullivan, and Mark G. Frank, "A Few Can Catch a Liar," *Psychological Science* 10 (1999): 263–265.

xiv *2 million Americans:* Gale, in *Polygraph Test,* 7. For numbers of polygraphers in U.S. and abroad, see Gordon H. Barland, "The Polygraph Test in the USA and Elsewhere," in Gale, *Polygraph Test,* 75. The only countries outside the U.S. to use the polygraph to any degree have a close national-security relationship with America—such as Israel and Japan.

xiv *from only 35 percent:* Office of Technology Assessment, *Scientific Validity of Polygraph Testing.*

xiv *53 percent:* David Thoreson Lykken, "The Case against Polygraph Testing," in Gale, *Polygraph Test,* 117. See also Lykken, *A Tremor in the Blood.*

### PART 1. THE ATHENS OF THE PACIFIC

1 Gertrude Stein, *A Novel of Thank You* (New Haven, Conn.: Yale University Press, [1925–1926], 1958), ch. CXCIII.

### CHAPTER 1. "SCIENCE NABS SORORITY SNEAK"

3 *"Her eyelids drooped":* Dashiell Hammett, *The Maltese Falcon* (New York: Random House, [1929], 1992), 89.

4 *"one of these big"* to *"very nervous type":* BPHS/64024: Fisher, 3/31/21–4/14/21. See also *BG,* 3/31/21.

5 *"eliminate all personal factors":* J.L., "Modification," 392.

6 *"If a lie":* Michel de Montaigne, *The Complete Essays,* trans. M. A. Screech (London: Penguin, 1991), 35.

7 *"a preliminary":* BPHS/64024: J.L., 3/30/21.

8 *"Do you like college?":* BPHS/64023: J.L., "Questions Used," [4/21]; a slightly different version appeared in J.L., "Modification," 397.

8 *"uncomfortable and painful":* J.L., "Modification," 393.

9 *"We forcibly prevented":* BPHS/64024: Larson, "Report," 5/7/21.

9 *"broke down":* BPHS/64024: Larson, "Report," 5/7/21.

11 *"Beyond my expectation":* JLP/7: J.L. to Reg Manning, [early 1960s].

11 *"It was an odd way":* George Haney quoted in JLP-BG: unidentified news clipping, James Craig, "Stories," n.d.

11 *"Fixing the":* *SFE,* 8/9/22.

12 *"pure hooey":* JLP: J.L. to W.M., 4/7/36.

12 *"might have been reacting":* J.L. in *Detroit News,* 1/23/48. See also David Redstone, "The Case of the Dormitory Thefts," *Reader's Digest* 51 (12/47): 18–21; JLP/7: Redstone to J.L., 4/10/47, 6/30/47, 7/16/47; J.L. to Redstone, 7/3/47; J.L., "Modification," 394; author interviews with Bill Larson, 9/11/2003, 5/12/2005.

13 *"My first knowledge"* to *"nervous breakdown":* BPHS/64023: Graham, as recorded by Fisher, 5/3/21; Graham (in Kansas) to J.L., 5/24/21.

13 *"all the indications":* J.L., "Modification," 398.

13 *"I am very sorry":* BPHS/64024: J.L. to Graham, [5/24].

14 *"Dr. Larson is":* BPHS/64023: Graham to A.V., [1922].

14 *"Listening in on":* A.V., "Caught by Lie Detector," *LAT,* 7/2/22.

15 *"I sometimes feel":* LKP-DoD: Richard North to Harold Morrison, 1/4/[68?].

## CHAPTER 2. POLICING THE POLIS

17 *"In a certain sense":* Hans Gross, *Criminal Psychology: A Manual for Judges, Practitioners, and Students,* trans. Horace M. Kallen (Boston, Mass.: Little, Brown, [1898], 1918), 474.

17 *"the most significant":* Joseph G. Woods, "Introduction," in Vollmer, *Law Enforcement in Los Angeles,* no page.

17 *"the genius and altruism":* J.L., *Single Fingerprint,* frontispiece.

17 *"First of all":* JLP/8: A.V. to J.L., 2/11/24.

18 *"mental acuity":* *BG,* 4/10/05, quoted in Carte and Carte, *Police Reform,* 19.

21 *"You're not to judge":* William Dean, in *August Vollmer: Pioneer in Police Professionalism* (Berkeley: Regional Oral History Office, Bancroft Library, University of California, 1972–1983), 1:1.

21 *90 percent of the "problem":* A.V., "Predelinquency," *JCLC* 14 (1923–1924): 279–283. Elisabeth Lossing, "The Crime Prevention Work of the Berkeley Police Department," in *Preventing Crime: A Symposium,* ed. Sheldon and Eleanor Glueck (New York: McGraw-Hill, 1936), 237–263.

21 *On his watch:* Anne Roller, "Vollmer and His College Cops," *The Survey* 62 (1929): 305. See also Julia Liss and Steven Schlossman, "The Contours of Crime Prevention in August Vollmer's Berkeley," *Research in Law Deviance and Social Control* 6 (1984): 79–107.

22 *"All three had":* William Dean, in *August Vollmer: Pioneer,* 1:1.

22 *"strike any person":* John Holstrom, in *August Vollmer: Pioneer,* 1:24.

22 *"mildly": Berkeley Record,* 6/13/06, quoted in Lawrence M. Friedman and Robert V. Percival, *The Roots of Justice: Crime and Punishment in Alameda County, California, 1870–1910* (Chapel Hill: University of North Carolina Press, 1981), 219.

23 *"Were you off"* to *"I don't need":* JLP/2: J.L., "Faked Holdup" (partial typescript), n.d. See also JLP/7: "Hoist on His Own Petard," n.d.

24 *"fearing," he recalled:* JLP: J.L., "Some past biographical notes," n.d.

24 *Larson created:* J.L., "Heredity in Finger-Prints" (M.A. thesis, Boston University, 1915).

24 *Ph.D. in physiology:* J.L., "On the Functional Correlation of the Hypophysis and the Thyroid . . ." (Ph.D. dissertation, University of California, Berkeley, 1920).

25 *"You go to":* J.L., *Lying,* 305. JLP/2: J.L., "Faked Holdup," n.d.

25 *"Mrs. Simons accused":* JLP/2: J.L., "Code Numbers: Berkeley Police Records," n.d.

25 *"Friend Larson":* BPHS/72848: "Battery" [W.B.] to J.L., 7/10/22. See also J.L., *Lying,* 290–294.

26 *"disturbances due":* J.L., *Lying,* 329–331.

26 *"very smooth":* BPHS/72828: "Larceny," 7/16/22.

27 *"Have you ever": BG,* 1/12/23.

27 *In two years:* J.L., "Polygraph and Deception," 26–27.

27 *"'lieing machine'": Officer 444* (Ben Wilson Production, Goodwill Pictures, 1926).

27 *"most convincing case":* Vollmer in *SFE,* quoted in J.L., *Lying,* 369. For Hahn case, see *BG,* 12/22/22; BPHS, J.L., "Case Book," [1921–1923].

## CHAPTER 3. A WINDOW ON THE SOUL

29 *"Then the Officer":* Franz Kafka, *In the Penal Colony,* trans. Ian Johnston [1919], (2003), http://www.mala.bc.ca/~johnstoi/kafka/inthepenalcolony.htm.

30 *"The Invincible Determination": SFE,* 5/20/1887, quoted in David Nasaw,

*The Chief: The Life of William Randolph Hearst* (Boston, Mass.: Houghton Mifflin, 2000), 77–78.

30 *"William A. Hightower is unshaken": SFE,* 8/17/21.

30 *"My dreams are": SFC,* 8/17/21.

31 *"Nothing could have been": SFCP,* 8/17/21.

31 *"100% accurate":* J.L., "Scientist Explains Features of Test," *SFCP,* 8/17/21.

31 *"no question":* A.V., "Proof Conclusive, Vollmer's Verdict," *SFCP,* 8/17/21.

31 *"the Lie Detector":* By the time the Hightower case was resolved, the term "lie detector" had become current, although I have not been able to locate the first citation. The first citation in *The Oxford English Dictionary* is Larson's later use in 1922; see J.L., "Berkeley Lie Detector." See also John Bruce, "The Flapjack Murder," in *San Francisco Murder,* ed. Joseph Henry Jackson (New York: Duell, Sloan, and Pearce, 1947): 213–242; J.L., *Lying,* 369–373.

31 *"No longer can we": SFE,* 6/20/22.

32 *"Haven't you enough?": SFE,* 5/31/22.

32 *"My daddy loved": SFCP,* 6/5/22.

32 *"a combination of a radio": SFE,* 6/10/22.

33 *"hands were clean": SFC,* 6/10/22.

33 *"hasty examination":* J.L., "Lie Test Clears Wilkens, Declares Larson," *SFE,* 6/10/22.

33 *"run for its money": SFCP,* 6/10/22.

33 *"doubtful":* BPP/10: J.L. to A.V., 4/20/27. See, however, how A.V. backed J.L. in *BG,* 6/10/22; J.L., *Lying,* 378–381.

35 *The next day, John Larson:* International Association of Chiefs of Police, *Proceedings, Twenty-Ninth Convention* (1922), 2:214–219.

35 *"would not be countenanced":* Captain Duncan Matheson, quoted in J.L., "Psychology in Criminal Investigation," *Annals of the American Academy of Political and Social Science* 146 (1929): 258–268, quotation, 263. There is some ambiguity about why Duncan objected to the tests. For police rumors, see J.L., *Lying,* 409.

35 *"How the Electric":* J.L., "The Polygraph and Deception," *Welfare Magazine* 18 (5/27): 646–669.

36 *"I am going to take": BG,* 6/26/23.

36 *truth serum was unethical:* BPP/10/1: J.L. to House, 6/11/23. JLP/8: J.L. and Dwight S. Beckner, "The Dangerous Truth about 'Truth Sera,'" n.d. For ban, see Missouri case, *State v. Hudson,* 289 S.W. 920 (1926).

36 *"are so striking":* J.L., "Berkeley Lie Detector," 628.

## CHAPTER 4. MONSTERWORK AND SON

39 "POLONIUS: *See you now*": *Hamlet,* II: i.

39 *In July 1922:* The most reliable account of *Frye* is J. E. Starrs, "'A Still-Life Watercolor': *Frye v. U.S.,*" *Journal of Forensic Science* 27 (1982): 684–694; but it suffers from inaccuracies. Frye's own version of events can be found in his pardon plea of fifteen years later; see NARA, RG204/56: Frye, "Application for Executive Clemency," 7/21/36.

41 *"weak and dangerous":* Ulpian, *Digest,* 48.18.1.23, quoted in Edward Peters, *Torture* (Oxford: Basil Blackwell, 1985), 34.

41 *"on the rack":* Gottfried Leibniz, quoted in Peter Pesic, "Wrestling with Proteus: Francis Bacon and the 'Torture' of Nature," *Isis* 90 (1999): 81–94, quotation, 82.

42 *"the nature of things":* Bacon, *The Great Instauration,* in *Works,* ed. James Spedding (London: Longmans, 1857–1874), 4:29.

42 *"to extremities":* Bacon, *Works,* 6:726.

42 *"When bodies":* Bacon, *Works,* 4:199–200.

42 *"By the laws of England":* Bacon, *Certaine Consideration Touching the Better Pacification and Edification of the Church of England* (London: Tomes, 1604), no page.

42 *"make windows":* Bacon, "Certain Observations, upon a Libell," [1592], in Bacon, *Resuscitatio* (London: Griffin, 1657), 127–128.

42 *"beyond a reasonable doubt":* Barbara J. Shapiro, *Beyond Reasonable Doubt and Probable Cause: Historical Perspectives on the Anglo-American Law of Evidence* (Berkeley: University of California Press, 1991); Langbein, *Torture and the Law.*

43 *The institution of the jury:* George Fisher, "The Jury's Rise as Lie Detector," *Yale Law Journal* 107 (1997): 575–713.

44 *"the common liar":* William L. Foster, "Expert Testimony—Prevalent Complaints and Proposed Remedies," *Harvard Law Review* 11 (1897): 169–186, quotation, 169.

45 *"We must bring":* Hugo Münsterberg, *On the Witness Stand: Essays on Psychology and Crime* (New York: Doubleday, 1908), 122–123.

45 *"[O]ur pulses strapped":* Grace Hollingswoth Tucker, "The Gods Serve Hebe," *Radcliffe Quarterly* (10/33): 192–204, quotation, 202.

45 *"the ablest":* William James to Henry James, 4/11/1892, in James, *Correspondence,* ed. Ignas K. Skrupskelis and Elizabeth M. Berkeley (Charlottesville: University Press of Virginia, 1992–2004), 2:217. William James, "What Is an Emotion?" (1884). For James's revisions, see James, "The Physical Basis of Emotion," *Psychological Review* 1 (1894): 516–529.

46 *"unquestioned first":* WJP: William James to Hugo Münsterberg, 5/15/1892.

46 *"Now I am yours":* WJP: Hugo Münsterberg to William James, 3/19/97.

46 *"automaton":* Gertrude Stein, "In a Psychological Laboratory," 12/19/1894, Gertrude Stein Papers, Yale University, quoted in Coventry Edwards-Pitt, "Sonnets of the Psyche: Gertrude Stein, the Harvard Psychological Lab, and Literary Modernism" (Senior Thesis, History of Science Department, Harvard University, 1998), 98.

46 *"Her record":* Ibid.

47 *"scientific conscience":* Münsterberg, *Witness Stand,* 137–171.

47 *"pierce his mind"* to *"not the slightest":* HMP: Münsterberg, "Experiments with Harry Orchard," [July 1907].

48 *"Monsterwork":* "'I Can Tell If You're a Liar!'" *NYT,* 9/15/07. See also Margaret Münsterberg, *Hugo Münsterberg: His Life and Work* (New York: Appleton, 1922), 149.

48 *"yellow psychology":* Charles Moore, "Yellow Psychology," *Law Notes* (10/07): 125–127. See also Wigmore, "Muensterberg."

48 *"Dr. Münsterberg can have":* Quoted in Burnham, *Superstition,* 94.

48 *"My nerves protest":* HMP: Münsterberg, "Experiments with Harry Orchard," [1907].

48 *"the methods of experimental":* Münsterberg, *Witness Stand,* 108–109.

48 *"truth-compelling machines": NYT,* 9/11/07.

48 *"There will be no jury": NYT,* 9/10/11.

49 *"This study":* W.M. quoted in *Boston Evening Standard,* 2/26/15.

49 *"the girl from Mt. Holyoke":* W.M., *Lie Detector Test,* 47–48.

49 *96 percent of the liars:* W.M., "Systolic Blood."

49 *100 percent of the liars:* W.M., "Physiological Possibilities." See also NRCP: W.M., "Report on Deception Tests," [10/17].

49 *"with very strong consciousness":* NRCP: W.M. to Yerkes, 1/21/18.

49 *"The factor of voluntary":* NRCP: W.M. to Yerkes, 2/23/18.

50 *"always proven entirely":* NRCP: Herbert Langfeld, 10/8/17.

50 *74.3 percent of the time:* NRCP: W.M. and John E. Anderson, "Deception Test," 12/10/18.

51 *"a matter of common knowledge"* to *"is what the jury is for":* NARA/RG21/38325: McCoy, *U.S. v. Frye,* 12–13, 18, 12. *Washington Daily News,* 7/20/22.

52 *"If such tests":* Zachariah Chafee, Jr., "The Progress of the Law, 1919–1921," *Harvard Law Review* 35 (1921–1922): 309.

52 *"Just when a scientific": Frye v. United States,* 293 Fed. 1013 (12/3/23).

52 *"a thing for the laboratory":* McCormick, "Deception Tests."

53 *"prevent detection":* Dorritt Stumberg, "A Comparison of Sophisticated and Naïve Subjects by the Association-Reaction Method," *American Journal of Psychology* 36 (1925): 88–95, quotation, 95.

53 *Marston himself, working with his wife:* W.M., "Sex Characteristics," 387–419; W.M., "Systolic Blood Pressure and Reaction Time Symptoms," 143–179.

54 *California State Medical Association:* "The Lie Detector," *California State Journal of Medicine* 20 (July 1922): 218.

## CHAPTER 5. THE SIMPLE HOME

55 *"A detective official":* Dashiell Hammett, "From the Memoirs of a Private Detective," *Smart Set* (3/23).

55 *"Send him down":* LKP/3: C.K. to L.K., 3/23/25. See also CKP/c5: C.K., "August Vollmer: Friends Bearing Torches," [1931–1932].

55 *"with the affection":* LKP/2: A.V. to Viola Stevens, 4/21/50.

55 *But there was another episode:* Warren Olney III, *Law Enforcement and Judicial Administration in the Early Earl Warren Era,* Earl Warren Oral History Project (Berkeley, Calif.: Bancroft Library, 1981), 63–65.

56 *"snake charmer and rotten liar":* JLP-BG: Waterbury to J.L., 12/28/35. See Waterbury's version in JLP/8: Frank Waterbury to J.L., 8/30/57, 9/16/57.

56 *"easy":* LKP/1: L.K., "The Lie Detector," *Cub Tracks* (4/26): 105–107, 138–140, quotation, 105.

57 *"Do you love":* E.K., *Lie Detector Man,* 1.

57 *"I imagine I was afraid":* LKP/3: E.K. to L.K., 12/3/44.

58 *"could fill stadiums":* E.K., quoted in C.K., *Simple Home,* xxxi.

58 *"I crave the truth":* Charles Keeler, "The Truth," *The Victory: Poems of Triumph* (New York: Gomme, 1916), 35.

58 *"simple and genuine"* to *"create it":* C.K., *The Simple Home,* xvi, xlv.

58 *"that you would grow":* LKP/3: C.K. to L.K., 8/17/30.

59 *"courage and initiative":* CKP/3: C.K. to L.K., 1/29/36.

59 *"indoctrinated":* CKP/9: C.K., "Cosmic Religion for the Upbringing of Children," n.d.

59 *"love, truth and beauty":* C.K., *An Epitome of Cosmic Religion* (Berkeley, Calif.: At the Sign of the Live Oak, 1925), 9.

60 *"the greatest single":* LKP/3: C.K. to L.K., 1/20/[35?].

60 *"the most beautiful": BG,* 4/29/21.

61 *"This is my land":* ADMP: Agnes de Mille, "[Keeler]," [1980s].

61 *"the best of all":* Agnes de Mille, "Do I Hear Violins?" *NYT Sunday Magazine,* 5/11/75.

61 *"My God":* ADMP: Agnes de Mille, "[Keeler]," [1980s].

## CHAPTER 6. POISONVILLE

63 *"And don't kid"*: Dashiell Hammett, *Red Harvest* (New York: Random House, [1929], 1992), 119.

63 *"battle politicians"*: *LAT*, 8/2/23.

63 *"honeycombed with crooks"*: *Los Angeles Record*, 6/11/23.

64 *"Is it true"*: G. A. Briegleb, R. P. Shuler, in *LAT*, 6/14/23.

64 *"semi-intelligent"*: *LAT*, 6/11/23.

64 *"WHO RUNS"*: Editorial, *LAT*, 8/18/23.

65 *"Business has continued"*: Editorial, *LAT*, 4/25/22.

65 *"divorce"*: *LAT*, 8/4/23.

65 *"Somehow one feels"*: *LAT*, 10/13/23.

66 *"hub of the police wheel"*: *LAT*, 9/29/23.

66 *"I am going to strip"*: *Los Angeles Record*, 10/13/23.

66 *"pressure"*: A.V., in [Wickersham], *Report on Police*, 14:61.

66 *"foolish worship"*: *LAT*, 2/12/24.

67 *"a modified, simplified"*: *LAT*, 1/25/24.

67 *"morbid, painted-up"*: LKP/3: L.K. to C.K., 11/28/23.

67 *"first victim"*: *LAT*, 1/25/24.

67 *"giant negro"*: *LAT*, 3/2/24.

67 *"great inward emotion"*: *LAT*, 2/1/24.

68 *"It seems to me"*: JLP/7: L.K. to J.L., 4/22/24.

68 *"You did right"*: JLP/7: J.L. to L.K., 5/5/24.

68 *"uncanny"*: *LAT*, 2/11/24.

69 *"the inhumane methods"*: *LAT*, 3/2/24.

70 *"In the solemn secrecy"*: JLP/7: Sloan to J.L., 6/26/24, provides an insider's account; see also *LAT*, 7/11/24.

70 *"mechanical truthseeker"*: *LAT*, 4/11/24.

71 *"inhuman third degree"*: *LAT*, 7/22/24.

72 *"the hounds"*: AVP: A.V. to Millicent Gardner Fell, [1923–1924].

72 *the results of his IQ survey:* A.V., in [Wickersham], *Report on Police*, 14: 59–60; Woods, "Progressives," 176–190.

72 *"tired, jaded"*: *Los Angeles Record*, 6/14/24.

72 *"The first of September"*: Alfred E. Parker, *Crime Fighter: August Vollmer* (New York: Macmillan, 1961), 164; see also *LAT*, 7/2/24.

72 *"Sometimes a frame-up"*: *LAT*, 8/9/24.

72 *"a cave man"*: *LAT*, 8/9/24.

72 *"If such a test"*: *LAT*, 8/9/24.

72 *"either mentally unbalanced"*: *LAT*, 8/8/24.

73 *"diseased"*: J. J. Mart, quoted in Vollmer, *Law Enforcement in Los Angeles*, 116.

73 *"The Chinese have":* Clyde Plummer, quoted in Vollmer, *Law Enforcement in Los Angeles,* 172.

73 *"poorly assimilated":* R. Lee Heath, quoted in Vollmer, *Law Enforcement in Los Angeles,* 163.

73 *"[I]t is my opinion":* A.V. to Guy E. Marion, 6/3/33, quoted in Carte and Carte, *Police Reform,* 61.

## CHAPTER 7. "SUBJECTIVE AND OBJECTIVE, SIR"

75 *"[A]nd even psychology":* F. Scott Fitzgerald, *This Side of Paradise,* ed. James L. West III and Lynn Seltzer (New York: Simon and Schuster, [1920], 1998), 80.

75 *"He is a practical":* CKP/3: C.K. to Rieber, 9/9/24.

76 *"the billiard ball tops":* LKP/2: Heck [Ralph Brandt] to L.K., [1924].

76 *"If study interferes":* Percy Marks, *The Plastic Age* (Carbondale: Southern Illinois University Press, [1924], 1980), 135.

76 *"And what are the symptoms":* LKP/3: L.K. to "Youngster," 1/27/26.

76 *"Rattlesnake":* LKP/2: Alyson to L.K., n.d.

76 *"sweetest when stewed":* LKP/3: E.K. to L.K., 9/26/27.

76 *A notebook:* LKP-DoD: L.K., "Physiological and Psychological Influences on Quantitative Blood Pressure," April 1925–June 1930.

78 *"let [the courts] come":* JLP/8: J.L. to L.K., 5/5/24.

78 *"Yes, I am ashamed":* LKP/3: L.K. to A.V., 5/20/25.

78 *"You might not":* JLP/7: J.L. to L.K., 3/23/27.

78 *"After all your kindnesses":* JLP/8: L.K. to J.L., 5/1/27.

79 *"Recording Arterial Blood Pressure":* LKP/3: C.K. to L.K., 2/20/30.

79 *"great scientific interest":* LKP/1: Eckhoff (White, Prost, and Fryer) to L.K., 8/6/29.

79 *"Apparatus for Recording":* L.K., "Apparatus," Patent, No. 1,788,434 (1/31/31).

80 *"turn[ed] out machines":* LKP/1: L.K. to W. J. Foster, 4/23/31.

80 *"Emotograph":* LKP/3: C.K. to L.K., 2/7/30.

80 *"Respondograph":* JWP: L.K. to John Wigmore, 8/12/29.

81 *"commercial and purely mechanical":* JLP-BG: J.L. and Robert Borkenstein, "Present Status of Lie Detector . . . ," [late 1950s].

83 *all but two out of sixty:* LKP/3: L.K. to C.K., 2/3/26.

83 *According to Applegate:* Author interviews with Penelope Eckert (10/19/2003) and Thomas "App" Applegate (1/18/2004).

83 *"We bought some":* KAP: K.A. to parents, 7/26.

83 *"the lie detector man":* KAP: K.A. to parents, 2/18/26.

84 *"Nard Keeler is the pleasantest":* KAP: K.A. to parents, 2/18/26.

84 *"Friday night":* KAP: K.A. to Clare Applegate, [1926].

84 *"an aloof dignity":* E.K., *Lie Detector Man,* 24.

84 *"startlingly brilliant":* LKP/3: L.K. to C.K., 3/5/33.
84 *"He is over six":* KAP: K.A. to parents, 12/26/24.
84 *"get it through":* KAP: K.A. to parents, 6/21/26.
85 *"Keeler has built":* JLP/8: A.V. to J.L., 5/1/29.
85 *"You're the only":* JLP/7: L.K. to J.L., 12/3/27.

## PART 2. IF THE TRUTH CAME TO CHICAGO

87 *"Law should be":* William T. Stead, *If Christ Came to Chicago!* (Chicago, Ill.: Laird and Lee, 1894), 354.

## CHAPTER 8. THE CITY OF CLINICAL MATERIAL

89 *"He that has eyes":* Sigmund Freud, "Fragment of an Analysis of a Case of Hysteria," [1901], *The Standard Edition of the Complete Psychological Works* (London: Hogarth, 1953–1974), 7:77–78.
89 *Boarding the "el":* JLP/2: J.L., "Faked Hold-Up," and JLP/8: J.L. to Charles Sloan, 7/2/24.
89 *"Nothing at home":* Clifford Shaw, *The Natural History of a Delinquent Career* (Chicago, Ill.: University of Chicago Press, 1931), 62–63. One of "Blotzman's" favorite books was *Sister Carrie,* and his description enthusiastically echoes her arrival.
90 *"playing the loop":* BPP/10: J.L. to A.V., 8/1/24.
90 *"the disease is ushered":* J.L. and A. Walker, "Paranoia," 353.
91 *"city fronting a lake":* BPP/10: J.L. to A.V., 4/1/26.
91 *"has any [city]":* BPP/10: J.L. to A.V., 8/9/24.
91 *"[T]he cops use":* BPP/10: J.L. to A.V., 12/12/24.
91 *"rubber hose":* BPP/10: J.L. to A.V., 10/17/24.
91 *"not Utopian":* J.L., "Present Police," 253.
91 *"I sure would like":* BPP/8: J.L. to A.V., 8/9/24.
93 *"swift implacable justice":* Robert E. Crowe, quoted in Hal Higdon, *Leopold and Loeb,* 64.
93 *"The mawkish sympathy":* Henry Barrett Chamberlin, "Crime as a Business in Chicago," *Bulletin of the Chicago Crime Commission* (10/1/19): 3–4.
93 *Larson first explained:* J.L., "The Use of the Polygraph in the Study of Deception," *Welfare Magazine* 18 (May 1927): 646–669, see 665. Keeler describes the method in similar terms in County of Green Lake Circuit Court: *State of Wisconsin v. Tony Grignano and Cecil Loniello* (2/35), 405–457.
94 *"comes hard and late":* G. Stanley Hall, "Children's Lies," *Pedagogical Seminary* 1 (1911): 211–218, quotation, 217.
94 *"precocious":* G. Stanley Hall, *Adolescence: Its Psychology* (New York: Appleton, 1904), 1:xvi; for deception, see 2:366.

94 *Studies seemed to show:* Hugh Hartshorne and Mark A. May, "Studies in Deceit," in *Studies in the Nature of Character* (New York: Macmillan, [1928], 1930), 1:245, 271. Lloyd Lewis and Herman Justin Smith, *Chicago: The History of Its Reputation* (New York: Harcourt, Brace, 1929), 491.

95 *"excessive and notorious":* William Healy and Mary Tenney Healy, *Pathological Lying, Accusation, and Swindling: A Study in Forensic Psychology* (Boston, Mass.: Little, Brown, [1915], 1922), 5.

95 *"clandestine sexual habits":* William Healy, *Honesty: A Study of the Causes and Treatment of Dishonesty among Children* (Indianapolis, Ind.: Bobbs-Merrill, 1915), 100.

95 *"impartial administration":* BPP/10: J.L. to A.V., 8/9/24.

95 *Healy, back in town:* William Healy, *People v. Leopold and Loeb,* 8/4/24, at http://www.leopoldandloeb.com/drhealy.html. See the court-ordered evaluation, Karl M. Bowman, "Richard Loeb," 6/24/24, 76–78, at http://homicide.northwestern.edu/crimes/leopold/.

96 *"good stuff on them":* BPP/10: J.L. to A.V., 9/18/24.

97 *Between 1923 and 1927:* J.L. and G. W. Haney, "Cardio-Respiratory Variations," 1054–1059.

97 *"showing up some of the methods":* BPP/10: J.L. to A.V., 5/31/26.

97 *"run the roost"* to *"It seems these women":* BP/10: J.L. to A.V., 8/9/24.

98 *"colossal joke":* LJP/7: J.L. to A.V., [6/26].

98 *"than this baby here":* BPP/10: J.L. to A.V., 7/27/27.

98 *"Dose guys were kidding":* JLP/2: J.L., "Faked Holdup."

98 *"Any honorable work":* BPP/10: J.L. to A.V., 6/14/26.

98 *"I own the police":* Torrio, quoted in Kenneth Allsop, *The Bootleggers and Their Era* (New York: Doubleday, 1961), 61.

98 *"Nobody knows": CT,* 11/4/23.

99 *"nearly had a hemorrhage":* BPP/10: J.L. to A.V., 5/26/26.

99 *"loaded at both ends":* BPP/10: J.L. to Herman Adler, 5/24/26.

99 *"wrong and a liar":* BPP/10: J.L. to A.V., 6/14/26.

99 *"caveman tactics":* JLP/7: A.V. to J.L., 6/18/26.

99 *"ditch":* BPP/10: J.L. to A.V., 8/9/24.

99 *"scatter-brained":* BPP/10: J.L. to A.V., 6/8/26.

99 *"Your reactions of necessity":* JLP/7: A.V. to J.L., 6/7/26.

100 *Larson had vilified Adler:* LJP/7: J.L. to A.V., [6/26].

100 *"natural inclination":* BPP/10: J.L. to A.V., 6/21/27.

100 *"I at least have":* BPP/10: J.L. to A.V., 7/27/27.

101 *"autopsies on old records":* JLP-JH: Adolf Meyer to J.L., 6/29/29.

101 *"something in his personality":* JLP-JH: Margaret Larson to Adolf Meyer, "Private Note," 6/23/29.

## CHAPTER 9. MACHINE V. MACHINE

103 *"There ain't a police"*: Raymond Chandler, *The Long Goodbye,* in *Later Novels and Other Writings* (New York: Library of America, [1953], 1995), 455.

103 *"The prisons are"*: LKP-DoD: L.K., "Day by Day around a Jail," 1929.

104 *"a nice kindly"*: LKP/3: L.K. to Aunt Kit [Gorrill], 7/23/29.

104 *"will be treated"*: LKP-DoD: L.K., "Day by Day around a Jail," 1929.

104 *Keeler now wanted:* LKP-DoD: L.K., "Merit System," [August 1929].

104 *"corrupt cesspools"*: LKP-DoD/B36: L.K., "Day by Day around a Jail."

104 *"But all this is about"*: LKP/3: L.K. to C.K., 7/18/29.

105 *"I hate the impudence"*: Walter Lippmann, "The Great Confusion: A Reply to Mr. Terman," *New Republic* 33 (1923): 146.

106 *African-American community:* Walter White, "People and Places," *Chicago Defender,* 12/25/43.

107 *"the man's conflicts"*: LKP-DoD: L.K., "Merit System," [8/29].

107 *"Damn rotten"*: LKP/3: L.K. to Aunt Kit [Gorrill], 7/23/29.

107 *"higher type of man"*: CKP/7: L.K. to C.K., n.d. [August–September 1929].

107 *"truth finder"*: LKP/3: C.K. to L.K., 8/17/30.

107 *"the soul to its lair"*: LKP/3: C.K. to L.K., 7/21/30.

107 *"a rescitation"* to *"get my wrath"*: BPP/10: L.K. to A.V., 4/17/30.

107 *"subject A"*: Harold Lasswell, "Certain Prognostic Changes during Trial (Psychoanalytic) Interviews," *Psychoanalytic Review* 23 (1936): 241–247, quotations, 245–246.

108 *"is a heavy one"*: Chafee, in [Wickersham], *Report on Lawlessness,* 11:126; see also 136.

108 *"the greatest blow"*: International Association of Chiefs of Police, 1932, quoted in Leo, "The Third Degree," 55.

109 *"The rack and torture chamber"*: *Brown v. State of Mississippi,* 297 U.S. 278 (1936).

109 *"Here's the best"*: Chafee, in [Wickersham], *Report on Lawlessness,* 11: 130–31.

110 *"smashed the machine"*: *NYT,* 11/23/29.

110 *"I know what"*: J.L., *Lying,* 386.

110 *"TOP BLEW OFF"*: E.K., *Lie Detector Man,* 60.

110 *"the machine will tell"*: J.L., *Lying,* 387.

110 *"more in keeping"*: *NYT,* 11/24/29.

111 *"one of the outstanding"*: Chafee, in [Wickersham,] *Report on Lawlessness,* 11:151.

111 *"laboratory"*: Newman F. Baker and Fred Inbau, "The Scientific Detection of Crime," *Minnesota Law Review* 17 (1932–1933): 602–625, quotation, 613.

112 *"fractures of the cervical"*: *CT,* 11/1/29. See also CCCP: Probate/151387, "In Matter of Anna Gustafson," 6/3/30; Keeler, "Canary."

112 *"This looks as if": CT,* 7/9/30.
113 *"Judge Horner doesn't":* BPP/10: L.K. to A.V., 8/17/30.
113 *And in Wichita:* LKP-DoD: Thomas Jaycox to L.K., 1/20/39.
113 *"She cared more":* LKP/3: C.K. to L.K., 6/13/30.
113 *"on your trail":* LKP/3: C.K. to L.K., [8/29].
114 *"[W]e know that":* LKP/2: C.K. to K.A., 8/13/30.
114 *"wayward husbands":* LKP/3: E.K. to L.K., 8/19/30.
114 *"So This Is Love!": Chicago Herald and Examiner,* 6/1/32.
114 *"infinitely more baffling":* Ernst P. Boas and Ernst F. Goldschmidt, *The Heart Rate* (Springfield, Ill.: Thomas, 1932), 100. See FBI: E. P. Coffey to Edwards, 11/13/35.
116 *"outside the pale":* Rex Collier, "Uncle Sam's Scientists Turn Detective," *Sunday Star* (Washington, D.C.), 9/14/30.
116 *"If there is ever":* Wigmore, *Anglo-American System of Evidence* (1923), 2:237–238.
116 *"As an axe leaves":* Wigmore, *Anglo-American System of Evidence,* 1:544.
116 *"My former skeptical":* JWP: John Wigmore to L.K., 8/8/29.
117 *"to further my knowledge":* LKP/2: Chester Gould, "Application," 11/8/32.
118 *"We take our place":* BPP/10: K.A. to A.V., 5/12/31.
118 *"I don't know": CT,* 12/30/33.

## CHAPTER 10. TESTING, TESTING

119 *"She frowned":* Dashiell Hammett, *The Glass Key* (New York: Random House [1931], 1989), 143.
119 *"the problems of":* JLP/1: J.L. to Douglas Kelley, 12/26/50, describing these efforts retrospectively.
120 *"shouted loudly":* BPP/10: L.K. to A.V., 4/17/30. For the case, see *Appleton Post-Crescent,* 4/1/30–4/12/30. See also LKP-DoD: E.K., "The Hidden Truth: The Black Creek Robbery," [1950].
120 *"doggone John":* BPP/10: L.K. to A.V., 4/17/30.
121 *"artificial":* AVP/18: J.L. to A.V., 1/5/[31], though the letter says 1930.
121 *"fishing expeditions":* JLP/7: J.L. to L.K., 10/31/30.
121 *"infallible":* AVP/18: J.L. to A.V., 1/22/31.
121 *"our young enthusiast":* BPP/10: J.L. to A.V., 4/28/31.
122 *"best modification":* JLP/7: J.L. to L.K., 12/21/31.
122 *"This sort of thing":* BPP/10: J.L. to A.V., 7/2/31.
122 *"unethical":* JLP/7: J.L. to Charles DeLacy, 2/5/41.
122 *"You have never":* JLP/8: J.L. to L.K., 10/4/32.
122 *"to every Tom":* JLP/7: J.L. to L.K., 12/21/31.
122 *"control the instrument":* LKP/2: L.K. to C. Wilson, 7/20/31.

122 *"prostitution and promiscuous use":* JLP/8: L.K. to J.L., 12/28/31.
123 *"tossing about among":* LKP/3: L.K. to A.V., 12/17/29.
124 *What he found shocked:* LKP/2: L.K. to Newman Baker, 7/21/33. See also LKP/2: L.K. to Henry Scarborough, 6/10/31.
124 *"Under the circumstances":* LKP/2: L.K. to E. R. Naugle, 6/3/41.
125 *"At this point":* ADMP: Agnes de Mille, "Keeler," [1980s]; the interrogation took place on 10/31/40.
125 *a clip of 75 percent:* L.K., "A Method for Detecting Deception," *Insurance Claim Journal* 3 (1932): 3–6, 25–26.
126 *After all, the courts: Commonwealth v. Hipple,* 333 Pa. 33 (1939). See also *People v. Becker,* 300 Mich. 562 (1942).
126 *"Don't be a darn": Topeka Daily Capital,* 5/27/33. See L.K., "Scientific Methods."
126 *privately surveyed:* LKP-DoD: Keeler survey of 1938–1940.
127 *"if he had to choose":* LKP-DoD: James Byars to L.K., 1/5/38.
127 *An officer with the Michigan:* L.K., "The Lie-Detector Proves Its Usefulness," *Public Management* 22 (1940): 163–166; reprinted in *Polygraph* 23 (1994): 181–185.
127 *"we rely on the Polygraph":* LKP-DoD: Don Kooken to L.K., 2/28/39.
127 *"legal fictions":* Raymond Moley, "The Vanishing Jury," *Southern California Law Review* 2 (1928): 97–127, quotation, 125. See also George Fisher, *Plea Bargaining's Triumph: A History of Plea Bargaining in America* (Stanford, Calif.: Stanford University Press, 2003).
128 *"The lie detector has not:"* Douglas Kelley, in NARA/RG326/149: AEC, "In the Matter of 'Lie-Detector' Panel Meeting," 1/24/52, 92.
128 *"Two Simple Ways":* Kenneth Murray, "Two Simple Ways to Make a Lie Detector," *Popular Science* 128 (5/36): 63, 98–99.
128 *"LIE!":* David Simon, *Homicide: A Year on the Killing Streets* (Boston, Mass.: Houghton Mifflin, 1991), 204. For recent practice and the laugh, see *U.S. v. Scheffer,* 523 U.S. 303 (1998), oral argument, 11/3/97, at www.oyez.org.
128 *Psychological studies suggest:* Edward E. Jones and Harold Sigall, "The Bogus Pipeline: A New Paradigm for Measuring Affect and Attitude," *Psychological Bulletin* 76 (1971): 349–364.
129 *the nation's top examiners:* Personal communication, DoD polygraph operator, Fort McClellan, Alabama, 6/98.
130 *"sweetheart": CT,* 5/23/31.
130 *"one hundred percent innocent": CT,* 5/13/31.
130 *Keeler expressing his willingness: Vidette-Messenger* (Valparaiso), 5/21/31.
130 *"any fantastic device": CT,* 5/21/31.
130 *"used as a testing room": CT,* 5/22/31.
130 *"Such a step": WP,* 5/21/31.
131 *"I'm sure he was":* BPP/10: L.K. to A.V., 5/26/31. See JLP-BG: J.L. to Crumpacker, 5/27/31; J.L., "The Detection of Lying," *Police 13-13* (7/32): 22.

131 *"Thought-Wave Detector": Look,* 1/4/38, 29.

131 *"Despite his mumbo-jumbo":* JLP/7: J.L. to Claude Broom, 6/25/63.

132 *"Diogenes searched":* CCCA/35943: Orlando Scott, business card, National Detection of Deception Laboratories.

132 *"in the doghouse":* JLP/7: J.L. to W.M., 10/27/37.

133 *"all scientists"* to *"We have made":* BPP/10: J.L. to A.V., 10/9/31.

133 *"probing for complexes":* J.L. and Haney, "Cardio-Respiratory Variations," 1051.

133 *"My own opinion":* JLP/5: A.V. to J.L., 10/1/31.

134 *"I flatly disagree":* BPP/10: J.L. to A.V., 10/9/31.

134 *"Golly, I'm sorry":* LJP/7: L.K. to J.L., 10/22/28.

134 *"While this work":* AVP/18: J.L. to A.V., 8/19/32.

134 *"breach of faith":* JLP/7: A.V. to J.L., 6/16/31.

134 *"source book":* BPP/10: J.L. to A.V., 4/28/31.

134 *"savored of cheap publicity":* Meyer, quoted in JLP-JH: J.L. to Adolf Meyer, 10/24/31.

135 *"Each case is a clinical":* J.L., "Detection," *Police 13-13,* 23.

135 *"ultra-conservative":* Harold Burtt, "*Lying and Its Detection,* by John A. Larson," *New York University Law Review* 10 (1932–1933): 563–565.

135 *"published every slanderous":* AVP/17: L.K. to A.V., 9/19/32.

135 *"It is my opinion":* AVP/42: A.V. to L.K., 10/4/32.

136 *"Our good friend John":* AVP/17: L.K. to A.V., 3/19/34.

136 *"I am glad to hear":* LKP/2: J.L. to L.K., 10/27/32.

137 *"inseparably linked":* UICA/Century of Progress/1-13301: C. Goddard to F. R. Moulon, 6/3/33.

137 *"unconscious schooling":* "Preliminary Report of the Science Advisory Committee," 6/31, quoted in John E. Findling, *Chicago's Great World's Fairs* (Manchester: Manchester University Press, 1994), 93.

137 *"features which might":* UICA/Century of Progress/1-13301: C. Goddard to F. R. Moulon, 6/3/33.

137 *"half-witted":* BPP/16: J.L. to Wiltberger, 3/1/32. *LAT,* 12/6/31.

137 *"overt activities":* AVP/17: L.K. to A.V., 3/19/34.

137 *"a more sympathetic":* UICA/Century of Progress/1-13303: L.K. to Lenox Lohr, 12/4/33.

138 *"snowed under":* JLP-BG: L.K. to J.L., 4/12/35.

138 *"People . . . made the farcical assumption":* Philip Wylie, *Generation of Vipers* (New York: Rinehart, 1942), 37.

## CHAPTER 11. TRACES

139 *"I was fascinated":* Dashiell Hammett, *The Thin Man* (New York: Random House, [1933], 1972), 4.

139 *"fantastic pajamas":* LKP-DoD: L.K., "Calendar," 1931.

140 *"If I had a hundred":* KAP: K.A. to parents, 3/31/32.

140 *"I miss the grand":* BPP/10: L.K. to A.V., 8/1/30.

140 *"He certainly has a gift":* KAP: K.A. to Clare Applegate, 5/1/32.

140 *"Nard is down":* LKP/2: K.A. to C.K., 10/21/30.

140 *"how bride and groom":* LKP-DoD: L.K., "Calendar," 1931.

140 *"man in a hotel":* KAP: K.A. to Clare Applegate, 7/30/32.

141 *"Can you tell":* K.A., "Tattle-Tale Tallies," 324.

142 *"Mrs. Keeler, [a] tall": Chicago Herald and Examiner,* 10/29/30.

142 *"to a nauseatingly":* KAP: K.A. to Clare Applegate, 11/9/32.

142 *"HOW SCIENCE WARS": New Orleans Times-Picayune,* 7/24/38.

143 *"I'd look at some": Chicago Herald and Examiner,* 10/5/39. See also Harrison T. Carter, "Kentucky's Hill-Billy Mail Order Swindle," *Inside Detective* (7/39): 14–19, 59–60.

143 *"[t]he ballot":* LKP-DoD: L.K. to Lorenzo Buckley, 9/25/34. See also Landesco, "Organized Crime," in *Illinois Crime Survey,* 1017–1021.

143 *In 1934, in one recount in Cook County:* K.A., "Tattle-Tale Tallies." See also Harold F. Gosnell, *Machine Politics: The Chicago Model* (Chicago, Ill.: University of Chicago Press, 1937), 87; *People ex rel. John S. Rusch v. Louis Greenzeit,* 277 Ill. App. 479 (1934); K.A., "Election Dispute."

143 *By 1940: NYT,* 11/10/40.

144 *"the extent that modern": Benton Evening News,* 3/23/36.

144 *Keeler tallied:* LKP/3: L.K. to C.K., Labor Day, 1935.

145 *"You can go out":* ISA/36-3000: *People v. John Mitchell McDonald* (3/16/36–3/21/36), 100. All quotations are from the trial. See also LKP/s2: "Lie Detector Traps Bombers," *Chicago American,* [1936].

145 *"Oh, Pete Benetti":* ISA/36-3000: Marion Hart, quoted in *People v. John Mitchell McDonald* (3/16/36–3/21/36), 862.

146 *"reign of terror": Benton Evening News,* 3/23/36; see also 3/20/36, 3/21/36.

146 *"corroborative evidence":* L.K. quoting judge in AVP/17: L.K. to A.V., 2/12/35. See also County of Green Lake Circuit Court: *State of Wisconsin v. Tony Grignano and Cecil Loniello,* 2/1/35–2/8/35, 405–457; AVP/17: L.K. to A.V., 9/19/32; and *Portage Daily Register,* 2/2/35–2/9/35.

146 *"It means that the findings": New York Post,* 2/26/35.

147 *"aura of near infallibility": U.S. v. Alexander,* 526 F.2d 161 (8th Cir. 1975).

147 *Keeler looked forward:* LKP: L.K., "Talk to Sigma [Xi], Mayo Clinic, Rochester, Minn., 1/18/34; L.K., "The Jury System Should Be Abolished," [1930].

148 *"incompetent":* Fred E. Inbau, "Case against the Polygraph," *American Bar Association Journal* 51 (1965): 857.

148 *"I would like": WP,* 12/12/35.

148 *"All scientists":* CKP/17: L.K. to C.K., 1/27/36.
149 *"as deeply as a son":* CKP/3: C.K. to L.K., 1/29/36.
149 *"which I should have controlled"* to *"I wanted publicity":* LKP/3: L.K. to C.K., 2/14/36.
149 *"doubts and suspicions":* LKP/3: C.K. to L.K., 2/20/36.
150 *"might"* to *"the mute lines of doom": CT,* 3/2/37. See also *Forverts,* 3/3/37. ISA/Pardon and Parole Board: "Rappaport," 10/22/36–2/25/37.
150 *Larson wrote:* J.L., *CDN,* 3/5/37. See also JLP/8: Verne Lyon to J.L., 3/4/37.
150 *"I can't afford":* Thomas B. Littlewood, *Horner of Illinois* (Evanston, Ill.: Northwestern University Press, 1969), 146–147.
151 *"Whatever happens":* LKP/3: L.K. to C.K., 3/5/33.
151 *"My sympathetic nervous system":* AVP/17: L.K. to A.V., 2/26/37. See also FGP/18: Dr. O. H. Horrall to Leon Green, 4/2/37.
152 *"I've been trying":* LKP/3: C.K. to L.K., 2/24/35.
152 *"his nervous system":* KAP: K.A. to Clare Applegate, 8/2/37.
152 *"I suppose it sounds":* AVP/3: L.K. to A.V., 9/17/37.
152 *one nonnegotiable condition:* FIP: Fred Inbau to Charles Wilson, 6/9/38; FGP/17: Leon Green to Walter Dill Scott, 6/14/38; author interview with Fred Inbau, 9/5/96.
153 *"until Mr. Keeler":* K.A. quoted in FGP/17: Fred Inbau to Leon Green, 7/18/38.
153 *"As long as I":* KAP: K.A. to Arthur Applegate, [10/38].

## CHAPTER 12. A SCIENCE OF THE SINGULAR

155 *"It was well said":* Edgar Allan Poe, "The Man of the Crowd," *The Complete Tales* (New York: Modern Library, [1850], 1938), 475.
156 *"Something in his personality":* JLP-JH: Margaret Larson to Adolf Meyer, "Private Note," 6/23/29.
156 *"Has the 'Lie Detector' Failed?":* Charles D. Delacy, [J.L.], "Has the 'Lie Detector' Failed?" *Police 13-13* (1/34): 5–6, 13.
156 *"does not apply":* Keeler, quoted in *Kansas City Daily Star,* 5/28/33.
157 *"Lady": CT,* 4/21/34.
157 *"pseudo-epilepsy": CT,* 4/27/34.
157 *"His answers all": CT,* 4/24/34. See CCCA/32514: "Summary of *People v. Allen R. Hammel alias Burt Armstrong,*" 6/13/34, 6/14/34, 6/15/34.
157 *"believes what he": Chicago Herald,* 4/24/34.
157 *"true as truth":* JLP/2: Hammel case file, 2/17/35.
158 *"neurotic and overactive":* JLP/2: J.L., "In Larson's book," n.d.
158 *So Leopold published:* William F. Lanne [Nathan Leopold], "Parole Prediction and a Science," *JCLC* 26 (1935): 377–400. Lloyd E. Ohlin and Richard A. Lawrence [Nathan Leopold], "A Comparison of Alternative Methods of

Parole Prediction," *American Sociological Review* 17 (1952): 268–274. See also UCA/Burgess/12: Nathan Leopold to E. Burgess, 2/8/34, 3/21/52.

158 *"precocious, egocentric"*: JLP-BG: J.L. and Alan Canty, "With Special Reference to Tattoo Symbolization," [1940s].

159 *By the late 1930s:* JLP-BG: J.L. to Bernard Diamond, 11/26/38. See also J.L., "Parole Prediction," *Police 13-13* (12/31): 10–13, 24–25.

159 *"no significant disturbances"*: JLP/2: J.L., "Faked Holdup," n.d.

159 *"narrow minded"*: J.L. in *CT,* 1/30/36.

159 *"Jesus God"*: J.L. quoting Day in *CDN,* 1/29/36.

159 *"abnormal proposals": CDN,* 1/29/36.

160 *"to reward": CDN,* 1/30/36.

160 *"Can we forget":* Bundesen, quoted in Littlewood, *Horner,* 180. See J.L., in *CT,* 2/15/36.

160 *"Poor old John"*: AVP/17: L.K. to A.V., 3/25/36.

160 *"false claims"*: JLP/8: Claude Broom [on behalf of J.L.] to J. Edgar Hoover, [1936].

161 *"same as used"*: J.L. quoted in JLP/8: W.M. to J.L., 6/30/36.

161 *"You have a strange"*: FBI: Verne Lyons to J.L., 5/6/41.

161 *"You didn't think"*: JLP-BG: J.L. to John Favill, 5/17/38.

161 *"revealed disturbances"*: JLP/8: J.L., "Sketches: Torso Case, Cleveland," [1938].

161 *"The crime"*: JLP-BG: J.L and A. Canty, "[M.O.]," c. 1940.

162 *"feebleminded, alcoholic"*: JLP-BG: J.L. and Alan Canty, "With Special Reference to Tattoo Symbolization," [1940s].

162 *"psychopathology, schizophrenia"*: JLP/6: J.L., "Dactyloscopy, Dermatoglyphics, and Psychobiology," talk presented at Ortho-Psychiatric Annual Meeting, 1937. See also *LAT,* 9/28/30.

162 *"magpie den"*: JLP/8: A.V. to J.L., 1/21/24.

162 *"no one has invented"*: J.L. et al., "The 'Lie Detector,' Oversold and Rashly Operated," *Police Journal* 25 (11/9/39): 13.

162 *"a method in this apparent"*: JLP-JH: J.L. to Adolf Meyer, 5/1/47.

## CHAPTER 13. FIDELITY

163 *"Janet Henry laughed"*: Hammett, *Glass Key,* 179.

164 *"a scientific aid"*: UCA/Burgess/193: Burgess, ["Review of O. Scott's Volume on the Lie Detector"], [1930s].

164 *Even in Canada: Toronto Star,* 3/16/35.

164 *"violation of the conscience"*: F. E. Louwage, "A Belgian Police Official's Impressions of His Travels in the U.S.A.," *JCLC* 42 (1951–1952): 237–239. And see "Scientific Detection of Lies," *International Criminal Police Review* 6 (August–September 1951): 230–233.

165 *"Interior language":* André Maurois, *The Thought-Reading Machine,* trans. James Whitall (New York: Harper, 1938), 216–217.

166 *"How would you":* *CT,* 10/19/24. See also *LAT,* 2/18/23.

166 *"For if all men":* Molière, *Misanthrope,* V:i: "Et si de probité tout était revêtu, / Si tous les coeurs étaient francs, justes, et dociles, / La plupart des vertus nous seraient inutiles." Author's translation.

166 *"prostituting the field":* LKP/2: L.K. and Alex Gregory, "Agreement," 6/3/44.

166 *"100 percent accuracy":* *WP,* 9/26/38.

167 *"go ahead and sell":* LKP/2: L.K. notes on Inman to L.K., 8/13/48.

167 *"Multiple choice":* LKP-DoD: [Keeler, Inc.], "Final Examination," n.d.

167 *"No. 26: Discover":* Baltasar Gracián y Morales, *A Truthtelling Manual and the Art of Worldly Wisdom,* ed. Martin Fischer (Springfield, Ill.: Thomas, [1653], 1945), 36.

167 *Keeler's success was:* J. P. McEvoy, "I Cannot Tell a Lie," *Forbes* (1941); reprinted as "The Lie Detector Goes into Business," *Reader's Digest* 38 (2/41): 69–72.

168 *"the third degree":* William Scott Stewart, "How to Beat the Lie Detector," *Esquire* (11/41): 35, 158, 160.

168 *"women's clubs":* McEvoy, "I Cannot Tell a Lie."

168 *"integrity, intentions":* CCCA/35943: Scott, "Summer Bulleton [*sic*]," n.d.

168 *"Little Brainwave":* CCCA/35943: Scott, "Ef You Don't Watch Out!" Christmas card, 1941.

169 *"Personal note to self":* LKP/1: L.K., "Be Friendly," n.d.

171 *"The Girl on the Case": Chicago Herald and Examiner,* 10/3/39–10/13/39.

171 *"Kay was temperamental":* Viola Stevens, quoted in E.K., *Lie Detector Man,* 143.

172 *"You're both too":* LKP/3: E.K. to L.K., 8/31/39.

172 *"It was the core":* ADMP: Agnes de Mille, "[Keeler]," [1980s].

172 *"it does make life":* KAP: K.A. to parents, 5/2/38.

172 *"Rosenfeldt":* LKP/2: L.K. to [Aunt] Betsy [Bunnell], 10/16/40.

172 *"Maybe that's because":* KAP: K.A. to Clare Applegate, 11/9/32.

172 *"You're voting":* ADMP: Agnes De Mille, "[Keeler]," [1980s].

172 *"I've made so many":* KAP: K.A. to Clare Applegate, 5/11/40.

173 *"cracked up":* AVP/17: L.K. to A.V., 5/3/41, but refers to 1940.

173 *"Kay drops in":* CKP/7: L.K. to Ormeida Keeler, 12/19/40.

173 *"not quite normal":* A.V., quoted in LKP/3: E.K. to L.K., 3/12/41.

173 *"[He told me] the same":* ADMP: Agnes de Mille, "[Keeler]," [1980s]. Author interview with Theodore "App" Applegate, 1/18/2004.

174 *"abandonment":* LKP/2: L.K. to [Uncle] Sterling [Bunnell], 5/15/41. See also CCCP/41S7328: *Katherine Keeler v. Leonarde Keeler,* "Report," 6/18/41.

174 *"Sure my machine":* *WP,* 10/5/46.

174 *"new birthday":* LKP/2: L.K. to Ormeida Keeler, 6/25/41.
174 *"I have a hunch":* LKP/3: L.K. to E.K., 6/21/41.
175 *"[M]arriage, even where":* LKP-DoD: [Clare Applegate] to L.K., 12/8/44.
175 *"those who wished":* L.K., quoted in Agnes de Mille to L.K., 9/4/41.
175 *"I think so well":* ADMP: Agnes de Mille to L.K., 9/4/41.
175 *"And I think you":* ADMP: Agnes de Mille to L.K., 9/14/[41].
176 *"I thoroughly agree":* ADMP: L.K. to Agnes de Mille, 9/8/41, emphasis in original.
176 *"master of the lie detector":* LKP-DoD: Cecil Martwell [Kentucky] to L.K., 5/14/47.
176 *"coincidences":* LKP-DoD: [Nutter Fort, W.V.] to L.K., n.d.
177 *"delicate needles"* and *"You have all"* and *"This is just as":* *CT,* 7/1/41. See also *CDN,* 7/1/41; *NYT,* 7/1/41; *LAT,* 7/1/41.
178 *"Mr. Keeler has":* Cook County Coroner's Office/18451: "Inquest on the Body of Carl M. Anderson," 7/1/41. See also E.K., *Lie Detector Man,* 147.

## PART 3. TRUTH, JUSTICE, AND THE AMERICAN LIE DETECTOR

179 *"POPE (exhausted)":* Bertolt Brecht, *Life of Galileo,* ed. Eric Bentley (New York: Grove, Weidenfeld, [1940], 1966), 110.

### CHAPTER 14. A LIE DETECTOR OF CURVES AND MUSCLE

181 *"STEVE: Well perhaps":* W.M., "Wonder Woman versus the Prison Spy Ring," *Wonder Woman,* No. 1 (Summer, 1942). Differs slightly from WMP-SI: W.M., "Wonder Woman," Original Script, Quarterly #1C, p. 5.
182 *"truth telling would be":* W.M., *Lie Detector Test,* 146.
183 *Marston's theory of human emotions:* W.M., *Emotions,* 113, 137, 369–373.
183 *"From these studies":* W.M., *Emotions,* 300.
184 *"A freshman girl":* W.M., *Emotions,* 312.
184 *When he polled college:* WMP: W.M., "Blondes, Brunettes and Redheads," 120.
184 *"He had a family":* Les Daniels, *The Life and Times of the Amazon Princess, Wonder Woman* (San Francisco, Calif.: Chronicle, 2000), 28, 31.
185 *Sanger was Olive:* Olive Byrne Richard, "My Aunt Margaret," in "Our Margaret Sanger," Ellen Watumull, ed., at www.nyu.edu/projects/sanger/our_ms_richard.htm.
185 *"the country will see":* *NYT,* 11/11/37.
185 *"The air was thick":* *NYT,* 1/31/28.
186 *"[b]londes prefer gentlemen":* WMP: [W.M.], "Blondes, Brunettes and Redheads," 426.

186 *"to apply psychology": NYT,* 12/25/28.
186 *"Motion pictures are emotion": LAT,* 1/11/29.
186 *"No other organization": New York Evening Post,* 12/28/28.
187 *"shaped by the demands":* Münsterberg, *Photoplay,* 41.
187 *"The horror which":* Münsterberg, *Photoplay,* 53.
187 *"a big money saver":* Gorham Munson, "Try Living, Try Loving, Try Laughing! Says William Moulton Marston," *Practical Psychology* (5/37): 13–20. See section on censorship in Pitkin and Marston, *Sound Pictures,* 52–81.
188 *"the best known psychologist":* WMP: Unidentified newspaper clipping, 12/18/29.
188 *"like a living, breathing":* BPP/10: L.K., "Summary of Activities," 1931.
188 *"emotional . . . dead spots": LAT,* 12/6/31. See also *Time,* 12/14/31; Leon Meehan, "Frankenstein," *Motion Picture Herald,* 11/14/31.
188 *"Now I know":* Gregory Wm. Mank, "Production Background," in *Frankenstein,* ed. Philip J. Riley (Absecon, N.J.: MagicImage Filmbooks, 1989), 40.
188 *"Frankenstein merely":* Mank, "Production Background," 37.
188 *"[W]e actors experienced":* Mank, "Production Background," 37.
189 *"spine-chilling": NYT,* 12/5/31.
189 *"grouchy and irritable":* "New Facts about Shaving," *Time,* 10/17/38; "Lie Detector 'Tells All,'" *Life,* 11/21/38, 12/19/38; "New Facts about Shaving," *Saturday Evening Post,* 10/8/38.
190 *"inducement":* JLP-BG: J.L. to E. P. Jordan, 12/12/38. See also JLP/2: J.L. draft report for Gillette, [1938]; and FBI: John Haus to J. Edgar Hoover, 7/13/39.
190 *"Try Living": LAT,* 11/7/37.
190 *"original, oldest":* www.discprofile.com/whatisdisc.htm, registered on 8/22/2006. See W.M., "This Test Reveals Your Secret Self," *Your Life* (9/39): 23–39.
190 *"twists, repression":* W.M., *Lie Detector Test,* 138.
190 *"higher emotions": LAT,* 1/25/39.
190 *For the readers: Look* (12/6/38), 16–17.
190 *"Healthy love":* W.M., *Lie Detector Test,* 119–120.
191 *"mayhem, murder":* Stirling North in *Chicago Daily News,* quoted in Olive [Byrne] Richard, "Don't Laugh at the Comics," *Family Circle* (10/25/40): 10–11, 22.
191 *"jealous, mercenary":* W.M., "What Comics Do to Your Children," *Your Life* (10/39), 80–89.
191 *"fundamental emotional":* Olive [Byrne] Richard, "Don't Laugh at the Comics."
191 *"Homeric inheritance":* W.M., "Why 100,000,000 Americans Read Comics," *American Scholar* 13 (1943–1944): 1–10, quotation, 6.

191 *"blood-curdling masculinity":* W.M., "Why 100,000,000 Americans Read Comics," 8.

191 *"Give them an alluring":* W.M., "Why 100,000,000 Americans Read Comics," 9.

192 *"woman's love charm":* WMP-SI: W.M. to Coulton Waugh, 3/5/45.

192 *"the growth in the power":* WMP-SI: W.M. to Sheldon Mayer, 4/12/42.

192 *"I want you":* WMP-SI: W.M. to Sheldon Mayer, 2/23/41.

192 *"Frankly":* WMP-SI: W.M. to Coulton Waugh, 3/5/45.

192 *"or lack of it":* WMP-SI: Josette Frank to M. C. Gaines, 2/17/43.

192 *"cult of force":* Walter Ong, quoted in *Time,* 10/22/45.

193 *"normally":* Frederic Wertham, *Seduction of the Innocent* (New York: Rinehart, [1954]), 235.

193 *"I am one of those":* WMP-SI: [J.J.] to W.M., 9/9/43.

193 *less "sexy":* WMP-SI: Dorothy Roubicek to M. C. Gaines, 2/19/43.

193 *"Superman without muscle":* WMP-SI: W.M. to M. C. Gaines, 9/15/43.

193 *"even a wee bit":* WMP: Annie Dalton Marston to W.M., 8/13/42.

193 *"The only hope":* WMP-SI: W.M. to M. C. Gaines, 2/20/43.

194 *"The Battle for Woman Kind": Wonder Woman,* No. 5 (June–July 1943).

## CHAPTER 15. ATOMIC LIES

197 *"GUILDENSTERN":* Tom Stoppard, *Rosencrantz and Guildenstern Are Dead* (New York: Grove, 1967), 17.

197 *"wrapt up here":* Laurence Sterne, *The Life and Opinions of Tristram Shandy* (London: Penguin, [1759–1767], 1987), 96.

197 *"Adam, Adam":* Ludovico à Páramo, *De Origine et Progressu Officcii Sanctae Inquisitionis* (Madrid, 1598), cited in Kimberly Lynn Hossain, "Was Adam the First Heretic? Luis de Páramo, Diego de Simancas, and the Origins of Inquisitorial Practice," *Archive for Reformation History* [2006]. For God as "Inquisitore maraviglioso," see John Tedeschi, *The Prosecution of Heresy* (Binghamton, N.Y.: Medieval and Renaissance Texts, 1991), 131.

198 *"But we abominate":* Francis Bacon, *The New Atlantis,* in *The Philosophical Works* (London: Midwinter, 1737), 3: 298.

199 *One historian of science:* Peter Galison, "Removing Knowledge," *Critical Inquiry* 31 (2004): 229–243.

200 *"subversives":* EFMP: E. F. McDonald to Captain E. B. Brown, 12/19/45.

200 *"was getting along beautifully":* KLP/2: L.K. to Sherman and Ruth Jennings, 11/13/42.

200 *"Of course, the 210":* LKP-DoD: L.K. to Captain James Delaney, 9/21/42.

201 *"potential menace":* J. Edgar Hoover, "Enemies at Large," *American Magazine* (4/44): 17, 97–100, quotation, 97.

202 *last mass execution: NYT,* 8/26/45. Richard Whittingham, *Martial Justice: The Last Mass Execution in the United States* (Annapolis, Md.: Naval Institute Press, [1971], 1997).

202 *"The Americans want":* Helmut Hörner, *A German Odyssey,* quoted in Ron Robin, *The Barbed-Wire College: Reeducating German POWs in the United States during World War II* (Princeton, N.J.: Princeton University Press, 1995), 37. See also Judith Gansberg, *Stalag: U.S.A.: The Remarkable Story of German POWs in America* (New York: Crowell, 1977), 61–62.

202 *"paranoid emotional core":* Richard M. Brickner, *Is Germany Incurable?* (Philadelphia, Pa.: Lippincott, 1943), 263.

202 *"Democracy":* Henry W. Ehrmann, "An Experiment in Political Education," *Social Research* 14 (1947): 304–320, quotation, 319.

202 *"militarists":* NARA/RG389/439A/48: Maxwell McKnight, "Memorandum for Chief, Field Service Branch," 6/7/45.

202 *"sold someone a bill":* NARA/RG389/413.6/1643: Maxwell McKnight to Colonel Powell, 7/27/45.

203 *"pro-communist in scope":* Robin, *Barbed-Wire College,* 143.

203 *Keeler found that 36 percent:* NARA/RG389/319.1/1627: L.K., "Report [POW Screening Project]," 9/12/45.

203 *"those with a sincere":* NARA/RG389/439A/40: R. W. Pierce, "Report on Screening of German Prisoners," 8/30/45.

204 *"We often were told":* NARA/RG389/1629: Siegfried Cammann to [Intelligence Officer], 10/12/45.

204 *"a habit not desirable":* NARA/RG389/1629: Alpheus Smith to Edward Davidson, 11/9/45.

204 *While only 6 percent:* Robin, *Barbed-Wire College,* 162–163. See also Arthur L. Smith, Jr., *The War for the German Mind: Reeducating Hitler's Soldiers* (Providence, R.I.: Berghahn, 1996).

205 *Oak Ridge was a city:* James Overholt, ed., *These Are Our Voices: The Story of Oak Ridge, 1942–1970* (Oak Ridge, Tenn.: Children's Museum of Oak Ridge, 1987). Russell B. Olwell, *At Work in the Atomic City: A Labor and Social History of Oak Ridge, Tennessee* (Knoxville: University of Tennessee Press, 2004).

206 *"insofar as possible"* and *"stolen product":* NARA/RG326/148: AEC, "Use of Lie Detector at AEC Installations," 3/24/53; with information on Keeler's work of 1946. For an explanation of the "thefts," see NARA/RG326/149: AEC, "In the Matter of 'Lie-Detector' Panel Meeting," 1/24/52, p. 42. For a description of the Oak Ridge program by someone with access to the private records of the examiners, see John G. Linehan, "The Oak Ridge Polygraph Program," *Polygraph* 19 (1990): 131–137.

206 *"heavily infested":* J. Parnell Thomas and Stacy V. Jones, "Reds in Our Atom-Bomb Plants," *Liberty* (6/21/47): 15, 90–93. See local and national coverage:

*Knoxville Journal,* 6/5/47; *WP,* 7/12/46, 6/5/47, 7/15/47. For context, see Wang, *Age of Anxiety,* 149–180; Richard G. Hewlett and Oscar E. Anderson, Jr., *A History of the United States Energy Commission,* vol. 2, *Atomic Shield,* ed. Richard G. Hewlett and Francis Duncan (Berkeley: University of California Press, 1990), 92–93.

206 *By the early 1950s:* NARA/RG326/149: AEC, "In the Matter of 'Lie-Detector' Panel Meeting," 1/24/52.

207 *"Q: Have you belonged":* LKP-DoD: Arnold Cohen, "Lecture," 2/12/54.

207 *"friends or relatives":* NARA/RG326/148: AEC, "Use of Lie Detector at AEC Installations," 3/24/53.

207 *"liberal thoughts":* NARA (Southeast)/RG326: F. P. Callaghan to J. S. Denton, 5/28/51.

207 *That the test uncovered:* National Academy of Sciences, ed. Fienberg, *Polygraph,* 48.

208 *"on the ball"* to *"so to speak":* NARA/RG340/1: Paul V. Trovillo, "Report on a Survey of . . . Polygraph Program at Oak Ridge," 4/14/51.

208 *"mental relief from worry":* ORP: J. C. Franklin to Carroll Wilson, 10/10/47.

208 *By 1952 some 50,000:* R. Chatham, "Public Employee Screening," in "Selected Papers on the Polygraph," *Conference on Criminal Interrogation and Lie Detection,* New York University, 11/8/1952 (N.p.: Board of Polygraph Examiners, 1956), 46. See also *Oak Ridger,* 4/14/53.

209 *"almost universal":* Dr. Richard Meiling, "Military Medical Problems," 6/27/51, quoted in Advisory Committee on Human Research Experiments, *Final Report* (Washington, D.C.: GPO, 1998), ch. 10, http://www.eh.doe.gov/ohre/roadmap/achre/.

209 *"atomic support":* Alfred Hausrath et al., "Troop Performance on a Training Maneuver Involving the Use of Atomic Weapons," Operation Research Office, Johns Hopkins University, ORO-T-0, (3/52), 53–61. For an account based on an unpublished manuscript by Chatham and Trovillo, see John G. Linehan, "Historical Note: Polygraph Research Study of Fear in a Field Situation, Review of a 1951 Study," *Polygraph* 25 (1996): 152–158. See also Howard L. Rosenberg, *Atomic Soldiers: American Victims of Nuclear Experiments* (Boston, Mass.: Beacon, 1980), 41.

209 *"men standing straight":* Paul Boyer, *By the Bomb's Early Light: American Thought at the Dawn of the Atomic Age* (Chapel Hill: University of North Carolina Press, 1994), 300.

210 *fears of 46 percent:* Irving Gitlin, "Radio and Atomic-Energy Education," *Journal of Educational Sociology* 22 (1949): 327–330.

210 *"others"* and *"lousy":* NARA/RG340/1: Trovillo, "Report on a Survey."

210 *"raising a fuss":* UCA/AORES/6: AORES, "It Is Time to Protest about Security," 9/7/47. See Truman, *NYT,* 9/14/48.

210 *"40 percent of senior physicists and chemists":* UCA/AORES/50: R. W. Stoughton to Senator Bourke Hickenlooper, 9/17/48.

210 *"an instrument of third-degree":* "'Lie Detector' Doesn't," *Science News Letter* 49 (3/30/46): 207.

211 *"an unfavorable and sometimes":* NARA/RG326/148: AEC, "Use of Lie Detector at AEC Installations," 3/18/53.

211 *"violent opposition"* to *"assur[ing] that":* NARA/RG326/149: AEC, "In the Matter of 'Lie-Detector' Panel Meeting," 1/24/52.

211 *In December 1951:* Anthony Leviero, "U.S. Tests Staff by Lie Detectors," *NYT,* 12/20/51. See also the follow-up in *WP,* 12/21/51.

211 *"un-American":* Wayne Morse, "Use of Lie Detectors," *Congressional Record* 98(7), 1/17/52, pp. 258–262. Leviero, "Morse Denounces," *NYT,* 1/18/52.

211 *"the examiner draws":* Fred Inbau, in NARA/RG326/149: AEC, "In the Matter of 'Lie-Detector' Panel Meeting," 1/24/52.

212 *"In many respects":* NARA/RG59/16/ConferenceReport: Peter Regis, "Conference of Regional Security Supervisors," 4/53.

## CHAPTER 16. PINKOS

215 *"Such a hypothetical":* Isaac Asimov, *Second Foundation,* in *The Foundation Trilogy* (Garden City, N.Y.: Doubleday, 1953), 208.

216 *"cover":* DDRS: Colonel S. Edwards to Director of CIA, "Project Bluebird, Memorandum," 4/5/50; Sidney Gottlieb, "Prospectus for Continuation of Research Project II on LSD," 6/9/53. See also John Marks, *The Search for the "Manchurian Candidate": The CIA and Mind Control* (New York: Times Books, 1979); McCoy, *Question of Terror.*

217 *Popular books:* Edward Hunter, *Brainwashing: The Story of the Men Who Defied It* (Farrar, Straus & Cudahy, 1956). For Hunter's affiliation with the CIA and the origins of the term "brainwashing," see Marks, *Search for the "Manchurian Candidate,"* 125–126. Even books with a different political orientation made comparable claims; see Joost A. M. Meerloo, *The Rape of the Mind* (Cleveland, Ohio: World, 1956).

217 *As one British author:* William Sargant, *Battle for the Mind: A Physiology of Conversion and Brain-Washing* (Garden City, N.Y.: Doubleday, 1957), 192–196.

217 *"we begin palely":* Joseph and Stewart Alsop, "Lie-Detector Piling Up Records," *WP,* 2/21/54.

217 *"An illusion can become":* George Orwell, "England, Your England," in *Inside the Whale and Other Essays* (Harmondsworth, England: Penguin, [1941], 1957), 71.

218 *Keeler's former partner:* George Haney, "Report on . . . Polygraph . . . of

Korean Nationals and Communist Chinese," Operations Research Office, ORO-S-85, 11/14/50, archived at http://antipolygraph.org/read.shtml.

219 *"playing it cool":* Albert D. Biderman, *March to Calumny: The Story of American POW's in the Korean War* (New York: Macmillan, 1963), 43–45. Compare this with Eugene Kinkead, *In Every War but One* (New York: Norton, 1959).

219 *Dean had been the first:* William F. Dean, *General Dean's Story* (New York: Viking, 1954), 130–131.

219 *"red herring": NYT,* 8/6/48.

219 *"whichever one of you":* Representative F. Edward Hebert, in Special Subcommittee of the U.S. House Un-American Activities Committee (HUAC), "Hearings Regarding Communist Espionage in the U.S. Government," 8/16/48, at http://www.law.umkc.edu/faculty/projects/ftrials/hiss/hiss.html.

220 *"CHAMBERS: Yes":* HUAC, "Hearings Regarding Communist Espionage," 8/7/48.

220 *"fly out and assist":* RNP: L.K. to Richard Nixon, 8/14/48.

220 *"the outstanding man":* HUAC, "Hearings Regarding Communist Espionage," 8/16/48.

220 "not, *however, a 'lie detector'":* Alger Hiss to J. Parnell Thomas, 8/18/48; in *New York Herald Tribune,* 8/20/48. Also in Alger Hiss, *In the Court of Public Opinion* (New York: Knopf, 1957), 420–421, emphasis in original. See also *New York Herald Tribune,* 8/17/48; Sam Tanenhaus, *Whittaker Chambers: A Biography* (New York: Random House, 1997), 253–254, 265, 586; Allen Weinstein, *Perjury: The Hiss-Chambers Case* (New York: Knopf, 1978), 21–40.

220 *"In the opinion": NYT,* 8/22/48. See also *New Yorker,* 9/4/48.

221 *"82 percent accurate": WP,* 8/23/48.

221 *In later years Nixon:* Richard M. Nixon, *Six Crises* (Garden City, N.Y.: Doubleday, 1962), 17–18, 29.

221 *"I don't know anything":* NARA/NixonWhiteHouseTapes/545-3: Nixon, 7/24/71.

221 *Now he offered: NYT,* 3/22/54. But see also *WP,* 6/2/54.

222 *"security risks": NYT,* 3/1/50.

222 *"the sexual perverts":* Guy George Gabrielson, quoted in *NYT,* 4/19/50.

223 *The Kinsey Report:* Alfred Kinsey, Wardell Pomeroy, and Clyde Martin, *Sexual Behavior in the Human Male* (Philadelphia, Pa.: Saunders, [1948], 1949).

223 *"Is he?":* Cheever, quoted in Johnson, *Lavender,* 53–55.

223 *"90 twisted twerps":* Jack Lait and Lee Mortimer, *Washington Confidential* (New York: Crown, 1951), 90.

223 *"When the son of the Democrat senator":* Johnson, *Lavender,* 140–141.

223 *"peculiar lips":* Blevins in NARA/RG59/1953–60/b12/BureauSecurity: C. M. Dulin to Michael J. Ambrose, 4/16/53.

223 *"quick test":* [Hoey Hearings], U.S. Senate, Investigation Subcommittee, Committee on Expenditures in the Executive Departments, Executive Sessions, "Report of Proceedings," 7/26/50, 2256.

224 *"Most authorities":* [Hoey Hearings], 2.

224 *Nonetheless Senator Karl Mundt:* [Hoey Hearings], 2292.

224 *"mannerisms and appearance":* Johnson, *Lavender,* 73.

224 *"Unless scientific tests":* NARA/RG59/b2/1E1: D. L. Nicholson and J. R. Ylitalo, "Review of 176 Closed Miscellaneous-M Files," 8/27/51.

224 *"ardent Catholicism":* NARA/RG59/b12/1E1: "Memo to Scott McLeod," 4/20/53.

224 *"hobbies, associates":* Johnson, *Lavender,* 73.

224 *"jelly hand shake":* Dean, *Imperial Brotherhood,* 116.

224 *"Look into his period":* NARA/RG59/b12/1E1: "Memo to Scott McLeod," 4/20/53.

224 *80 percent confessed:* NARA/RG59/1953-60/b16: Peter Regis, "Conference of Regional Security Supervisors," 4/53.

225 *"clear their name":* NARA/RG59/1953–60/b12/Bureau of Security: Thurston Morton to Senator Olin Johnston, 6/30/55, "Polygraph," [1955].

225 *"just happened to remember":* Douglas Kelley in NARA/RG326/149: AEC, "In the Matter of 'Lie-Detector' Panel Meeting," 1/24/52, p. 84.

225 *"without mentioning sexual perversion":* LKP-DoD: [Keeler, Inc.], "Final Examination," n.d.

226 *"anguish, revulsion":* Thayer diaries, 3/23/53, quoted in Dean, *International Brotherhood,* 111, 132, 136.

226 *"beat":* NARA/RG59/1953–60/b12/BureauSecurity: Thurston Morton to Senator Olin Johnston, 6/30/55, and attached memo, "Polygraph," [1955].

227 *In all, some 1,000:* Johnson, *Lavender,* 166.

227 *"More than anything":* Hoover, introduction to Fendall Yexra and Ogden R. Reid, "The Threat of Red Sabotage," reprinted from the *New York Herald Tribune,* 1950.

227 *"psychological aid":* See *WP,* 6/8/53; Hoover, quoted in DDRS: Dwight D. Eisenhower and John Foster Dulles, "Memorandum of Telephone Conversation," 3/17/53; FBI: E. H. Winterrow to D. M. Ladd, Re McCarthy, 5/13/49.

227 *"experimental":* J. Edgar Hoover, *Scientific Methods of Crime Detection in the Judicial Process* (Washington, D.C.: GPO, 1935), 18.

227 *Initially skeptical:* LKP-DoD: E. P. Coffey to Clyde Tolson, "Polygraph Study," 1935.

228 *"I personally would not":* Albert Deutsch, *The Trouble with Cops* (New York:

Crown, 1954), 171. See also *WP*, 6/8/53. See also FBI: J. Edgar Hoover, note on E. A. Tamm to E. P. Coffey, 9/24/45; FBI: Clyde Tolson to J. Edgar Hoover, 3/7/50. Athan Theoharis, *J. Edgar Hoover, Sex, and Crime: A Historical Antidote* (Chicago: Ivan Dee, 1995).

## CHAPTER 17. DEUS EX MACHINA

229 *"He was of no"*: Chandler, *Long Goodbye*, 732.

229 *"No one has ever"*: LKP-DoD: [Waukegan, Ill.] to L.K., 3/10/46.

230 *"behind the scenes"*: Gardner, "Two Men Wait," *Argosy* (1/51): 35.

230 *"greatest murder mystery"*: Gardner, "My Most Baffling Murder Case," *Mercury Mystery Book Magazine* 4 (1/58): 106–114, quotation, 106.

230 *"the most intense drama"*: Gardner, *New York Journal American*, 10/19/43.

231 *"turn around the bad"*: Alfred de Marigny, *A Conspiracy of Crowns: The True Story of the Duke of Windsor and the Murder of Sir Harry Oakes* (New York: Crown, 1990), 89. See also Marshall Houts, *King's X: Common Law and the Death of Sir Harry Oakes* (New York: Morrow, 1972).

231 *"just as Perry"*: Gardner, *New York Journal American*, 10/18/43.

231 *"THE MAN WHO"*: Gardner, *New York Journal American*, 10/19/43.

231 *"grandstanding"*: JLP/7: J.L. to C. Goddard, 11/10/43.

232 *"The Magic Lie Detector"*: Alva Johnston, *Saturday Evening Post* (4/22/44): 26–27, 63; (4/29/44): 20,101–102; (4/15/44): 9–11, 72.

232 *"Lie Detector Clears Joe"*: *CDT*, 12/4/44. See also ISA/Pardon and Parole Board/Majczek: L.K., "Joseph Majczek, case #3749," 11/27/44; John J. McPhaul, *Deadlines and Monkeyshines: The Fabled World of Chicago Journalism* (Englewood Cliffs, N.J.: Prentice-Hall, 1962), 191–202.

233 *"FIRST REPORTER"*: Jay Dratler, *Call Northside 777* (Twentieth Century Fox, 1947); phrasing differs slightly from revised shooting script, 9/13/47. See also Jennifer Mnookin and Nancy West, "Theaters of Proof: Visual Evidence and the Law in *Call Northside 777*," *Yale Journal of Law and the Humanities* 13 (2001): 329–390.

234 *"[A] lawyer would ask"*: *CT*, 2/10/48.

234 *"brain-wave detector"*: *Chicago Herald American*, 2/6/48.

234 *"the entire truth"* and *"didn't believe"*: *Chicago Daily Sun and Times*, 2/9/48.

234 *Then Theodore Marcinkiewicz:* CCCP/31684: W. J. Devereux and D. I. McCain, "Theodore Marcinkiewicz," 2/3/48. *Chicago Daily Sun and Times*, 2/24/50.

235 *"a militant body"*: Gardner, "A Dead Man Works for Justice," *Argosy* (5/50): 80.

235 *"Remember"*: Gardner, "Was Gross Railroaded?" *Argosy* (9/49): 79.

235 *"We had to know"* and *"[W]hen Keeler"*: Erle Stanley Gardner, *The Court of*

*Last Resort* (New York: William Sloane, 1952), 24, 87. See also Dorothy B. Hughes, *Erle Stanley Gardner: The Case of the Real Perry Mason* (New York: Morrow, 1978).

236 *"still-waters-run-deep": CDN,* 3/19/48.

236 *"good ones":* EFMP: E. F. McDonald to L.K., 7/10/44.

236 *"My darling":* LKP/2: Marie Trolle to L.K., [1943].

236 *"After all, if the":* LKP/2: Marie Trolle to L.K., 2/27/44.

237 *"Oh hold me":* LKP/2: Sarah Elizabeth Rodger, "And If I Cry Release," [1944]; also Rodger, *If I Cry Release* (New York: Doubleday, 1940), 2–3.

237 *"one you can":* LKP/2: Rodger to L.K., Friday, [1944].

237 *"I just don't know":* LKP/2: Rodger to L.K., 8/44.

237 *"stunning young woman": CT,* 7/31/42.

237 *"tall and poised": CT,* 8/9/42.

238 *"so take heart":* LKP/2: Franja Hutchins to L.K., 5/1/44.

238 *"We cannot love":* LKP/2: Franja Hutchins to L.K., 4/18/45.

238 *"My hobby is collecting"* and *"Many times":* LKP/2: Anon., "A Tribute to Leonarde Keeler," *International Association for the Detection of Deception,* [1949–1950].

239 *According to the head:* ADMP: Agnes de Mille, "[Keeler]," [1980s].

239 *In her diaries:* ADMP: Agnes de Mille, "[Keeler]," [1980s]. De Mille recorded the episode a few weeks after the events, and wrote several versions over the years.

240 *"bitter memory":* L.K. in LKP/2: Jesse and Mildred Bollman to L.K., [1944].

241 *"Kay loved to live":* LKP-DoD: Claire Applegate to L.K., 12/8/44. See also TWU: Adolph Kurek, "Report of Major Accident," 11/27/44.

241 *Total income:* LKP/2: "Leonarde Keeler, Inc., Balance Sheet," 3/31/50.

241 *"complete wash-out":* LKP-DoD: L.K. to James Elder, 10/10/45.

241 *"on a wharf":* LKP/2: L.K. to E.K., 7/25/49.

241 *His sisters divided:* CCCP/49P8223: Leonarde Keeler probate, 12/27/50.

242 "INVENTOR OF LIE": *CT,* 9/21/49; *NYT,* 9/21/49.

242 *"true or only made up":* Quesalid, quoted on p. 150 of Harry Whitehead, "The Hunt for Quesalid: Tracking Lévi-Strauss' Shaman," *Anthropology and Medicine* 7 (2000): 149–168, which recounts the true story of Quesalid, a half-English, half-Tlingit man also known as Bruce Hunt, who married into the Kwakiutl group and long served as an interpreter for the anthropologist Franz Boas.

242 *"Quesalid did not":* Claude Levi-Strauss, "The Sorcerer and His Magic," in *Structural Anthropology,* trans. Claire Jacobson and Brooke Grundfest Schoepf (New York: Basic Books, [1949], 1963), 1:180.

## CHAPTER 18. FRANKENSTEIN LIVES!

243 "*FRANKENSTEIN: They?*": Garrett Fort and Francis Edwards Faragoh, *Frankenstein,* ed. Philip J. Riley (Absecon, N.J.: MagicImage Filmbooks, 1989), 80.

243 "*I regarded him*": LKP/2: A.V. to Viola Stevens, 4/21/40.

243 "*high school kid*" to "*always refused*": JLP-BG: J.L. and Robert Borkenstein, "Present Status of Lie Detector . . . ," [late 1950s].

244 "*flamboyant self-confidence*": JLP/7: J.L., typescript attached to letter of J.L. to Claude Broom, 6/25/63.

244 "*invented*": JLP-BG: J.L. to Lucile Burke, 6/6/51.

244 "*many years*": AVP/18: J.L. to A.V., 6/2/51.

244 "*Maybe I have*": JLP/8: A.V. to J.L., 6/12/51, emphasis in original.

244 "*America's greatest cop*": Albert Deutsch, *Collier's,* 2/3/51.

246 "*their disorganized*": JLP-JH: J.L. to Adolf Meyer, 5/1/47.

246 "*awful mess*": JLP-JH: J.L. to Adolf Meyer, 9/22/47.

246 "*Reactograph*": *Logansport Pharos Tribune,* [January 1950].

246 *A glowing write-up:* Albert Q. Maisel, "Scandal Results in Real Reform," *Life* (11/12/51): 140–154.

246 "*[Larson] is one*": JLP/7: Warren Pugh, "Re: John A. Larson," 7/26/55.

247 "*fifteen-to-twenty hour days*": JLP-JH: Frank D. Tikalsky, 4/4/62. LeRoy A. Stone, "Using the Polygraph to Detect Lying and Deception: The Hoax of the Century, *Journal of Forensic Psychonomics* (2003), at www.home.earthlink.net/~lastone2/forensicarticle6.htm.

247 "*Sometimes his idealism*": JLP-JH: R. F. Borkenstein, 12/3/56.

248 "*whipped into shape*": JLP/2: J.L., "Beginning the . . . ," [1960s].

248 "*magnum opus*": JLP/7: J.L. to Joseph Kubis, 4/10/61.

248 *In the 1940s his manuscript:* JLP-BG: Mary Irwin (U. Chicago Press) to J.L., 3/7/40.

248 "*By the 1950s, it had*": JLP/2: J.L., "[Summary of] *Psychobiology of Detection . . . ,*" n.d.

248 "*Our text from the book*": JLP-BG: J.L. to Alexander Moris, 3/5/57.

248 "*Alpha and Omega*" and "*the entire work*":JLP/4: J.L. "Corrections made [by] Larson, John [to] Holographic Will," 12/15/64–12/16/64.

248 "*cheese-cake type*": JLP/2: J.L., "[Summary of] *Psychobiology of Detection . . . ,*" n.d.

248 *9,000 pages:* JLP-JH: J.L., "Rational[e]," [1962].

249 "*key to this puzzle*": BPHS: Robert Borkenstein to Lucile Burke, 5/24/71.

249 "*Beyond my expectation*": JLP/7: J.L. to Reg Manning, n.d. [early 1960s].

249 "*fostered a Frankenstein's*": JLP-BG: J.L. to Robert Borkenstein, 6/18/51.

## CHAPTER 19. BOX POPULI

251 *"Nobody will ever":* Hammett, "Letter to the Editor," *Black Mask* (1/1/24), 127, in *Selected Letters of Dashiell Hammett,* ed. Richard Layman (Washington, D.C.: Counterpoint, 2001), 24.

252 *"twentieth-century witchcraft": NYT,* 5/25/73.

252 *And O. J. Simpson: Sharon Rufo, et al. v. Orenthal James Simpson,* Superior Court of the State of California for the County of Los Angeles, 11/25/96, at www.courttv.com/casefiles/simpson/transcripts/nov/nov25.html.

252 *"Q: "Did President":* "Hannity and Colmes," *Fox News Network,* 3/9/2005.

252 *Representative Gary Condit: NYT,* 7/21/2001; William Saletan, "Polygraph Theism," at www.Slate.com, 7/20/2001; Gallup Poll, www.gallup.com, 7/11/2001.

253 *"cautioning against advertising":* Wladimir Eliasberg, "Forensic Psychology," *Southern California Law Review* 19 (1945–1946): 362. See also *NYT,* 6/25/44; JLP-BG: [J.L.] to Lowell Selling, 11/10/44.

253 *"interior domain": NYT,* 4/11/58.

253 *"voluntarily":* Paul Falick, "The Lie Detector and the Right to Privacy," *Case and Comment* (September–October 1968): 36–44. Yet even a critic who attacked the lie detector as scientifically dubious accepted it for police interrogation; see Jerome H. Skolnick, "Scientific Theory and Scientific Evidence: An Analysis of Lie Detection," *Yale Law Journal* 70 (1961): 694–728.

254 *"invasion of the mind":* Rep. Neal Gallagher (D., N.J.) quoted in *Detroit Free Press,* 4/7/64.

254 *By the mid-1980s, twenty-two:* James A. Suffridge, "The Silent Assault on the Right to Privacy," *AFL-CIO: American Federationist* (8/65), no pagination. Lois Recascino Wise and Steven J. Charvat, "Polygraph Testing in the Public Sector: The Status of State Legislation," *Public Personnel Management* 19 (1990): 381–390.

254 *eighty cases a month:* FBI: White to Tamm, 1/3/61.

254 *"useful tool":* "Editorial: The Polygraph," *American Bar Association* 51 (5/65), no page numbers.

254 *some 1,000 operators:* Alan F. Westin, *Privacy and Freedom* (New York: Atheneum, 1967).

254 *one-fourth of all U.S.:* Paul R. Sackett and Phillip J. Decker, "Detection of Deception in the Employment Context: A Review and Critical Analysis," *Personnel Psychology* 32 (1979): 487–506.

255 *"up to his keister":* Jack Brooks, "Polygraph Testing: Thoughts of a Skeptical Legislator," *American Psychology* 40 (1985): 348–354.

255 *"the minute": NYT,* 12/20/85.

256 *twice as many polygraphers:* Edward E. Cureton, "A Consensus as to the Validity of Polygraph Procedures," *Tennessee Law Review* 22 (1953): 728–742. See also Alder, "Untruth," 32–33.

257 *"the jury is the lie detector":* Thomas, citing *U.S. v. Barnard,* 490 F.2d 907 (1973), in *U.S. v. Scheffer,* 523 U.S. 303 (1998), emphasis by Thomas. For jurors' attitudes, see Carol Hubbard et al., "Brief for the Respondent in the Supreme Court of the United States . . . ," *U.S. v. Scheffer,*" 8/97, reprinted in *Polygraph* 26 (1997): 150–183; Stephen C. Carlson et al., "The Effect of Lie Detector Evidence on Jury Deliberations: An Empirical Study," *Journal of Police Science and Administration* 5 (1977): 148–154; Alan Markwart and Brian E. Lynch, "The Effect of Polygraph Evidence on Mock Jury Decision-Making," *Journal of Police Science and Administration* 7 (1979): 324–332.

257 *The muddled ruling:* New Mexico continues to be the only state to allow the polygraph in criminal cases without stipulation; see *Lee v. Martinez,* 2004 NMSC 27; 96 P.3d 291 (2004)—now with the assent of the U.S. Supreme Court; see *Associated Press,* 7/15/2004. As of 2003, the Fifth and Ninth Circuits have left the use of the polygraph up to the discretion of the district judge, though not many trial judges have allowed the evidence; see *U.S. v. Prince-Oyibo,* 320 F.3d 494 (2003).

258 *For conservatives:* The conservative justices who upheld the per se exclusion voted four years later to affirm the power of law enforcement officials to require prisoners to take polygraph exams as part of a sex abuse treatment program—even though answering the questions honestly might force a prisoner to incriminate himself; see *David R. McKune et al. v. Robert G. Lile,* 536 U.S. 24 (2002).

258 *For liberals:* The liberal justices were prescient in expecting *Scheffer* to be cited principally as allowing the state to place reasonable limits on the defendant's ability to request evidence; see *U.S. v. Cianci,* 378 F.3d 71 (2004).

258 *"Like most junk science":* Aldrich Ames to Steven Aftergood (FAS), [11/28/2000], at www.AntiPolygraph.org.

258 *"not deceptive":* Department of Justice, "Bellows Report," *Attorney General's Review Team on the Handling of the Los Alamos National Laboratory Investigation* (Washington, D.C.: Department of Justice, 5/2000), 632, at http://www.usdoj.gov/ag/readingroom/bellows.htm.

259 *"found to be deceptive":* James Risen and Jeff Gerth, "China Stole Nuclear Secrets . . . ," *NYT,* 3/6/99. See also Dan Stober and Ian Hoffman, *A Convenient Spy: Wen Ho Lee and the Politics of Nuclear Espionage* (New York: Simon and Schuster, 2001), 173–174; Wen Ho Lee, with Helen Zia, *My Country versus Me* (New York: Hyperion, 2001), 23–29.

259 *accused Lee twenty-six times:* Carol Covert and John Hudenko, interview of Wen Ho Lee, 3/7/99 (FBI/004868-004950), at www.wenholee.org.

259 *"stonewalled":* James Risen, "U.S. Fires Researcher . . . ," *NYT,* 3/9/99.

259 *"not uncommon": WP,* 1/8/2000.

259 *"would never pass muster":* Alan P. Zelicoff, in Society of Professional Scientists and Engineers, "DOE Public Hearings on the Proposed Polygraph Regulations," in Livermore, 9/14/99; Albuquerque, 9/16/99; Los Alamos, 9/17/99; Washington, D.C., 9/22/99, at www.spse.org/Polygraph_hearing.html. See also "Polygraphs: Worse Than Useless," *WP,* 5/27/2003; Steven Aftergood, "Polygraph Testing and the DOE National Laboratories," *Science* 290 (11/3/2000): 939–940; "Outspoken Nuclear Scientist 'Forced Out' over Polygraph Row," *Nature* 428 (3/18/2004): 243.

259 *For more than five years, some 20,000:* DOE, "Counterintelligence Evaluation Regulations," *Federal Register* 70, no. 5 (1/7/2005). See *Albuquerque Journal,* 1/10/2005; "Editorial," *Science* 300 (5/23/2003): 1201. See also *The Scientist,* 3/26/2003. For news that the screening program would end in October 2006, see Aliya Sternstein, "DOE Reduces Use of Polygraph," at www. fcw.com, 10/16/2006.

260 *When Abdallah Higazy: NYT,* 6/29/2002, 8/6/2002, 8/16/2002, 10/29/2002. *New York Law Journal,* 12/2/2002. James B. Comey (DOJ) to Judge Jed S. Rakoff, 10/31/2002, at www.AntiPolygraph.org. See also *Abdallah Higazy v. Millennium Hotel et al.,* 346 F. Supp. 2d 430 (2004).

260 *Documents seized by U.S. forces:* "Devices Used in Interrogation," from *Mawu'at al-jihad* (*Encyclopedia of Jihad*), [2002], at www.AntiPolygraph.org. "Usturah jahaz kashf al-kidhb" ("The Myth of the Lie Detector"), *Al-Fath Magazine* 1 (December 2004): 19–23, translation and text at www.AntiPolygraph.org.

260 *As the historian Alfred McCoy:* McCoy, *Question of Torture.* For the standard CIA interrogation manual of the cold war, see CIA, "Kubark Counterintelligence Interrogation" (July 1963) at www.gwu.edu/~nsarchiv/NSAEBB/NSAEBB122. For an overview of the current debate on the American use of torture, see Sanford Levinson, ed., *Torture: A Collection* (Oxford, UK: Oxford University Press, 2004).

262 *"Some examinees were familiar":* Clint Richter, quoted in Carolyn Collins, "OSI Provides the Truth in Baghdad," *Global Reliance* (March–April 2004), at http://antipolygraph.org/articles/article-044.shtml. See the denunciations of CIA torture in a Web-published op-ed on 2/24/2006 by the former CIA polygrapher John F. Sullivan at http://blogsisters.blogspot.com/2006/02/posting-what-mainstream-wont-publish.html.

262 *televsion dramas like* 24: Adam Green, "Television; Normalizing Torture, One Rollicking Hour at a Time," *NYT,* 5/22/2005.

262 *intentionally inflicted on a person":* UN Convention Against Torture, 1984, quoted in McCoy, *Question of Torture,* 100.

262 *senior aide Abu Zubaydah:* Dan Eggen and Dafna Linzer, "Secret World of Detainees Grows More Public," *WP*, 9/7/2006. Mark Mazzetti, "Questions Raised about Bush's Primary Claims," *NYT*, 9/8/2006. David Johnston, "At a Secret Interrogation, Dispute Flared Over Tactics," *NYT*, 9/10/2006.

262 *McCain detainee amendment:* Josh White, "New Rules of Interrogation Forbid Use of Harsh Tactics," *WP*, 9/7/2006. Adam Liptak, "As Bush Proposes Freer Hand," *NYT*, 9/8/2006.

263 *Paul Ekman has compiled:* Ekman, *Telling Lies.* Compare Ekman's claims with those in other articles in Granhag and Strömwall, eds., *Detection of Deception*, especially Bella M. DePaulo and Wendy L. Morris, "Discerning Lies from Truth: Behavioral Clues to Deception and the Indirect Pathway of Intuition," 15–40. David Livingstone Smith, *Why We Lie: The Evolutionary Roots of Deception and the Unconscious Mind* (New York: St. Martin's, 2004).

264 *99.9 percent: Economist* (7/10/2004): 71–72.

264 *"TIME 100": Time*, 11/26/2001. See also *NYT*, 10/9/2001.

264 *"art rather than science":* J. P. Rosenfeld, in U.S. GAO, *Investigative Techniques: Federal Agency Views on the Potential Application of "Brain Fingerprinting,"* GAO-02-22 (10/2001), 18. For the views of Farwell's mentor, see Emanuel Donchin, in U.S. GAO, *Investigative Techniques*, 14–15; for the CIA's views, see p. 9.

265 *"stolen":* D. D. Langleben et al., "Brain Activity during Simulated Deception: An Event-Related Functional Magnetic Resonance Study," *NeuroImage* 15 (2002): 727–732.

265 *"interesting scenarios":* G. Ganis et al., "Neural Correlates of Different Types of Deception: An fMRI Investigation," *Cerebral Cortex* 13 (2003): 830–836.

266 *"blobology":* Russell A. Poldrack and Anthony D. Wagner, "What Can Neuroimaging Tell Us about the Mind?" *Neuroimaging and Mind* 13 (2004): 177–181. Sean Spence, cited in Helen Pearson, "Lure of Lie Detectors Spook Ethicists," *Nature* 441 (6/22/2006): 918–919.

266 *"media information":* Daniel Langleben, U.S. patent #20050154290, filed 6/15/01.

## EPILOGUE

269 *"If the fixture":* Sterne, *Tristram Shandy*, 96–98.

# Selected Bibliography

## Primary Sources

Adler, Herman, and John Larson. "Deception and Self-Deception." *Journal of Abnormal Psychology* 22 (1928): 364–371.

Borkenstein, R. F., and John A. Larson. "The Clinical Team Approach," *Academy Lectures on Lie Detection,* ed. V. A. Leonard. Springfield, Ill.: Thomas, 1957, vol. 1, pp. 11–20.

Cannon, Walter B. *Bodily Changes in Pain, Hunger, Fear, and Rage: An Account of Recent Researches into the Function of Emotional Excitement.* New York: Appleton, 1915.

Inbau, Fred. *Lie Detection and Criminal Investigation.* Baltimore, Md.: Williams and Wilkins, 1942, 1948, 1953. Reissued as Fred Inbau and John E. Reid. *Criminal Investigation and Confessions.* Baltimore, Md.: Williams and Wilkins, 1962, 1967, 1986, 1989, 2001, 2005.

James, William. "What Is an Emotion?" *Mind* 9 (1884): 188–205.

Keeler, Charles Augustus. *Evolution of the Colors of North American Land Birds.* San Francisco: California Academy of Sciences, [1893].

———. *The Simple Home,* intro. Dimitri Shipounoff. Santa Barbara, Calif.: Peregine Smith, [1904], 1979.

Keeler, Katherine [Applegate]. "A Study of Documentary Evidence in Election Frauds: Tattle-Tale Tallies." *JCLC* 25 (1934–1935): 324–337.

———. "Documentary Evidence Involved in an Election Dispute." *JCLC* 27 (1936–1937): 249–262.

Keeler, Leonarde. "A Method for Detecting Deception." *American Journal of Police Science* 1 (1930): 38–51.

———. "The Canary Murder Case." *American Journal of Police Science* 1 (1930): 381–386.

———. "Apparatus for Recording Arterial Blood Pressure; U.S. Patent, No. 1,788,434." Filed 7/30/25. Granted 1/31/31; reprinted in *Polygraph* 23 (1994): 128–133.

———. "Scientific Methods of Crime Detection with a Demonstration of the Polygraph." *Kansas Bar Association Journal* 12 (1933): 22–31.

Larson, John Augustus. "Modification of the Marston Deception Test." *JCLC* 12 (1921): 390–399.

———. "The Berkeley Lie Detector and Other Deception Tests." *Reports of the American Bar Association* 47 (1922): 619–628.

———. "Cardio-Pneumo-Psychogram in Deception." *Journal of Experimental Psychology* 6 (1923): 420–454.

———. "Present Police and Legal Methods for Determination of the Innocence or Guilt of the Suspect." *JCLC* 16 (1925): 219–271.

———. *Lying and Its Detection; A Study of Deception and Deception Tests,* with George W. Haney and Leonarde Keeler, intro. August Vollmer. Chicago, Ill.: University of Chicago Press, 1932.

Larson, John Augustus, and G. W. Haney. "Cardio-Respiratory Variations in Personality Studies." *American Journal of Psychiatry* 11 (1932): 1035–1081.

Larson, John Augustus, and A. Walker. "Paranoia and Paranoid Personalities: A Practical Police Problem." *JCLC* 14 (1923–1924): 350–375.

Marston, William M. "Systolic Blood Pressure Symptoms of Deception." *Journal of Experimental Psychology* 2 (1917): 117–163.

———. "Physiological Possibilities . . . in the Deception Test." *JCLC* 11 (1921): 551–570.

———. "Systolic Blood Pressure and Reaction Time Symptoms of Deception and Constituent Mental States." Ph.D. dissertation, Harvard University, 1921.

———. "Sex Characteristics of Systolic Blood Pressure Behavior." *Journal of Experimental Psychology* 6 (1923): 387–419.

———. *Emotions of Normal People.* London: Kegan Paul, 1928.

———. *The Lie Detector Test.* New York: R. R. Smith, 1938.

Marston, William M., C. Daly King, and Elizabeth H. Marston. *Integrative Psychology: A Study of Unit Response.* New York: Harcourt, Brace, 1931.

Macdonald, Dwight. "The Lie-Detector Era." *Reporter* (6/8/54): 10–18; (6/22/54): 22–29.

McCormick, Charles T. "Deception Tests and the Law of Evidence." *California Law Review* 15 (1926–1927): 484–503.

Meyer, Adolf. "The Scientific Study of Behavior and Its Support," *Collected Papers of Adolf Meyer,* vol. 4, pp. 132–137. Institute for Juvenile Research (1/26). Baltimore, Md.: Johns Hopkins University Press, 1950–.

Münsterberg, Hugo. *On the Witness Stand; Essays on Psychology and Crime.* New York: McClure, 1908.

———. *The Photoplay: A Psychological Study,* [1916]. Reprinted as *The Film: A Psychological Study,* ed. Richard Griffith. New York: Dover, 1970.

Pitkin, Walter B., and William M. Marston. *The Art of Sound Pictures.* New York: Appleton, 1930.

Shils, Edward A. *The Torment of Secrecy: The Background and Consequences of American Security Politics,* intro. Daniel P. Moynihan. Chicago, Ill.: Dee, [1956], 1996.

Vollmer, August. "The Policeman as Social Worker." *Proceedings of the International Association of Chiefs of Police* (1919): 32–38.

———. "Annual Address of the President." *Proceedings of the International Association of Chiefs of Police* (1922), 1: 11–15; reprinted, *JCLC* 13 (1922): 251–257.

———. *Law Enforcement in Los Angeles,* intro. Joseph G. Woods. LAPD Annual Report for 1924. New York: Arno, 1974.

———. "The Chicago Police," *The Illinois Crime Survey,* ed. John Henry Wigmore. Chicago: Illinois Association for Criminal Justice, 1929, pp. 357–372.

———. *The Police and Modern Society.* Berkeley: University of California Press, 1936.

[Wickersham Commission]. U.S. National Commission on Law Observance and Enforcement. *Report on Lawlessness in Law Enforcement,* vol. 11, ed. Zechariah Chafee, Jr. *Report on Police,* vol. 14, ed. August Vollmer. Washington, D.C.: U.S. GPO, 1931.

Wigmore, John Henry. "Professor Muensterberg and the Psychology of Testimony." *Illinois Law Review* 3 (1909): 399–445.

———. *A Treatise on the Anglo-American System of Evidence in Trials at Common Law,* 2nd ed. Boston, Mass.: Little, Brown, 1923.

## Secondary Sources

Alder, Ken. "The Honest Body: The Polygraph Exam and the Marketing of American Expertise." *Historical Reflections* 24 (1998): 487–525.

———. "A Social History of Untruth: Lie Detection and Trust in Twentieth-Century America." *Representations* 80 (2002): 1–33.

Bok, Sissela. *Lying: Moral Choice in Public and Private Life.* New York: Vintage, 1989.

Bond, Charles F., Jr., and Bella M. DePaulo. "Accuracy of Deception Judgments." *Personality and Social Psychology Review* 10 (2006): 214–234.

Brooks, Peter. *Troubling Confessions: Speaking Guilt in Law and Literature.* Chicago, Ill.: University of Chicago Press, 2000.

Bunn, Geoffrey C. "The Hazards of the Will to Truth: A History of the Lie Detector." Ph.D. dissertation, York University, 1997.

———. "The Lie Detector, *Wonder Woman* and Liberty: The Life and Work of William Moulton Marston." *History of the Human Sciences* 10 (1997): 91–119.

Burnham, John C. *How Superstition Won and Science Lost: Popularizing Science and Health in the United States.* New Brunswick, N.J.: Rutgers University Press, 1987.

Carson, John. *The Measure of Merit: Talents, Intelligence, and Inequality in the French and American Republics, 1750–1940.* Princeton, N.J.: Princeton University Press, 2006.

Carte, Gene E., and Elaine H. Carte. *Police Reform in the United States: The Era of August Vollmer, 1905–1932.* Berkeley: University of California Press, 1975.

Danziger, Kurt. *Constructing the Subject: Historical Origins of Psychological Research.* Cambridge: Cambridge University Press, 1990.

Dean, Robert D. *Imperial Brotherhood: Gender and the Making of Cold War Foreign Policy.* Amherst: University of Massachusetts Press, 2001.

DePaulo, Bella, et al. "The Accuracy-Confidence Correlation in the Detection of Deception." *Personality and Social Psychology Review* 1 (1997): 346–357.

Dror, Otniel E. "The Scientific Image of Emotion: Experience and Technologies of Inscription." *Configurations* 7 (1999): 355–401.

Ekman, Paul. *Telling Lies: Clues to Deceit in the Marketplace, Politics, and Marriage.* New York: Norton, 1985.

Fass, Paula. "Making and Remaking an Event: The Leopold and Loeb Case in American Culture." *Journal of American History* 80 (1993): 919–951.

Foucault, Michel. *History of Sexuality,* vol. 1, *An Introduction,* trans. Robert Hurley. New York: Vintage, 1990.

Gale, Anthony, ed. *The Polygraph Test: Lies, Truth, and Science.* London: Sage, 1988.

Golan, Tal. *Laws of Nature and the Laws of Men.* Cambridge, Mass: Harvard University Press, 2004.

Granhag, Pär Anders, and Leif A. Strömwall, eds. *The Detection of Deception in Forensic Contexts.* Cambridge, UK, and New York: Cambridge University Press, 2004.

Hacking, Ian. *Rewriting the Soul: Multiple Personality and the Sciences of Memory.* Princeton, N.J.: Princeton University Press, 1995.

Hale, Matthew. *Human Science and Social Order: Hugo Münsterberg and the Origins of Applied Psychology.* Philadelphia, Pa.: Temple University Press, 1980.

Higdon, Hal. *The Crime of the Century: The Leopold and Loeb Case.* New York: Putnam, 1975.

Hyde, Alan. *Bodies of Law.* Princeton, N.J.: Princeton University Press, 1997.

Johnson, David K. *The Lavender Scare: The Cold War Persecution of Gays and Lesbians in the Federal Government.* Chicago, Ill.: University of Chicago Press, 2004.

Keeler, Eloise. *The Lie Detector Man: The Career and Cases of Leonarde Keeler.* N.p.: Telshare, 1983.

Langbein, John H. *Torture and the Law of Proof: Europe and England in the Ancien Régime.* Chicago, Ill.: University of Chicago Press, 1977.

Leo, Richard A. "From Coercion to Deception: The Changing Nature of Police Interrogation in America," in *The Miranda Debate: Law, Justice, and Policing,* ed. Richard A. Leo and George C. Thomas III. Boston, Mass.: Northeastern University Press, 1998, pp. 65–74.

——— "The Third Degree and the Origins of Psychological Interrogation in the United States," in *Interrogations, Confessions, and Entrapment,* ed. Daniel Lassiter. New York: Kluwer, 2004, pp. 37–84.

Leys, Ruth. "Types of One: Adolf Meyer's Life Chart and the Representation of Individuality." *Representations* 34 (1991): 1–28.

Lykken, David Thoreson. *A Tremor in the Blood: Uses and Abuses of the Lie Detector.* New York: McGraw-Hill, 1981.

McCoy, Alfred W. *A Question of Torture: CIA Interrogation, from the Cold War to the War on Terror.* New York: Henry Holt, 2006.

National Academy of Sciences, Committee to Review the Scientific Evidence on the Polygraph, Stephen E. Fienberg, chair. *The Polygraph and Lie Detection.* Washington, D.C.: National Academy of Sciences Press, 2003.

Office of Technology Assessment, U.S. Congress. *Scientific Validity of Polygraph Testing: A Research Review and Evaluation—A Technical Memorandum.* Washington, D.C.: GPO, 1983.

Paul, Annie Murphy. *The Cult of Personality: How Personality Tests Are Leading Us to Miseducate Our Children, Mismanage Our Companies, and Misunderstand Ourselves.* New York: Free Press, 2004.

Potter, Claire Bond. *War on Crime: Bandits, G-Men, and the Politics of Mass Culture.* New Brunswick, N.J.: Rutgers University Press, 1998.

Ruth, David E. *Inventing the Public Enemy: The Gangster in American Culture, 1918–1934.* Chicago, Ill.: University of Chicago Press, 1996.

Segrave, Kerry. *Lie Detectors: A Social History.* Jefferson, N.C.: McFarland, 2004.

Sokal, Michael M., ed. *Psychological Testing and American Society, 1890–1930.* New Brunswick, N.J.: Rutgers University Press, 1987.

Stearns, Peter N. *American Cool: Constructing a Twentieth-Century Emotional Style.* New York: New York University Press, 1994.

Stearns, Peter N., and Jan Lewis, eds. *An Emotional History of the United States.* New York: New York University Press, 1998.

Tanenhaus, David Spinoza. *Juvenile Justice in the Making.* New York: Oxford University Press, 2004.

Wang, Jessica. *American Science in an Age of Anxiety: Scientists, Anticommunism, and the Cold War.* Chapel Hill: University of North Carolina Press, 1999.

Willrich, Michael. *City of Courts: Socializing Justice in Progressive Era Chicago.* Cambridge: Cambridge University Press, 2003.

Winter, Alison. "The Making of 'Truth Serum.'" *Bulletin of the History of Medicine* 79 (2005): 500–533.

Woods, Joseph Gerald. "The Progressives and the Police: Urban Reform and the Professionalization of the Los Angeles Police." Ph.D. dissertation, UCLA, 1973.

# Acknowledgments

Growing up, as I did, in the 1970s, the glory days of Californian authenticity, I have only gradually come to appreciate the democratic virtues of cognitive privacy. But interior privacy is of little value without the camaraderie and intellectual engagement of family, friends, and colleagues. The following people read the manuscript as a whole or in part, and I am grateful that they shared their comments with me: John Carson, Peter Gaffney, Andrew Nelson, Ted Porter, Bronwyn Rae, and Mike Sherry. I would also like to thank those people who influenced my thinking while I was writing this book; sometimes an oblique comment had a greater impact than either of us realized at the time: Francesca Bordogna, Bob Brain, Stephen Fienberg, Mike Fortun, Dario Gaggio, John Lear, Sarah Maza, Jock McLane, Joel Mokyr, Ed Muir, Shobita Parthasarathy, Jessica Riskin, Patrick Singy, John Tresch, and Rob Warden.

Some of the ideas in this book were first tested in presentations for faculty seminars at the following institutions: University of California-Berkeley, University of Chicago, Harvard University, American Bar Foundation, Max Planck Institute for the History of Science, École des Mines de Paris, Oregon State University, University of California-San Diego, Princeton University, École des Hautes Études en Sciences Sociales, and the Newberry Library. I would like to thank both the hosts and the participants. And last but not least, this book was written with the financial support of the National Science Foundation (grant SBR-9710438); the National Endowment for the Humanities at the Newberry Library; the American Bar Foundation; and the Weinberg College of Arts and Sciences, Northwestern University. In my "Note on Sources" I thank the people who provided the inky traces on which this book is based. Here I would like to extend my appreciation to Bill Larson and Penny Eckert for letting me share their family memories; and to Michael Holland, for service above and beyond. I would also like to thank the undergraduate research assistants who hauled so many books and papers from the library for me: Stas Rosenberg, Rebecca Rogalski, Rashaun Sourles, Hannah Nam, and William Slaughter.

Bruce Nichols has been a scrupulous and generous editor; his engagement with the manuscript made this a stronger book. As always, I am deeply indebted to the judgment and advocacy of Christy Fletcher and her colleagues at Fletcher and Perry. Finally, I thank my wife, Bronwyn, and daughter, Madeleine, for refraining from telling me the truth about what they thought of my spending all my time in the attic working on this book—maybe because they tell me the truth often enough. At least it sure *feels* like the truth.

# Index

# Photo Credits

**1.** Courtesy Bill Larson; **2.** Courtesy of Texas Woman's University, The Woman's Collection; **3.** Courtesy of The Bancroft Library, University of California, Berkeley; **4.** Jaquet Apparatus, Catalogue, 1073, late nineteenth century, from The Virtual Laboratory, http://vlp.mpiwg-berlin.mpg.de, Max Planck Institute for the History of Science, Berlin; **5.** Münsterberg lab in 1893, courtesy Harvard University Archives; **6.** Marston/Lampe family; **7.** Bettmann/CORBIS; **8.** Courtesy of the Berkeley Police Department Historical Unit; **9.** *San Francisco Examiner*, August 9, 1922; **10.** *San Francisco Call and Post*, August 17, 1921; **11.** Bettmann/CORBIS; **12.** Courtesy of The Bancroft Library, University of California, Berkeley; **13.** *Newsweek*, March 13, 1937, p. 34; **14.** Courtesy of The Bancroft Library, University of California, Berkeley; **15.** Bettmann/CORBIS; **16.** Tribune Media Services, Inc. All Rights Reserved. Reprinted with permission; **17.** Tribune Media Services, Inc. All Rights Reserved. Reprinted with permission; **18.** E. Keeler, *Lie Detector Man*, 144; **19.** Applegate family; **20.** John Larson, *Lying*, figs. 47, 48; **21.** Christian A. Ruckmick, *The Psychology of Feeling and Emotion* (New York: McGraw-Hill, 1936), 287; **22.** Marston/Lampe family; **23.** Marston/Lampe family; **24.** News clipping, unknown source, dated August 11, 1929; **25.** *Life* magazine, December 19, 1938; permission Proctor & Gamble; **26.** *Look* magazine, December 6, 1938, pp. 16–17; **27.** Charles Moulton [William Marston], "Wonder Woman Versus the Prison Spy Ring," *Wonder Woman*, No. 1 (Summer 1942); **28.** Alva Johnston, "The Magic Lie Detector," *Saturday Evening Post*, April 15, 1944, p. 9; **29.** *Chicago Herald American*, February 7, 1948, p. 15; **30.** *Liberty* magazine, June 21, 1947, p. 15; **31.** CORBIS; **32.** The Herb Block Foundation, 1954; **33.** Hulton-Deutsch Collection/CORBIS